Biotic Stress and Yield Loss

Biotic Stress and Yield Loss

Edited by
Robert K. D. Peterson
Leon G. Higley

CRC Press
Boca Raton   London   New York   Washington, D.C.

Library of Congress Cataloging-in-Publication Data

Biotic stress and yield loss / edited by Robert K.D. Peterson and Leon G. Higley.
    p. cm.
  Includes bibliographical references (p.).
  ISBN 0-8493-1145-4 (alk. paper)
  1. Agricultural pests. 2. Crop losses. 3. Crops—Ecology. I. Peterson, Robert K. D.
II. Higley, Leon G.
SB601 .B47 2000
626′.6—dc21
                                                   00-049829
                                                          CIP

© 2001 by CRC Press LLC

No claim to original U.S. Government works
International Standard Book Number 0-8493-1145-4
Library of Congress Card Number 00-049829
Printed in the United States of America 1 2 3 4 5 6 7 8 9 0
Printed on acid-free paper

Dedication

We dedicate this book to Professor Larry P. Pedigo, whose outstanding mentoring and pioneering work on insect injury and yield loss inspired us to work in this area.

Preface

The idea for this book came to us after we organized a symposium at the 1996 meeting of the North Central Branch of the Entomological Society of America. The symposium was titled *Understanding Yield Loss from Insects* and its focus was on current knowledge of insect herbivores and their impact on plant fitness and yield loss. When we discussed the idea, we recognized it would be important to cover all biotic stressors (insects, plant pathogens, and weeds) because it is critical to compare and contrast plant responses to all biotic stressors if we are ever going to develop more encompassing understandings of plant stress. The participation of writers from many areas was essential for this project, and we are greatly appreciative of the contributions from our authors for this volume. We also appreciate and thank Marijean Peterson for her editorial reviews and assistance in manuscript preparation.

This book is a natural extension of a previous volume, *Economic Thresholds for Integrated Pest Management*, edited by Higley and Pedigo. Where that book focused on economic decision levels and pest management for insects, plant pathogens, and weeds, this book concentrates on plant physiological, developmental, growth, and yield responses to biotic stress. Most of the chapters discuss biotic stress primarily within an agricultural context. This largely reflects the current state of knowledge for plants and biotic stress. Much more is known about biotic injury and agricultural crops than biotic injury and wild plants within ecosystems. Many of the chapters also discuss plant response and yield loss to insect herbivores. This is partly a reflection of our backgrounds and disciplinary bias, but is also a reflection of the relative richness of entomological research efforts to characterize plant responses to insect injury.

This is not to imply that our understandings of biotic stress and plant response are adequate. To our knowledge, this is the first book to cover the topic of biotic stress and yield loss. Although one type of biotic stress—arthropod injury—may be better understood than the others, plant responses to biotic stresses continue to be largely ignored and poorly understood. Indeed, the term *plant stress* has been synonymous with the abiotic stresses—temperature, moisture, and mineral nutrition stress. We need to elevate our knowledge of biotic stress to the same level as abiotic stress if we are to meaningfully comprehend biotic stress and integrate both abiotic and biotic stress into a general understanding of plant stress.

One theme that emerges from many chapters in this volume is the need to regard stress as a general phenomenon affecting plants. Although there are profound differences between plants grown in agroecosystems and natural ecosystems, to us, one of the most appealing features of focusing on biotic stress is the potential for greater integration of plant ecophysiology. Many authors have recognized the schism between work on biotic stress in the basic and applied sciences (research on herbivory offers one striking example). As discussions in this book point out, it is time to look for commonalities and exploit the advantages offered by work in both systems. A similar appeal can be made for research that spans disciplines. Happily, over the 15 years we have been working in this area, we have seen disciplinary boundaries weaken. Such a change is hardly surprising, given the intrinsically interdisciplinary nature of work in ecophysiology, and is welcome as the best hope for improving our understandings of plant stress.

Finally, our dedication of this volume to Larry Pedigo is fitting on many levels. Both of us studied under Larry: Bob as an undergraduate and Leon for both his graduate degrees. Larry established a research program that not only offered students the opportunity to work on state-of-the-art issues in pest management, but also encouraged students to look beyond where we were in entomology and pest management and seek a different future. Our later, postgraduate research on photosynthesis, yield loss, and stress would not have been possible without Larry's early encouragement. And, we have had the wonderful opportunity to make the transition from students to collaborators, so that our continued research and writing with Larry has been one of the most valued aspects of our careers.

Looking beyond this personal connection, it was Larry, along with his student Jay Stone, who recognized the importance of yield loss and stress in defining economic damage. Stern and co-workers rightly receive credit for inventing the economic injury level (EIL) and key concepts in 1959 that lead to the development of integrated pest management. But, there were no calculated EILs until Stone and Pedigo in 1972 showed how to do it. Their seminal contribution is that they linked data on economics with data on yield loss from pests. In doing so they helped define what a pest is and showed how understandings of stress are as important as understandings of pest population biology. Starting with that 1972 paper, Larry's work with his students on EILs, defoliation, yield loss, and stress interactions has helped move pest management beyond pests and pest populations to questions of economics and of plant stress.

It is our hope that this book will continue that journey.

Robert K. D. Peterson
Dow AgroSciences
Indianapolis, IN

Leon G. Higley
University of Nebraska
Lincoln, NE

Editors

Robert K. D. Peterson, Ph.D., is a senior research biologist with Dow AgroSciences, where he currently is regulatory manager for biotechnology. He also is adjunct associate professor in the Department of Entomology at the University of Nebraska. His research interests include physiological responses of plants to biotic stress, agricultural technology risk analysis, economic decision levels, and pest management theory.

Leon G. Higley, Ph.D., is professor in the Department of Entomology at the University of Nebraska. His research interests include plant stress physiology, insect ecology, and pest management. He also is distance education coordinator for the Department of Entomology.

Contributors

G. David Buntin, Ph.D.
Georgia Experiment Station
University of Georgia
Griffin, GA 30223
Email:
    gbuntin@gaes.griffin.peachnet.edu

Michael D. Culy, Ph.D.
Dow AgroSciences
Indianapolis, IN 46268
Email: mdculy@dowagro.com

Kevin J. Delaney
Department of Entomology
University of Nebraska
Lincoln, NE 68583
Email: delaney@unlserve.unl.edu

Fikru J. Haile, Ph.D.
Dow AgroSciences
Fresno, CA 93706
Email: fhaile@dowagro.com

Leon G. Higley, Ph.D.
Department of Entomology
University of Nebraska
Lincoln, NE 68583
Email: lhigley1@unl.edu

Scott H. Hutchins, Ph.D.
Dow AgroSciences
Indianapolis, IN 46268
Email: shhutchins@dowagro.com

Stevan Z. Knezevic, Ph.D.
Department of Agronomy
Haskell Agricultural Laboratory
University of Nebraska
Concord, NE 68728
Email: sknezevic2@unl.edu

John L. Lindquist, Ph.D.
Department of Agronomy
University of Nebraska
Lincoln, NE 68583
Email: jlindquist@unl.edu

Tulio B. Macedo
Department of Entomology
University of Nebraska
Lincoln, NE 68583
Email: macedo@unlserve.unl.edu

Brian D. Olson, Ph.D.
Dow AgroSciences
Geneva, NY 14456
Email: bdolson@dowagro.com

Robert K. D. Peterson, Ph.D.
Dow AgroSciences
Indianapolis, IN 46268
Email: rkpeterson@dowagro.com

Stephen C. Welter, Ph.D.
Department of Environmental Science,
    Policy, and Management
University of California
Berkeley, CA 94720
Email: welters@nature.berkeley.edu

Contents

1 Illuminating the Black Box: The Relationship Between Injury and Yield

Robert K. D. Peterson and Leon G. Higley

CONTENTS

1.1 THE BLACK BOX

The observation that insects, weeds, and plant pathogens reduce crop yield undoubtedly predates recorded history. The concept that yield is progressively reduced with increasing numbers of pests is widely known. Modern science and agriculture have long recorded and studied the deleterious impact of biotic stressors on yield. The recognition that pest infestations reduced yields prompted early agriculturists to select and breed plants that survived after infestation. Thus, the fundamental concepts of biotic stress and yield loss have been known and acted upon for millennia.

What is known beyond the observation that biotic stressors impact yield? Surprisingly, very little. The relationship between biotic stress and yield remains poorly understood despite centuries of advances in science and agriculture. For most agricultural and natural systems, knowledge has not moved beyond quantitative

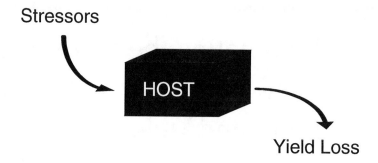

FIGURE 1.1 The black box reflects current understandings of the relationship between biotic stressors and plant yield.

descriptions of pest numbers and resulting yield loss. The plant has been relegated to the status of a black box in the overall equation (Figure 1.1). Pests approach the black box and feed on it. Yield is the endpoint—the thing of value that comes out of the black box. A great deal is known about the pests, their physiology, behavior, life history, and ecology. Similarly, much is known about quantifying plant yield and fitness. But, what about the plant itself—the black box? What happens between stressors and yield? In the absence of pests, we know a great deal about plant growth and development, especially for cultivated species. Advances in the past few decades have provided stunning insights into plant physiology and genetics. Therefore, we can conclude that we know a lot about plants, but very little about how plants respond to pest stress.

1.1.1 THE IMPORTANCE OF UNDERSTANDING RELATIONSHIPS BETWEEN PESTS AND HOSTS

Why is it important to understand how plants respond to biotic stress? After all, in agriculture are we not most concerned about the end result—yield loss—in crops, and how to prevent pests from reducing yield? Similarly, in natural systems can we not forget about plant response and simply measure fitness to understand plant-stressor interactions, evolution, and community ecology?

The short answer to the first question is that pest management rests on the principle of tolerating pests. (The long answer is the topic of the next chapter.) Unless we are willing to regard up to some density of pests as acceptable, then pest management is reduced to pest control—reducing pest populations to the maximum extent technically possible. Instead, if we accept managing pests as a more appropriate goal, then the question of biotic stress becomes very important. Through understanding biotic stress, particularly in economic terms from yield loss, it is possible to determine tolerable levels of pests (such as with the economic injury level). Additionally, it may be possible to increase plant tolerance, through plant breeding or cultural practices, as an alternative to mortality-based approaches.

In natural systems the question of plant responses to stress is no less important. Indeed, the discipline of physiological ecology exists as recognition of the importance of the physiological interaction of an organism with its environment. Although much work in physiological ecology focuses on abiotic stressors, the need for comparable understandings of the impact of biotic stress are increasingly clear. It is not possible to understand how competition, herbivory, or disease influences plant population biology without a detailed understanding of how competition, herbivory, and disease affect individual plants and plant processes. Consequently, both from applied and basic perspectives, understanding plant stress is a crucial objective.

1.1.1.1 Constraints on Understanding

Why do we know a great deal about plants, yet know so little about how plants respond to biotic stress? There are many reasons, which can be divided into two major categories: *disciplinary emphasis* and *practical limitations.* We have already stated that there is generally abundant knowledge about the pests that injure plants. The research area of insect–plant interactions typically focuses on how plants affect insect herbivores, not the reciprocal.[1, 2] Similarly, pathogen–plant interaction research largely focuses on the infection process rather than on disease physiology. The focus on stressors, as opposed to plant response, is largely the result of disciplinary emphasis because most scientists conducting research on biotic stress are trained in entomology, plant pathology, and weed science. Their research training, background, and interest are on the stressor. Many entomologists have only limited training in plant physiology, making research on the relationship between injury and yield difficult.[3]

Overcoming disciplinary limitations is necessary if we are to develop more encompassing knowledge of biotic stress and yield loss. This is certainly possible, given recent trends toward improving interdisciplinary research and unifying concepts in molecular biology and biotechnology. Interdisciplinary research conducted by teams of scientists is critical if we are to understand the entirety of pest impact on plant cells, organs, whole plants, and plant populations. However, interdisciplinary teamwork is not the only option. A pest scientist or ecologist with extensive interdisciplinary training in plant science certainly can investigate physiological responses to biotic stress.

There are several practical limitations to understanding how plants respond to biotic stress. The experimental techniques required to explore relationships between injury and yield require comprehensive knowledge of both plants and pests.[4] In many systems, this knowledge simply does not exist. Experimental procedures are difficult to employ where knowledge of basic biological processes is lacking and can lead to equivocal results. Quantifying injury and injury rates also is difficult. This is especially true for weeds, plant pathogens, and insects that remove plant assimilates with sucking mouthparts. (This will be discussed in more detail in a subsequent section.)

The primary constraint most likely is the influence of environment. Environmental conditions affect both plant responses to biotic stress as well as the stressors themselves.[4] Indeed, environmental conditions are the abiotic stressors and their impact on plants and biotic stressors is more well known than biotic stress. Environmental factors may include light penetration, water availability, temperature, and nutrient availability. Experiments designed to elucidate relationships between injury and yield must consider the confounding potential of the environment. (See Chapter 6 for a more thorough discussion of environmental influences on plant physiology.)

1.1.1.2   Illuminating the Black Box

How do we progress from the black-box approach to understanding how plants respond to biotic stress? We believe the answer lies within a physiologically based approach. Physiology provides a common language for characterizing plant stress and is essential for integrating understandings of stress.[1, 5] For the purposes of this discussion, and indeed for the entire book, we define plant physiology as all processes that determine plant growth, development, and yield.

To understand the physiology of plants in response to biotic stress, measurements from all levels of plant organization often are necessary. These levels include molecules, organelles, cells, tissues, organs, whole plants, and populations. Biotic stressors may impact plants at all of these levels of organization. Indeed, responses to biotic stress may be dramatically different at these varied organizational levels. Chapter 6 discusses plant organization levels in more detail, and other chapters in this volume present many details of physiological understanding for specific pests and specific systems.

Advances in instrumentation offer a prospect for greatly improved understandings of stress. One striking example is the growth in research on stressors and gas exchange in plants over the past 15 years, following from the development of portable infrared (IR) CO_2 analyzers. Molecular biology has already had a huge impact on our understanding of plant disease physiology, and the growth of molecular biology has the potential to lead to dramatically new understandings in other areas of stress physiology. Currently, we see many new insights into stress responses at the molecular and cellular level. Ultimately, integrating responses across all levels of plant organization, including populations, is needed. As molecular approaches become more commonplace and as instrumentation continues to improve, we can reasonably expect new understandings of stress to emerge from these new technologies.

1.2   UNIFYING UNDERSTANDINGS OF BIOTIC STRESS AND YIELD LOSS

The common language of physiology gives us a way to unify our knowledge of biotic stress and plant response. To integrate understandings of biotic stress, we need to rely on common concepts and appropriate terminology.[4] The distinction between injury, the action of a stressor on a plant, and damage, the response of the plant to injury, has been recognized for several decades.[6-8] Higley et al.[1] redefined the terms injury, damage, and stress, to align them better with physiological processes. *Injury* is a stimulus producing an abnormal change in a physiological process. *Damage* is a measurable reduction in plant growth, development, or reproduction resulting from injury. *Stress* is a departure from optimal physiological conditions. For example, a wilt pathogen causes injury through blockage of xylem vessels. Stress results in the form of a reduction in water potential. Damage is a reduction in fruit number, size, and quantity.[1] Higley and Peterson[4] stated, "This terminology provides a common linkage for addressing all types of stress (not only biotic stress)."

1.2.1 STRESS

Formal concepts of stress as a phenomenon are surprisingly recent. Higley et al.[1] discuss the evolution of stress concepts for plants, and we will briefly reiterate their points here. Originally, the term stress was associated with the mechanical concept of a force being applied to a body. From this application of force, some strain would result (like the bending of a metal bar). For plants, much of the literature on stress follows from these engineering notions of stress,[9, 10] and we speak of such things as elastic (reversible) or plastic (irreversible) plant stresses. Indeed, for many abiotic stresses the physical model of stress is appropriate, because such stresses are measurable in physical terms (e.g., water potentials).

An alternative concept for plant stress was offered by Higley et al.,[1] and we follow that proposal here. The key observation is that for many types of stresses (especially resulting from biotic agents) the action of the stress is not reducible to a simple physical measure. Indeed, we would argue that the mechanical view of stress is too limiting to be an appropriate view for plant stress. Does stress involve the response of an organism to a single stimulus in a stereotypic fashion or is stress a complex suite of responses to single or multiple stimuli? The answer to this question was offered in work by Seyle[11] investigating stress on humans. In a series of studies, Seyle demonstrated that a characteristic stress syndrome exists for humans, in response to various physical and psychological conditions. Following from his work, Seyle[11] defined stress as "the nonspecific response of the body to any demand placed on it." Although a single type of "nonspecific response" is not associated with plant stress, the notion of a "syndrome" or variety of effects does seem to fit many plant responses to biotic and abiotic conditions.

Higley et al.[1] argued that because plant stress involves a (typically) adverse reaction to environmental factors (biotic and abiotic), then we might define stress on the broad basis of that reaction. They proposed that stress be defined as "a departure from optimal physiological conditions." This is the definition we have used, but we recognize that it has difficulties. In particular, this definition seems to suggest that there is a recognizable set of optimum physiological conditions and that even transient departures from this optimum represent stress. Nevertheless, the emphasis of this definition on changing physiological conditions seems to us a more appropriate perspective than one of mechanical forces and strains.

1.2.2 INJURY

Why do we distinguish between *injury* and resulting yield loss vs. *pests* and resulting yield loss? The issue largely is one of accuracy and relevance. Many agricultural studies have simply related pest numbers to yield loss. These studies have been beneficial to pest management because they allowed for the calculation of some economic injury levels (EILs). However, we would argue that this is a black-box approach because it does not differentiate the pest from the plant response to the injurious action of the pest. Simply relating pest numbers to yield loss may mask physiological processes that more precisely describe the impact of pests on plants.[4] This has both applied and basic implications. From a pest management perspective, by not considering injury, the accuracy of the yield loss component of the EIL may be impaired. From an ecological perspective, by not considering injury the significance of biotic stressors on plant evolution and succession may be impaired. Further, we believe the ability to integrate biotic and abiotic stresses would be seriously impaired without considering a physiologically based understanding of injury.

So, what is injury? According to the definition above, injury is a stimulus producing an abnormal change in a physiological process. Because injury alters a plant's physiological process, injury necessitates recognition of the *processes* affected by the stressor. For example, Mexican bean beetle injury reduces leaf tissue and photosynthetic rates of remaining leaf tissue.[12] The feeding action of the pest is manifested in an alteration of a physiological process—photosynthesis. For most weeds, injury consists of a reduction in light and water availability. The presence of the weed, disease, or insect herbivore is not the injury, but rather it is its impact on physiological processes.

The identification and terminology of plant injury is critical if we are to move toward more comprehensive understandings of stress and yield loss. The injury guild concept has emerged as a unifying concept when characterizing and discussing injury. In a seminal paper, Boote[13] defined injury guilds by emphasizing the physiological responses of the plant to unique injury types. Boote suggested five injury guilds: stand reducers, leaf-mass consumers, assimilate sappers, turgor reducers, and fruit feeders. Pedigo et al.[8] proposed *plant architectural modifier* as an additional injury guild. Higley et al.[1] incorporated the six injury types, and suggested several more, into categories of physiological impact. These include: population or stand reduction, leaf-mass reduction, leaf photosynthetic-rate reduction, leaf senescence alteration, light reduction, assimilate removal, water-balance disruption, seed or fruit destruction, architecture modification, and phenological disruption.

The type of injury does not solely determine yield loss. Injury type coupled with the magnitude and duration of injury is an important determinant of yield loss. Considerations of magnitude and duration of injury necessitate dividing injury into two types: acute and chronic. Acute injury occurs over a relatively short time in which each unit of injury is discrete and stress can occur from the effect of one or a few units of injury. Chronic injury occurs over an extended time in which units of injury are indistinct and stress only occurs from the combined effect of many units of injury.[1] Injury from most insect defoliators is an example of acute injury, while injury from aphids, weeds, and diseases represents chronic injury.

The magnitude and duration of injury may be influenced by several abiotic and biotic factors. These factors are important to consider when determining injury rates for individual pests and pest populations. For insects, injury rates may be highly species-specific. Injury rates vary depending on the insect's developmental stage.[4] Early instars consume only a few percent of the total consumption. Consumption rates also may be temperature dependent. Crowding of pests, or density effects, can significantly impact consumption. Finally, the action of natural enemies, such as parasites and pathogens, can influence consumption rates.

What becomes evident rather quickly is that if injury guilds are based on physiological response, then pests of different species, phyla, and taxa may be placed into the same injury guild. Homogeneities in physiological response have been identified for different pest species. This has substantial practical advantage because by placing different pests into an injury guild, pest management programs can be developed for the entire injury guild, as opposed to managing individual pest species. Most of the research in this area has focused on the identification of injury types, the construction of injury guilds, and the development of multiple-species economic injury levels (EILs). (Chapter 6 on photosynthesis, yield loss, and injury guilds discusses this in more detail.)

1.2.3   DAMAGE

1.2.3.1   The Damage Curve

Injury produces stress, which results in physiological alterations in the host plant. But, how does injury relate to damage—the measurable reduction in plant growth, development, or reproduction? And, ultimately, how does injury relate to yield? The theoretical relationship between injury and yield is known as the *damage curve* (Figure 1.2). Tammes[6] first established the theoretical and empirical basis for the damage curve in 1961. However, nomenclature for the specific portions of the curve was not developed until 1986. Pedigo et al.[8] named segments of the curve that are indicative of unique types of response between injury and yield.

The segments and their definitions are

- *Tolerance:* No damage per unit injury; yield with injury equals yield without injury;
- *Overcompensation:* Negative damage (yield increase) per unit injury; curvilinear relationship, positive slope;
- *Compensation:* Increasing damage per unit injury; curvilinear relationship, negative slope;
- *Linearity:* Maximum (constant) damage per unit injury; linear relationship, negative slope;
- *Desensitization:* Decreasing damage per unit injury; curvilinear relationship, negative slope;
- *Inherent Impunity:* No damage per unit injury; yield with injury is less than yield with no injury, constant slope.

It is important to note that "not all plants display the entire array of responses, but all potential responses are encompassed by the damage curve and its

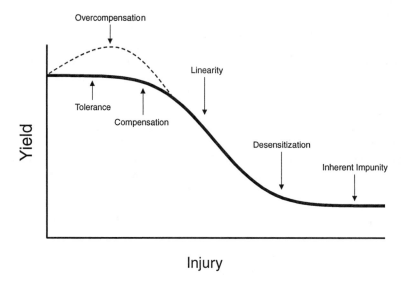

FIGURE 1.2   The damage curve.

components."[1] Despite poor understandings of the physiological mechanisms under-lying plant responses to biotic stress, the relationship between injury or numbers of pests and yield has been documented in hundreds of studies.[3, 4] Most studies have revealed one or, at the most, only a few portions of the damage curve. However, Delucchi[14] and Shelton et al.[15] observed all portions except overcompensation in response to insect injury. The linear portion of the damage curve has been observed most frequently probably because this level of injury and yield loss is of most prac-tical interest to agricultural researchers.[4] (See Higley and Peterson[4] for a thorough discussion of the portions of the damage curve, the studies that reveal its structure, and cases of inappropriate curve fitting.)

1.2.3.1.1   Factors influencing the damage curve
We briefly discussed above the influence of injury type, duration, and magnitude in determining plant responses to biotic stress. Ultimately, we are interested in how injury impacts plant fitness, or, in an agricultural context, yield. Consequently, Pedigo et al.[8] identified factors associated with biotic stress that influence yield. More specifically, they discussed five factors in relation to the damage curve: (1) time of injury, (2) plant part injured, (3) injury types, (4) intensity of injury, and (5) envi-ronmental effects.

    Time of injury refers to when injury is occurring in relation to plant growth and development. Typically, plants are more susceptible to injury during seedling and early reproductive than during vegetative and mature stages. Knowledge of the tim-ing of injury in relation to plant phenology is critical for both improved understand-ing of fitness and development of pest management decision criteria for individual plant developmental stages (plant stage-specific EILs).[16]

Plant part injured refers to the plant structure that is injured by the stressor. For insect herbivores, injury to yield-forming organs has been termed *direct injury.* Injury to nonyield-forming organs has been identified as *indirect injury.* Further distinctions can be made to more accurately characterize yield loss. For example, many insect herbivores prefer to feed on the upper one-third of a plant canopy. This spatial preference causes the plant canopy to respond differently to this injury than to injury to the lower or whole canopy.[17] Similarly, some ear-feeding insects of corn prefer to feed on developing kernels, while others prefer silks. Chapter 4 discusses this in greater detail.

Injury types and *intensity of injury* were discussed above. *Environmental effects* are critical to understand and quantify if we are to improve our knowledge of biotic stress and yield loss. Indeed, the role of the environment in influencing plants and biotic stressors should not be underestimated. The ability of environmental stress (abiotic stress) to interact with biotic stress is well known. Dramatic differences in plant response and yield occur when a plant is injured by insect defoliators and moisture stress in combination as opposed to each stress alone. Chapters 2 and 7 discuss stress interactions in more detail. Several chapters in this book discuss abiotic stress alone and in combination with biotic stress.

1.3 STRESS INTERACTIONS

A longstanding concern regarding characterizing stress in agricultural settings is the potential for interactions among stressors. In fact, such interactions are sometimes held as an insurmountable limitation to characterizing stress and yield loss. Obviously, as editors of a book on yield loss and stress, we do not subscribe to the view that stress and yield are so variable as to be unknowable. Indeed, considerable research on yield loss relationships demonstrates that such relationships can be reliably described.[18] But the observation that many factors influence plant responses to stress, as we outlined in Section 1.2.3.1.1, is clearly true.

The scientific meaning of the term *interaction* is of the dependence of one factor on another.[19] From this perspective, any factor that alters response of a plant to a given stressor represents an interaction. Thus, plant age, plant part injured, or environmental effects all might be considered interacting factors with a given stress. However, stress interactions more commonly refer to dependence of one stress on another. Stress interactions represent the potential relationships among different agents that produce stress in a plant.

Higley et al.[1] presented a classification scheme for different types of stress relationships. A key point in this scheme was the recognition that interaction might occur in two ways. One type of stress interaction involves changes in plant responses to stress when a second stress occurs. A second type of stress interaction can occur when one stress actually alters the incident of a subsequent stress. This latter form of interaction is possible if an initial stress impairs a resistance mechanism for a subsequent stress or if the initial stress somehow makes the environmental conditions more favorable for a subsequent stress. Higley et al. pointed out that the notion of dependency in response among stressors need not necessarily be associated with statistical

measures of interaction. For example, two stressors producing a common, but curvilinear, response in some measured plant response (such as yield) would not represent
an interaction (because each stressor is acting independently); however, because the
combined action of these factors is non-additive (curvilinear) it would appear as an
interaction in a statistical test.

In brief, the categories of stress relationships Higley et al. proposed and their definitions were

1. *Independence*—Plant responses to or occurrence of one stress are not
   influenced by another stress or stresses.
2. *Interaction (= Dependence)*—Plant responses to or occurrence of one
   stress are influenced by another stress or stresses.
   a. *Stress Response Interactions*—Plant responses to two or more stresses
      are greater or less than the sum of responses to the individual stresses.
      (Both biotic and abiotic stresses can produce stress response interac-
      tions. Stress response interactions indicate that physiological processes
      affected by the stresses are interrelated with respect to a measure of
      damage [such as seed yield or biomass production]).
   b. *Stress Incidence Interactions*—The occurrence of an initial stress
      changes the incidence of a subsequent stress. (The initial stress can be
      biotic or abiotic, but the subsequent stress is almost always biotic.
      Stress incidence interactions indicate that the initial stress impairs plant
      resistance mechanisms to the biotic agent, causing the subsequent
      stress or altering the environment for the second stress.)
3. *False Relationships*—A failure to recognize common injury by two different factors
   a. *False Independence*—Two stressors appear to be independent but actu-
      ally affect the same physiological process, with the sum of their injuries
      producing a linear damage response.
   b. *False Interaction*—Two stressors appear to interact but actually affect
      the same physiological process, with the sum of their injuries produc-
      ing a nonlinear damage response.

One motivation for such a classification system is that classifying interactions
may provide insights into the mechanisms behind a given interaction. This point
seems especially pertinent for stress incidence interactions, where the subsequent
occurrence of stress comes from some alteration of resistance or the environment.
The interaction between insect defoliation and increased weed competition (arising
from increased light penetration through a crop canopy) is one example of such an
interaction. Similarly, distinguishing between actual and false dependences among
stressors clearly is of value.

Nevertheless, both schemes for classifying stressors and the notion of stress
interactions itself are probably reflective of agricultural perspectives of stress. In natural systems, multiple stressors are constantly interacting with plants. Additionally,
plants in natural settings will face limited resources and, frequently, more interspe-

cific competition, while agricultural plants may face a more intense intraspecific competition. Work on multiple stressors in agricultural systems, under near optimum conditions and with high levels of experimental control, may provide good evidence of the potential influences of different stressors. A difficult challenge lies in seeing where, or if, these understandings can be translated to complex and variable natural systems.

1.4 CONCLUSIONS

Substantial progress has been made in recent years to understand how plants respond to insects, pathogens, and weeds. Additionally, we have learned a great deal about plant response in both natural and agricultural systems. A key to improving knowledge is to use plant physiology as a common language for explaining the relationships between plants and stressors. Future research must address questions framed around the concepts outlined above. A focus on and differentiation of stress, injury, and damage must occur before meaningful integration of biotic and abiotic stress can result.

REFERENCES

1. Higley, L. G., Browde, J. A., and Higley, P. M., Moving towards new understandings of biotic stress and stress interactions, in *International Crop Science I,* Buxton, D.R., Shibles, R., Forsberg, R. A., Blad, B. L., Asay, K. H., Paulson, G. M., and Wilson, R. F., Eds., Crop Science Society of America, Madison, WI, 1993, 749.
2. Welter, S. C., Responses of plants to insects: Eco-physiological insights, in *International Crop Science I,* Buxton, D. R., Shibles, R., Forsberg, R. A., Blad, B. L., Asay, K. H., Paulson, G. M., and Wilson, R. F., Eds., Crop Science Society of America, Madison, WI, 1993, 773.
3. Peterson, R. K. D., The status of economic-decision-level development, in *Economic Thresholds for Integrated Pest Management,* Higley, L. G., and Pedigo, L. P., Eds., University of Nebraska Press, Lincoln, 1996, 151.
4. Higley, L. G., and Peterson, R. K. D., The biological basis of the EIL, in *Economic Thresholds for Integrated Pest Management,* Higley, L. G., and Pedigo, L. P., Eds., University of Nebraska Press, Lincoln, 1996, 22.
5. Peterson, R. K. D., and Higley, L. G., Arthropod injury and plant gas exchange: current understanding and approaches for synthesis, *Entomol. (Trends Agric. Sci.),* 1, 93, 1993.
6. Tammes, P. M. L., Studies of yield losses. II. Injury as a limiting factor of yield, *Tijdschr. Plantenziekten,* 67, 257, 1961.
7. Bardner, R., and Fletcher, K. E., Insect infestations and their effects on the growth and yield of field crops: a review, *Bull. Entomol. Res.,* 64, 141, 1974.
8. Pedigo, L. P., Hutchins, S. H., and Higley, L. G., Economic injury levels in theory and practice. *Annu. Rev. Entomol.,* 31, 341, 1986.
9. Levitt, J., Stress concepts, in *Responses of Plants to Environmental Stress, vol. II. Water, Radiation, Salt, and Other Stress,* 2nd edition, Academic Press, New York, 1980, 3.
10. Bradford, K. J., and Hsiao, T. C., Physiological responses to moderate water stress, in Lange, O. L., Nobel, P. S., Osmond, C. B., and Ziegler, H., Ed., *Physiology and Plant*

Ecology III. Water Relations and Carbon Assimilation, Springer-Verlag, New York, 1982, 264.

11. Seyle, H., The evolution of the concept of stress, *Am. Sci.,* 61, 692, 1973

12. Peterson, R. K. D., Higley, L. G., Haile, F. J., and Barrigossi, J. A. F., Mexican bean beetle (Coleoptera: Coccinellidae) injury affects photosynthesis of *Glycine max* and *Phaseolus vulgaris, Environ. Entomol.,* 27, 373, 1998.

13. Boote, K. J., Concepts for modeling crop response to pest damage, ASAE Pap. 81-4007, American Society of Agricultural Engineers, St. Joseph, MI, 1981.

14. Delucchi, V., Integrated pest management vs. systems management, in *Biological Control: A Sustainable Solution to Pest Problems in Africa,* Yaninek, J. S., and Herren, H. R., Eds., International Institute of Tropical Agriculture, Ibadan, Nigeria, 1989, 51.

15. Shelton, A. M., Hoy, C. W., and Baker, P. B., Response of cabbage head weight to simulated Lepidoptera defoliation, *Entomol. Exp. Appl.,* 54, 181, 1990.

16. Peterson, R. K. D., Danielson, S. D., and Higley, L. G., Yield responses of alfalfa to simulated alfalfa weevil injury and development of economic injury levels, *Agron. J.,* 85, 595, 1993.

17. Higley, L. G., New understandings of soybean defoliation and their implications for pest management, in *Pest Management in Soybean,* Copping, L. G., Green, M. B., and Rees, R. T., Eds., Elsevier, London, 56, 1992.

18. Higley, L. G., and Pedigo, L. P., Eds., *Economic Thresholds for Integrated Pest Management,* University of Nebraska Press, Lincoln, Nebraska, 1996.

19. Sokal, R. R., and Rohlf, F. J., *Biometry,* 2nd edition, W.H. Freeman and Co., New York, 1981.

2 Yield Loss and Pest Management

Leon G. Higley

CONTENTS

2.1 INTRODUCTION

The focus of this chapter is to examine how understandings of yield loss apply to the practical problem of managing pests. Yield loss is only one expression of plant stress, so one cannot talk about yield loss in isolation. Even in agricultural systems, the affect of stressors on yield is not the only issue of concern; quality and plant longevity are as important as, if not more important than yield for some crops. However, for many practical issues yield is taken as the final arbiter of stress, and it is the parameter most closely tied to economic impacts of stress. So focusing on yield loss as an index of plant stress is appropriate in an agricultural context.

One of the most common issues relating to stressors and yield is crop loss assessment. As a *post-hoc* approach to evaluating yield reductions from pests, crop loss assessment is important for indemnifying producers for hail damage, drought, or certain types of pest injury. Crop loss assessment is useful in this context and as an indicator of the yield loss potential of various stressors, which is important in deciding to use preventive management tactics. Many simulation modeling approaches for plant stress may achieve the same goal, although models also can be extremely useful for decisions on therapeutic tactics. I believe it is a fair criticism to note that crop loss assessment focuses heavily on the results of stress rather than the process that leads to yield loss. This same criticism might be raised for some regression approaches for relating pest densities to yield loss (as are needed for economic injury levels, or EILs), but much of the work in this area clearly is directed at more than simple measures of loss. Teng[1] offers a good review of approaches to crop loss assessment.

Beyond crop loss assessment, what other areas are there where yield loss relates to management? I will answer that question in the next section, as we consider the broader issues of plant stress and their implications for pest management. My intent in this chapter is not to provide a comprehensive review or survey on the topic. Instead, I want to offer one perspective on how plant stress and yield loss are relevant to pest management. Other chapters in this book explore in detail specific areas pertinent to stress and management. Additionally, a substantial literature exists on EILs, and on the use of yield loss information in building pest management decision tools.[2, 3]

2.1.1 STRESS, YIELD, AND PEST MANAGEMENT

Characterizing how plants are stressed by abiotic and biotic factors is a challenging undertaking, not only from a technical standpoint, but also theoretically. What is stress? What is an appropriate yardstick for evaluating stress? Are there common responses to different stressors or are all agents unique? Why look at stress at all? Why not focus our attention on reducing the number of stressors (and avoid this confusion)?

Of course, focusing attention on reducing stressors is exactly what has been done for weeds, insects, and plant pathogens. Such an approach probably is easier than trying to elucidate the nuances of plant stress physiology, and eliminating stressors certainly eliminates stress. However, we cannot easily eliminate all stressors (especially abiotic stresses, although we try through practices such as irrigation), and we cannot circumvent some fundamental limitations in focusing on stressors. In particular, reducing weeds, insects, or pathogens through mortality-based tactics (the common approach) means that we also expose pest populations to serious selection pressures. It is hardly surprising that in the face of such selection pressures pests have evolved counter-responses.

Are there other reasons for focusing on stress? In articulating the principle of pest management, Geier and Clark[4] called for tolerating pests. How do we determine what is tolerable? The EIL, which requires an explicit characterization of stress (in terms of yield loss), is one answer. A second obvious question is whether we can improve the ability of plants to tolerate pests. We can through the mechanism of plant resistance called *tolerance,* but here also understanding stress is necessary to most effectively develop tolerant plants. Finally, the architects of modern pest management like Stern et al.[5] and Geier and Clark[4] were interested in applying basic understandings of ecological principles to practical questions of pest management. How might ecological insights into the nature of plant stress offer opportunities for use in pest management?

These three areas, the development of EILs, the development of tolerant plant genotypes, and the development of ecological theory on plant stress, are where understanding yield loss is essential for pest management. Yield loss is not the only manifestation of stress, and defining yield itself has many of the same pitfalls as trying to define stress. Nevertheless, organisms are plant pests (by definition) because they reduce plant utility, and the most common and most important measure of utility (at least for food crops) is yield.

Given the importance of yield, it seems logical to use yield as our yardstick for measuring stress. Unfortunately, even in highly controlled experimental settings direct relationships between the action of a stressor and reductions in yield may be obscure. This is not to say we cannot document direct relationships between stressors and yield loss (although some have argued we cannot), but rather to recognize that yields are endpoints influenced by many, many factors including the action of any given stressor. Establishing relationships between the actions of stressors and resulting yield loss are serious challenges to using yield loss information in pest management. Another challenge is the recognition that understanding how stressors reduce yield requires that we look at many plant responses beyond yield. Because stress (by most definitions) involves an alteration or impairment of physiology, these "plant responses beyond yield" are changes in plant physiology.

These arguments lead to what is an important observation. Both in its genesis and evolution, the link of pest management to ecology has been through population ecology. Although we have yet to see the emergence of a comprehensive application of population theory to the key practical question of how to minimize selection pressure in the application of mortality-based tactics, population ecology forms the foundation for most work in pest management. However, once we acknowledge the importance of understanding yield loss to pest management, we move into a new arena. Here, physiological ecology is the key underlying discipline. I believe this is an exciting prospect, because it suggests that as comprehensive understandings of plant stress physiology are developed, new opportunities will arise for mitigating the impact of plant pests.

As outlined in other chapters here on domesticated vs. wild plants (Chapter 10) and on plant stress in natural systems (Chapter 9), ecological and agricultural understandings on plant stress present opportunities and challenges. The barriers to unified approaches include differences between domesticated vs. wild plant species and differences between agroecosystems (resource rich, simplified monocultures) vs. natural ecosystems (resource limited, spatially and temporally diverse polycultures). Another key point is that in natural systems the factor of interest is fitness—how stressors alter the ability of individual plants to perpetuate their genotypes. In agroecosystems, the factor of interest is yield, although in some perennial systems plant longevity also is a goal. To argue that yield and fitness are equivalent is clearly wrong, but it is also wrong to argue that they are unrelated. From ecological and agronomic perspectives, we are rightly interested in the endpoints of stress: fitness and yield. But in looking for better synthesis between agronomic and ecological research on stress, it is the process of stress—the physiological mechanisms by which plants are affected by stressors—that seems to hold the most promise for more unified perspectives between agriculture and ecology.

2.2 TOLERANCE

Tolerance is one of the three recognized "mechanisms" or forms of host plant resistance. The other forms include antibiosis, plant traits that impair pest biology in some fashion, and antixenosis, plant traits that render plants less acceptable to pests as hosts. In contrast, tolerance represents the ability of plants to withstand pest attack without appreciable deleterious effects (such as yield loss). In the context of natural systems, tolerant plants can maintain fitness in response to pest injury.[6]

Of the three forms of plant resistance, tolerance is unique in that it involves only a response of the host, not of a pest. From this point, it follows that unlike other forms of resistance, tolerance does not represent a selection pressure on pest populations. In a sense, tolerance changes pest status by reducing the impact of pests on a host. Broadly speaking, tolerance operates through buffering capacities of plants for pest injury and through compensatory mechanisms to minimize physiological insults produced through injury. Of course, hidden in that simple statement are many mysteries regarding the mechanisms and genetics underlying tolerant responses of plants to injury. Nevertheless, tolerance might be regarded as the ultimate pest management tactic: an approach that presents no selection pressure on pests, no untoward consequences to the environment, and sustainability as a long-term solution to pest problems. And yet, as many authors have recognized, tolerance is the least used and understood of any form of host plant resistance.[7–10]

Despite its theoretical advantages, there are compelling reasons why tolerance has not been more widely used in host plant resistance. First, screening lines for tolerance is a difficult undertaking, given that in many circumstances plants must be grown to yield to demonstrate tolerance and that screening techniques must have sufficient resolution to identify tolerant genotypes. With these requirements it is difficult to construct a procedure for evaluating hundreds of lines (such as in initial screening efforts) for tolerance. Second, tolerance may operate only over some range of pest densities or only under specific environmental conditions (typically highly favorable conditions of water and nutrients). Third, the genetic basis of most forms of tolerance is held to be polygenic, which makes the prospect of incorporating tolerance into advanced breeding lines problematic at best. And fourth, tolerance does not reduce pest populations. In some instances, such as pest species able to move between and damage different crops (e.g., soybean looper moving between cotton and soybeans), tolerance in one crop could conceivably contribute to increased problems in another. Probably the more compelling objection, though, is that producers (and plant breeders) are used to management approaches that reduce pest numbers.

Despite these potential objections, tolerance is used. In particular, many disease resistant lines express tolerant traits. Similarly, some crops naturally exhibit high levels of tolerance to certain pests. For example, soybeans and alfalfa have striking abilities to tolerate insect defoliation.[11, 12] Besides genetic approaches to this issue, environmental conditions and crop management practices can greatly influence the ability of crops to withstand various stresses. For example, under conditions of high water availability, full bloom soybeans defoliated by 70% recovered all lost leaf area in about three weeks.[13]

Is it possible to develop a sufficient understanding of tolerance responses of plants that practical applications might follow? Certainly, Trumble et al.'s[14] review of plant compensation highlights many identified mechanisms of tolerance. Just as one example, compensatory regrowth, leaf area production, and delayed leaf senescence have been documented by a number of researchers as compensatory mechanisms for defoliation.[12, 14–17] Moreover, work by Haile et al.[13] with isogenic soybean lines demonstrated that increased tolerance to herbivory could be associated with single gene differences (in this study, a gene coding for different leaf morphologies).

In keeping with the theme of this book, better understandings of plant stress may offer new opportunities for developing more pest-tolerant crop varieties. Because tolerance operates through compensatory mechanisms, identifying such mechanisms might present one approach for screening germplasm for tolerance. Additionally, with the advent of transgenic crop improvement techniques, the potential for moving tolerance genes into elite lines is greatly improved. Beyond these technical issues, better understandings of tolerance are likely to be increasingly important with the need to develop more sustainable agricultural production systems.[18]

2.3   YIELD LOSS AND ECONOMIC THRESHOLDS

Perhaps the most compelling reason for considering yield losses from pests is that this information is essential for calculating EILs. As outlined in the preface, although Stern et al.[5] defined the EIL in 1959, it was not until 1972 that the procedures for calculating EILs were published by Stone and Pedigo.[19] The essential point was to define economic damage (yield losses equal to management costs) in calculable terms. Stone and Pedigo's goal was to define an EIL for the green cloverworm, *Plathypena scabra,* on soybean, but yield loss data were lacking. So, as a first approximation they used data from studies on yield losses from defoliation by hail. This problem of having appropriate yield loss data for specific insects on specific crops is a continuing impediment to the development of EILs.[3, 20]

The equation for the EIL presents costs of management and benefits of preventing yield loss. Drawing on previous definitions, Pedigo et al.[21] defined the EIL as:

$$EIL = C/VDIK \qquad [2.1]$$

where C = management costs per production unit, V = value per unit production, D = damage (yield loss) per unit injury, I = injury per pest, and K = proportion of injury prevented by management. The D variable represents yield loss, the mathematical relationship between pest injury and yield. Depending on the pest, it may not be possible to distinguish between D and I, so both are sometimes combined (D') into a term relating pest density to yield. Generally, relating pest numbers to yield loss is even more variable than relating injury to yield loss, but unless pest injury is quantifiable, D and I cannot be distinguished.

Elsewhere, colleagues and I have written extensively on issues surrounding the EIL and in particular the D variable, so there is no need to repeat that discussion here.[2, 20] Additionally, Chapter 3 in this volume speaks greatly to the issues associated with determining D. In short, determining D frequently comes down to experimental measures of yield loss and the use of regression procedures to describe the relationship. Because yield is a parameter influenced by many factors, getting good relationships between yield loss and injury can be challenging. Another key problem is that imposing and quantifying injury may be difficult in many systems. Finally, the use of regression offers many pitfalls for the unwary who do not try to relate regressions to underlying biological relationships (for instance, quadratic relationships may indicate a yield increase at high levels of injury that is not biologically possible).

Regression is an essential tool in establishing yield loss relations, but are other approaches possible to reduce the variability in our mathematical descriptions? For at least some plant–pest relationships, the answer to this question is a resounding yes. The key observation is that by identifying mechanisms underlying certain types of injury it may be possible to circumvent variability in yield responses. The best example of the power of more physiologically based explanations for yield loss comes from work on defoliation.

2.3.1 THE DEFOLIATION EXAMPLE

Since Stone and Pedigo's initial calculation of an EIL for a defoliating soybean pest,[19] improving characterizations of defoliation and yield loss was a significant research thrust. Unfortunately, results among researchers were highly variable and no clear explanation for the diversity in findings was evident. Debates involved questions of defoliation techniques, timing, environmental interactions, and cultivar differences. Through the 1970s and early 1980s workers examined many of these factors to identify which were most associated with variability in yield loss relationships, but no obvious solutions occurred. A change in outlook among soybean entomologists began in the mid-1980s and took hold by the end of the decade. That change was the recognition that reductions in yield following defoliation must follow from relevant physiological changes in the plant. While much effort was directed at relating leaf loss (often defined as percentage defoliation) to yield, the physiologically relevant issue was leaf tissue remaining in a defoliated canopy. More precisely, it was the capacity of this remaining canopy to intercept light and continue to photosynthesize that seemed most likely to be related to yield.

My involvement in what we called the defoliation–light interception hypothesis began with initial studies in 1987 and large, multi-state studies in the late 1980s and early 1990s. At the same time, in an exchange of letters in *Phytopathology,* Waggoner and Berger[22] emphasized the importance of remaining leaf area and Johnson[23] argued for the importance of light interception as driving yield in crops affected by foliar pathogens. From these beginnings, a growing body of evidence supports the contention that light interception of remaining leaves is a key determinant of yield in defoliated soybean (and likely many plant systems).[12, 13, 24–26]

The relevance of this example to EILs is that by refining how we characterize injury we have the potential to dramatically improve the accuracy of our pest management decisions. Also, we may be able to identify plant traits or cultural practices to increase the tolerance of plants to injury (in this example, finding ways to increase plant light interception efficiencies). Relationships between leaf area and light interception or between light interception and yield seem a long way from yield loss per pest and the relatively simple EIL equation. Let us look at how these can be merged into the EIL and how both injury-defined and pest-defined EILs might be developed.

2.3.2 STRESS-BASED AND INJURY-DEFINED EILS

The notion of a "stress-based EIL," as I envision it, is that the mathematical basis of yield loss portion of the EIL (the D variable) comes from relationships of physiologically significant parameters, rather than from yield loss-injury or yield loss-pest density regressions. Actually, the title "injury-defined EILs" is slightly misleading, because in principle the EIL always defines a level of injury sufficient to cause economic loss. In practice, however, pest densities are used as an index of injury, so my terms "injury-defined" and "pest-defined" really just refer to how we express the EIL (in terms of leaf loss or insect densities). As background for our calculations, Haile[26] obtained data in 1997 and 1998 on simulated insect defoliation to soybean, and defined strong linear relations between the leaf area index (LAI) (the ratio of leaf area to ground area) and intercepted photosynthetically active radiation (PAR). More specifically, the linear relationship held for LAIs below the critical LAI of 3.5. (The critical LAI for a crop is that LAI at which a canopy intercepts approximately 95% of all PAR.) Additionally, he observed another strong linear relationship between intercepted PAR (immediately post-defoliation) and yield. Haile's findings are consistent with a growing body of other research on soybean defoliation.

The following example is based on calculations with absolute yield, but similar calculations can be made for proportional yield. In the following equations, a1, b1 and a2, b2 are linear regression parameters. PAR refers to intercepted PAR after defoliation, and LAI refers to the LAI of plants after defoliation. As a first step, we need to relate LAI to yield because we want to define the EIL as leaf tissue lost, and we can convert lost tissue into an insect density by considering insect consumption rates:

$$YLD = a1 + b1*PAR, \text{ and}$$

$$PAR = a2 + b2*LAI, \text{ so}$$

$$PAR = (YLD-a1)/b1, \text{ and} \tag{2.2}$$

$$LAI = (PAR-a2)/b2;$$

combining the two previous equations yields

$$LAI = (((YLD-a1)/b1)-a2)/b2. \tag{2.3}$$

This explains the relationship of remaining LAI to yield, but we need to calculate how much leaf area would need to be removed to reach an economic level of yield loss. Without injury, we have what I am calling check yield (CYLD); this occurs at or above the critical LAI (what I have called CRTLAI), so

$$CRTLAI = (((((CYLD-a1)/b1)-a2)/b2) \text{ and} \tag{2.4}$$

at the EIL (GT = gain threshold)

$$EILLAI = (((((CYLD-GT-a1)/b1)-a2)/b2). \tag{2.5}$$

The gain threshold (defined originally by Stone and Pedigo[19]) is the amount of yield loss necessary to justify management; the GT is determined as GT = C/V, where C is the cost of management and V is the value of production. By subtracting CRTLAI-EILLAI, we get the LAI reduction necessary from a critical LAI to the EIL. This LAI reduction is an injury-defined EIL. For an EIL expressed in a pest density, the LAI is converted to a row-m basis and divided by the consumption rate per insect (e.g., green cloverworm larvae consume $0.00531 m^2$ leaf area per larva). If the initial LAI (ILAI) is greater than the critical LAI, the real EIL (in LAI or insects) represents the difference between the ILAI and CLAI, plus the adjustment for the GT below the CLAI. Mathematically, this is

$$EILLAI = ILAI - CRTLAI + (((((CYLD-GT-a1)/b1)-a2)/b2). \qquad [2.6]$$

Use of this approach requires an estimate of the starting canopy size (LAI) as well as estimates of pest densities. It has the great advantage over existing procedures in that it properly recognizes both the buffering capacity of soybean canopies for defoliation (because yield losses do not occur until the LAI is reduced below critical levels) and the compensatory responses of soybean to defoliation (through the descriptions of relationships between LAI and intercepted PAR and between intercepted PAR and yield). This relatively simple analysis also can serve as the basis for more detailed understandings. For instance, in looking at genotypic differences in soybean responses to defoliation, Haile et al.[13] observed that more tolerant genotypes had altered intercepted PAR and yield relationships. The more tolerant genotype proved to have a high canopy light extinction coefficient; in other words, in the more tolerant genotype soybean, leaf positions were altered so that the canopy intercepted more light with a given leaf area than other genotypes.

Instrumentation for evaluating crop canopy sizes is available and has been demonstrated to be suitable for use on insect defoliated crop canopies.[27] With such instrumentation or other approaches to estimating canopy size it will be possible to dramatically improve the accuracy of yield loss predictions for defoliation. Additionally, with new understandings of the importance of canopy size and light interception as the issues most responsible for driving yield loss, the possibility exists of using remote sensing techniques to evaluate the need for intervention in crops exposed to defoliating insects.

2.4 CONCLUSIONS

Elsewhere, my colleagues and I have argued that for most of its existence, pest management has been dominated by population issues, particularly the development of new approaches for imposing mortality.[2, 3] However, in relying on the principle of tolerating pests, pest management has long had an implicit dependence on understanding and defining plant stress. As a direction for more sustainable management through improved tolerance, better understandings of stress clearly have an important role. Similarly, as a direction for improving our decision tools, better understandings of stress are clearly essential. Too often advances in control tactics have been held as the ideal for advancing pest management. What I find most promising in our growing understanding of biotic stress in both agricultural and natural systems is the prospect for genuinely new applications and approaches for pest management.

REFERENCES

1. Teng, P. S., *Crop Loss Assessment and Pest Management,* American Phytopathological Society, St. Paul, MN, 1987.
2. Higley, L. G., and Pedigo, L. P., *Economic Thresholds for Integrated Pest Management,* University of Nebraska Press, Lincoln, 1996.
3. Peterson, R. K. D., The status of economic-decision-level development, in *Economic Thresholds for Integrated Pest Management,* Higley, L. G., and Pedigo, L. P., Eds., University of Nebraska Press, Lincoln, 1996.
4. Geier, P. W., and Clark, L. R., An ecological approach to pest control, in *Proc. Tech. Meeting Intern. Union Conser. Nature and Nat. Resources,* 8th, 1960, Warsaw, Poland, 1961, 10.
5. Stern, V. M., Smith, R. F., van den Bosch, R., and Hagen, K. S., The integrated control concept, *Hilgardia,* 29, 81, 1959
6. Rosenthal, J. P., and Kotanen, P. M., Terrestrial plant tolerance to herbivory, *Trends Ecol. Evol.,* 9, 145, 1994.
7. Smith, C. M., *Plant Resistance to Insects: a Fundamental Approach,* Wiley & Sons, New York, 1989.
8. Pedigo, L. P., and Higley, L. G., The economic injury level concept and environmental quality: a new perspective, *Am. Entomol.,* 38, 12, 1992.
9. Welter, S. C., Responses of plants to insects: eco-physiological insights, in Buxton, D. R., Shibles, R., Forsberg, R. A., Blad, B. L., Asay, K. H., Paulson, G. M., and Wilson, R. F., Eds., *International Crop Science I,* Crop Science Society of America, Madison, WI, 1993, 773.
10. Reese, J. C., Schwenke, J. R., Lamont, P. S., and Zehr, D. D., Importance of quantification of plant tolerance in crop pest management programs for aphids: greenbug resistance in sorghum, *J. Agric. Entomol.* 11, 252, 1994.
11. Hutchins, S. H., Buntin, G. D., and Pedigo, L. P., Impact of insect feeding on alfalfa regrowth: a review of physiological responses and economic consequences, *Agron. J.,* 82, 1035, 1990.
12. Higley, L. G., New understandings of soybean defoliation and their implications for pest management, in *Pest Management of Soybean,* Copping, L. G., Green, M. B., and Rees, R. T., Eds., Elsevier, London, 1992, 56.
13. Haile, F. J., Higley, L. G., Specht, J. E., and Spomer, S. M., Soybean leaf morphology and defoliation tolerance, *Agron. J.,* 90, 353, 1998.
14. Trumble, J. T., Kolondny-Hirsch, D. M., and Ting, I. P., Plant compensation for arthropod herbivory, *Annu. Rev. Entomol.,* 38, 93, 1993.
15. Ostlie, K. R., Soybean Transpiration, Vegetative Morphology, and Yield Components following Simulated and Actual Insect Defoliation, Ph.D. dissertation, Iowa State University, Ames, 1984.
16. Welter, S. C., Arthropod impact on plant gas exchange, in *Insect–Plant Interactions, vol. 1,* Bernays, E. A., Ed., CRC Press, Boca Raton, Florida, 1989, 135.
17. Peterson, R. K. D., Danielson, S. D., and Higley, L. G., Photosynthetic response of alfalfa to actual and simulated alfalfa weevil (Coleoptera: Cuculionidae) injury, *Environ. Entomol.,* 21, 501, 1992.
18. Welter, S. C., and Steggall, J. W., Responses of tomato to simulated and real herbivory by tobacco hornworm (Lepidoptera: Sphingidae), *Environ. Entomol.,* 20, 1537, 1993.
19. Stone, J. D., and Pedigo, L. P., Development and economic-injury level of the green cloverworm on soybean in Iowa, *J. Econ. Entomol.,* 65, 197, 1972.

20. Higley, L. G., and Peterson, R. K. D., The biological basis of the economic injury level, in *Economic Thresholds for Integrated Pest Management,* Higley, L. G., and Pedigo, L. P., Eds., University of Nebraska Press, Lincoln, 1996.
21. Pedigo, L. P., Hutchins, S. H., and Higley, L. G., Economic injury levels in theory and practice, *Annu. Rev. Entomol.,* 31, 341, 1986.
22. Waggoner, P. F., and Berger, R. D., Defoliation, disease, and growth, *Phytopathology,* 77, 393, 1987.
23. Johnson, K.B., Defoliation, disease, and growth: a reply, *Phytopathology,* 77, 1495, 1987.
24. Herbert, D. A., Mack, T. P., Backman, P. A., and Rodriguez-Kabana, R., Validation of a model for estimating leaf-feeding by insects in soybean, *Crop Prot.,* 11, 27, 1992.
25. Klubertanz, T. H., Pedigo, L. P., and Carlson, R. E., Reliability of yield models of defoliated soybean based on leaf area index versus leaf area removed, *J. Econ. Entomol.,* 89, 751, 1996.
26. Haile, F. J., Physiology of Plant Tolerance to Arthropod Injury, Ph.D. Dissertation, University of Nebraska–Lincoln, 1999.
27. Hunt, T. E., Haile, F. J, Hoback, W. W., and Higley, L. G., Indirect measurement of insect defoliation, *Environ. Entomol.,* 28, 136, 1999.

3 Techniques for Evaluating Yield Loss from Insects

G. David Buntin

CONTENTS

3.1 INTRODUCTION

Crop loss can be defined as the measurable reduction in quantity or quality of yield.[1] Quantifying the relationship between pest numbers and damage and crop loss is a vital component needed for the development of decision making rules in integrated pest management (IPM) programs. Optimal control decisions cannot be made without reliable crop loss estimates.[1] The economic injury level (EIL) is the basis of economic decision making in pest management and is defined as the lowest number of insects that will cause economic damage, which is the amount of injury that will justify the cost of control.[2] Economic damage is determined by the integration of control costs, commodity market value, and damage per insect. Damage per insect often is further divided into injury per insect and damage per unit of injury.[2] Injury focuses on the pest and is the effect of pest activity (usually feeding) on host physiology, whereas damage is the measurable crop loss as a result of plant response to pest injury. Control costs and commodity values are easily determined for most control measures and commodities, but quantifying the relationship between damage and insect numbers requires expensive detailed studies under representative field conditions. Achieving representative conditions often requires experiments at more than one location and in more than one year. This chapter reviews experimental approaches for quantifying the relationships between insect numbers or populations and crop yield loss.

0-8493-1145-4/01/$0.00+$.50

3.2   EXPERIMENTAL APPROACHES FOR QUANTIFYING YIELD LOSS RELATIONSHIPS

Experimental approaches for quantifying damage yield loss relationships of insects and plants can be broadly categorized as (1) observation of natural populations, (2) establishment of artificial populations, (3) manipulation of natural populations, and (4) simulation of damage by surrogate injury techniques. Each approach has advantages and disadvantages with the selection of methodology often being dictated by the nature of injury, relative impact of injury on plant loss, and typical abundance of pest populations relative to levels needed to cause economic damage. If pest populations normally occur at high levels or if multiple insecticide applications are needed during a season to prevent extensive damage, then manipulation of pest populations using insecticides is a viable approach. If, however, a pest is occasional and typically does not cause economic damage and require control, then this approach may not be technically feasible in the course of a few seasons. Instead, establishment of artificial infestations by infesting plots or simulation of insect injury would be a more reasonable experimental approach.

3.2.1   OBSERVATION OF NATURAL POPULATIONS

The simplest approach to quantifying the damage yield loss relationships is to observe natural infestations over a range of levels and to relate pest population numbers to crop yield or productivity. If a pest population typically occurs over a range of levels thought to be above and below levels needed to cause economic damage, then observation of natural populations and documentation of crop yield and quality are feasible. With this approach, a series of plots of fields is sampled to measure pest population levels and crop yield or production, and yield or crop productivity is quantitatively related to population level. For this approach to work, it is usually necessary to sample a large number of fields or sites in selected fields to generate enough data to provide a meaningful description of the yield loss relationship. Furthermore, the range of infestations must encompass the levels which cause economic damage to be applicable for developing decision rules for pest control.

For example, the level of economic damage of feeding scars on snap bean pods by western spotted cucumber beetles, *Diabrotica undecimpunctata undecimpunctata,* is established by processors at 1.5 scars per 100 pods. Weinzierl et al.[3] measured beetle populations using a sample unit of 10 sweeps and mean number of feeding scars per 100 pods in 50 snap bean fields before harvest. Pod scarring ranged from about 0.2 to 4.0 scars per 100 pods and was linearly related to beetle numbers. Using this regression they calculated an EIL of 4.1 beetles per 10 sweeps. Dutcher and All[4] used multiple regression procedures to relate girdling and number of feeding sites of the grape root borer, *Vitacea polistiformis,* to berry yield of individual grape vines. Story et al.[5] sampled cutworm infestations and numbers of cut plants at differing stages of larval and plant growth in 116 fields of field corn in three states over three years. They used these data to generate a model to predict yield loss based on the number of species and stage of cutworm, plant population, and expected yield in the absence of cutworm injury. Brown et al.[6] sampled aerial and root feeding (edaphic) populations of the wooly apple aphid, *Eriosoma lanigerum,* during years of high and low aphid infestation, and related aphid population level to yield and quality of individual apple trees. In a different variation, Hutchison and Campbell[7] related sugar beet productivity to damage by the sugarbeet root aphid, *Pemphigus betae,* by comparing sugar content and yield of uninfested plants and of plants within series of aphid foci with varying levels of root damage within the same field. Thus, yield loss of plants grown under similar field conditions could be compared from a number of fields representing a range of growing conditions. Their results demonstrate the importance of environmental conditions in that the relationship between yield loss and aphid damage was substantially different between years of normal and excessive rainfall, which prevented the calculations of a unified yield loss equation and EIL. They speculated that the additional precipitation ameliorated aphid-induced stress and prevented the expression of damage in infested plants.

Observing natural infestations has the advantage of simplicity. Infestations are natural with feral populations and without bias from cages or barriers, adverse effects of insecticides, or problems arising from manual infestation techniques. Phenology and spatial distribution of injury also are natural. Disadvantages are that the range of data may not encompass the critical levels where economic damage occurs, or data from a large number of fields or plot-years may be needed to generate an appropriate range of infestations levels, injury, and yield loss. Furthermore, this approach does not allow a direct experimental distinction between injury and damage per se.

3.2.2 ESTABLISHMENT OF ARTIFICIAL POPULATIONS

Creating a gradient of artificial populations is a useful approach for studying yield loss by occasional pests where the occurrence of damaging infestations is not reliable. This approach also is useful when a precise range and level of population densities and damage are desired. The procedure involves field collecting and/or rearing the insect and manually infesting plants in plots with a known number of individuals.[8-10] Depending on the insect's mobility, manually infested populations may be confined by some type of cage or barrier, or left unconfined. Unrestrained infestations are most desirable because this avoids the added complications of disturbance caused by cages and barriers (i.e., cage effects). Cages can alter the microclimate under the cage by reducing wind and photosynthetically active radiation (light) levels, increasing relative humidity, affecting (usually increasing) ambient and soil temperatures, and possibly reducing penetration of rain. Cage screen-mesh size should be as large as possible to minimize microclimatic changes while still confining the pest and excluding predators and parasitoids or other pest species.[11] Cages typically are used for a short time, 1 to 2 weeks, thereby assuming that any cage effects are transient and not substantial over an entire season. Nevertheless, caged and uncaged controls should be included as treatments, and data from the caged control rather than the uncaged control used to generate yield loss relationships. Comparison of caged and uncaged controls provides a measure of cage effects.

The mobility of a pest may vary from crop to crop; thus pests may require confinement in one crop but not in another. For example, manual infestations of small fall armyworm, *Spodoptera frugiperda,* larvae usually remain in the whorl of infested corn plants and do not typically require confinement.[12] Some plant-to-plant movement by larvae can occur, so typically a few border rows of uninfested corn are sufficient to prevent larval movement between plots. However, fall armyworms actively move on the soil surface in a bermudagrass pasture. Jamjanya and Quisenberry[13] used metal barriers to confine fall armyworm larvae in small plots of bermudagrass at varying densities to create a range of forage yield loss. In another example, Mailloux and Bostanian[14] confined nymphs of the tarnished plant bug, *Lygus lineolaris,* on strawberry plants using plastic barriers, whereas a similar study with adults would require enclosed cages. However, if natural enemy populations are present, nonmobile arthropods may need caging to reduce pest mortality that would occur in unconfined plots.

Typically, winged adults must be confined with cages to prevent their immediate movement out of the designated plots. Cages must be sized to be practical while enclosing a large enough area to provide meaningful results. Where crops are grown as separate plants such as in orchard crops, individual plants or (for large trees) individual branches can be caged. However, for crops grown as a continuous population of plants, including most agronomic and vegetable crops, a cage must enclose a representative portion of the plant populations or crop canopy rather than individual or small groups of plants. Indeed, cage studies that enclose single plants, a stem, or another portion of plants are of limited use in developing yield–loss relationships for many crops, because crop yield is determined on the basis of plant populations and not individual plants.[15] A common cage size for agronomic crops encloses about 2 m^2 of ground area with the height varying with crop height. In a typical cage study,[16, 17] yield loss by potato leafhopper, *Empoasca fabae,* was studied in alfalfa by caging plots with cages measuring 1 m wide by 2 m long by 1 m high and covered with Saran screening. Highly mobile potato leafhopper adults were collected from nearby fields using a D-Vac vacuum insect net, sorted and counted, and introduced into cages in predetermined numbers. Leafhoppers were allowed to feed and lay eggs on alfalfa plants for 14 days, after which cages were removed and populations of nymphs were quantified and allowed to feed up to harvest. Other crops where cages have been successfully used to study yield loss by arthropod pests include soybean,[18] grain sorghum,[11, 19] barley,[20] pinto bean,[21] and oilseed rape.[22, 23]

Wingless insects such as aphids or foliage-inhabiting lepidopteran larvae usually do not require confinement to prevent their movement from plots, although border plants between plants often are necessary to minimize movement to adjacent plants.[24] Ground barriers have been successfully used to confine wingless insects that have potential for movement such as ground-inhabiting lepidopteran larvae and beetles.[25–28] Barriers typically are metal or plastic, measuring from 0.3 m to 1 m wide with a portion of the barrier buried in the ground. Open barriers have the advantage of not restricting light, rainfall, or air movement in enclosed plots. However, barriers, especially if buried in soil, can disrupt root growth, cause soil compaction, and affect water movement in soil. Indeed, an impermeable barrier may affect water runoff and movement in soil, causing an enclosed plot to fill with water after a heavy rain. When barriers are buried in the ground, plots should be large enough to minimize soil compaction and disruption of plant roots during barrier installation. For example, Buntin and Pedigo[26] enclosed plots measuring 1.5 m by 4 m with 0.5-m high aluminum barriers to study yield loss in alfalfa by larvae of the variegated cutworm, *Peridroma saucia.* Barriers also should be tall enough to prevent plants from hanging over the top and allowing insects to escape. The top edge of barriers also may need coating with an insect adhesive such as Tanglefoot®* or Fluon®** to prevent escape by climbing insects. To study strawberry yield loss by the tarnished plant bug nymphs, Mailloux and Bostanian[14] enclosed a 0.5-m row of strawberry plants with a 0.3-m-high plastic strip which had the inside upper edge coated with a film of Fluon to prevent escape of plant bug nymphs.

Other difficulties can arise because artificial infestations may not mimic the phenology, injury, and population dynamics of natural infestations. Genetic shifts in laboratory-reared populations, especially those reared for several generations, may cause reared insects to behave differently and be less injurious than populations of feral insects. Furthermore, mortality from handling and transportation and from biotic and abiotic factors may be large in cages; thus, infestations must be monitored to document changes in pest numbers without excessive disturbance of small plots. The presence of herbivores or natural enemies after cages are established but before pest infestation may require removal of these insects manually or by using a broad spectrum insecticide with short residual activity.

Many of the problems of phenology disruption and/or cage effects with artificial infestations can be avoided by infesting plots with eggs or by manipulating adult populations to achieve a gradient of egg deposition in a series of plots. Parman and Wilson[29] studied yield loss by *Philaenus spumarius* nymphs in the spring growth cycle of alfalfa by caging various numbers of adults in the fall. This created a gradient of egg deposition that produced a range of nymphal infestations the following spring which did not need confinement. Gradients of damage to corn roots by corn rootworm, *Diabrotica* spp., larvae have been artificially created by infesting plants with different densities of laboratory-reared eggs.[30-33] In these examples, artificial infestations were created without environmental disruption or affecting plant growth by avoiding the need to confine infestations during the period of pest damage.

3.2.3   MANIPULATION OF NATURAL POPULATIONS

Manipulation of natural populations is most suitable for severe or perennial pests[34] whose infestations usually cause economic damage. Four general methods for modifying pest populations are to (1) reduce populations using natural enemies, (2) enhance populations using attractant baits or trap/cover crops, (3) reduce populations

---

*Registered trademark of The Tanglefoot Co., Grand Rapids, MI.
**Registered trademark of Northern Products, Inc., Woonsocket, RI.

using plant genotypes with varying levels of pest resistance, and (4) increase or reduce populations using insecticides. Sometimes these methods are combined to create a gradient of pest densities in a single study.

Natural enemies are challenging to use as a tool for manipulating pest populations in yield loss studies because of possible delays in pest reduction and inherent variability in levels of reduction, but mainly because of the logistics of obtaining enough natural enemies at the correct time. Hartstack et al.[35] used a series of field trials to examine damage to cotton by cotton bollworm, *Helicoverpa zea,* and tobacco budworm, *Heliothis virescens,* which included 16 fields where *Trichogramma* sp. and *Chrysopa* sp. had been released to reduce numbers of bollworm/budworm eggs and small larvae. Shipp et al.[36] used a combination of insecticide treatments and repeated introductions of the predatory mite, *Amblyseius cucumeris,* or hemipteran, *Orius insidiosus,* to create different densities and levels of injury by thrips to greenhouse-grown sweet peppers.

Baits can be used to attract and enhance pest populations. Meat-and-bone meal has been effectively used to attract adult egg-laying seedcorn maggots, *Delia platura,* to enhance larval damage to germinating seeds of soybean and field bean.[37, 38] Cancelado and Radcliffe[39] interplanted alfalfa to enhance potato leafhopper populations in nearby plots of potato to study yield loss in potato. Corn seedling injury by the southern corn rootworm, *Diabrotica undecimpunctata howardi,* is enhanced following no-till planting into hairy vetch as a winter cover crop as compared with a winter wheat cover crop or fallow.[40]

The use of plant resistance as a tool for manipulating pest populations is appealing because of simplicity of use and lack of problems associated with insecticides and artificial infestations. However, there are few examples of this approach because of the difficulty of obtaining agronomically comparable genotypes with varying levels of pest resistance. Ideally, genotypes should be isolines or near isolines with and without pest resistance. A series of winter wheat lines with varying levels of resistance to the Hessian fly, *Mayetiola destructor,* was used to relate Hessian fly populations to forage yield loss in winter wheat[41] and grain yield loss of winter barley.[42] The deployment of genetically modified crops with novel insect resistance, such as genes expressing toxins derived from *Bacillus thuringiensis* (Bt), should provide a powerful tool for measuring yield loss. Yield of a crop variety expressing a high dose of Bt toxin could be compared with a susceptible isoline or near-isoline planted over a range of growing conditions and pest infestation levels, thereby providing a direct measure of yield loss to a given level of infestation in the susceptible isoline. This approaches assumes the production of the Bt toxin by a plant does not cause a yield penalty in the resistant isoline. For insects that are difficult to control with insecticides, such as the European corn borer, *Ostrinia nubilalis,* in corn, comparison of Bt and susceptible isolines may provide the first true season-long assessment of a pest's impact on crop yield.

The most common method for manipulating pest populations is the use of insecticides either to directly reduce pest numbers or enhance them by eliminating natural enemies. Because of the difficulty of control, using pesticides to manipulate pest populations and injury may be the only feasible approach to studying damage yield loss relationships for some pests. When manipulating pest populations with insecticides, care must be taken to ensure that yield differences are being caused by the target pest and are not being confounded by the control of other arthropod infestations. Methods with insecticides are serial dilution from a standard rate, variable numbers of multiple insecticide applications, selective insecticides, series of treated and untreated plots, and comparison of control of different target populations.

Serial dilutions from a standard rate of an insecticide can create a gradient of pest densities where a single application effectively controls a pest. Wilson et al.[43] examined the yield response of oats to defoliation by the cereal leaf beetle, *Oulema melanopus,* using serial dilutions of malathion, which created a gradient of larval numbers and yield loss. Hintz et al.[44] conducted six trials using serial dilutions of heptachlor to quantify the relationship between alfalfa forage yield and larval numbers of the alfalfa weevil larvae, *Hypera postica.* The amount of insecticide applied over a season also can be varied to create a gradient of pest numbers and damage where repeated applications are needed to minimize pest damage. Naranjo et al.[45] generated economic injury levels for *Bemisia tabaci* biotype B (= *B. argentifolii*) on cotton by varying the number of weekly sprays of a mixture of fenpropathrin and acephate from 0 to 15 times per season. This created a range of whitefly densities and damage from which yield loss equations could be generated.

Selective insecticides and rates also can create a gradient of pest densities. Wilson et al.[46] modified spider mite populations in cotton by using several rates of dicofol or methyl parathion to reduce mite numbers and also used permethrin to eliminate predators, thereby enhancing mite numbers. Selective insecticides also can be used to examine the interactions of injury by two or more pests. Nault and Kennedy[47] studied the single and combined effect of two insects on potato yield by selectively controlling Colorado potato beetle, *Leptinotarsa decemlineata,* using acephate and controlling European corn borer using oxamyl. Esfenvalerate was used to control both pests. They found an absence of interaction between corn borer damage and beetle defoliation on potato yield. There are numerous examples of the use of selective insecticides to control an insect pest with the most notable examples being the use of *Bacillus thuringiensis* sprays to control lepidopterans[48] and the use of pirimicarb to selectively control aphids.[49]

Another method using insecticides is to compare pest numbers in untreated plots with the difference in yield between treated and untreated plots for a series of many paired plots. This method is commonly used by plant pathologists to measure crop losses by plant pathogens.[1] The method assumes that damage can be effectively eliminated using insecticides and that a range of populations causing economic and noneconomic damage can be included. Bechinski et al.[50] examined yield loss in sugarbeet caused by the sugarbeet root maggot, *Tetanops myopaeformis,* by comparing cumulative fly catches on sticky traps with yield loss estimates from each of 34 fields. Yield loss was determined from replicated plots untreated or treated with aldicarb at 10 days after peak catch. Butts and Lamb[51] used replicated paired untreated and treated plots at different stages of flowering and pod development to demonstrate that feeding injury by lygus bugs, *Lygus* spp., reduced canola grain yield. Butts et al.[52] also used trials of paired untreated and treated plots to show that Russian wheat aphid, *Diuraphis noxia,* feeding injury in autumn reduces stand and yield of winter wheat. Buntin[53] used 76 replicated trials over nine years to determine the relationship between Hessian fly infestations and grain yield loss of winter wheat by comparing yield difference between untreated and insecticide-treated plots of Hessian fly-susceptible (50 trials) and -resistant (26 trials) winter wheat. This provided a wide range of population levels from which yield loss equations could be generated for three measures of Hessian fly populations. Walker et al.[54] studied the cumulative effect of injury by the citrus bud mite on yield of lemon trees over four years by comparing untreated trees with treated trees where mites were suppressed with an acaricide whenever mite numbers exceeded a predetermined level. They did not find a significant yield difference until the third year of treatment, thereby documenting the cumulative impact of mite injury on lemon productivity.

A final method with insecticides is to compare a series of target population levels or economic thresholds where the pest is controlled with an insecticide application whenever populations reach a predetermined target density. Crop productivity measured as yield, product quality and/or marginal economic returns of each target threshold are compared for best return.[48, 55–65] The objective of this method is to select an optimal economic threshold for managing a pest rather than to specifically establish the relationship between pest damage and yield loss. Nevertheless, crop yield can be compared with pest numbers as measured by direct counts or intensity such as insect days to quantify yield loss.[66] Evaluation of target thresholds is a common method for severe pests where economic thresholds are low, as in many vegetable crops, and multiple pesticide applications normally are applied during a season. However, Welter et al.[67] controlled mite infestations after they reached target populations to study multiple year effects of Willamette spider mite, *Eotetranychus willamettei,* feeding injury on berry yield and quality of grapes. In situations where cosmetic damage is important, studies using target thresholds may select a range of thresholds all below the damage boundary where plant physiological injury occurs and yield loss can be measured.

Regardless of methodology, the main criticism of manipulating natural infestations with pesticides is that the pesticide may alter, usually suppress, plant physiological processes.[68] Numerous pesticides affect plant physiological processes such as leaf photosynthesis, stomatal conductance, and transpiration which may adversely affect plant growth and yield.[68, 69] Adverse effects on plant health typically occur with high dosages or after frequent use such as weekly or biweekly applications.[70, 71] Although insecticide effects on plant physiology usually are transient, a single application can reduce photosynthetic rates for several weeks.[72] Potential effects of an insecticide on crop physiological process should be evaluated or documented before being used to modify pest populations and damage in yield loss studies.

3.2.4 SIMULATION OF INSECT INJURY

In this approach, laboratory-derived consumption data are used to relate injury to pest numbers, and pest injury is related to crop loss by manually removing or injuring plant tissue to mimic actual tissue injury caused by a pest.[2, 34] Because the amount of injury can be precisely controlled and injury per pest can be separated experimentally from damage per unit injury, damage simulation often provides more flexibility in investigating damage-loss relationships than other approaches. Damage simulation most often is used to study crop response to defoliators, seedling cutters, flower or pod feedings with mandibulate mouthparts that chew plant tissue, or by insects that destroy apical meristems or flower buds regardless of mouthpart morphology. Because of technical difficulties, simulation has rarely been used to mimic injury by arthropods with piercing/sucking mouthparts.

Showers et al.[73] simulated corn seedling cutting by black cutworm, *Agrotis ipsilon,* by clipping plants at ground level with scissors or by manually digging and removing plants at several stages of corn seedling development. This provided a quantitative relationship between corn yield loss and seedling damage and stand loss by black cutworm. Terminal bud destruction of cotton by cotton bollworm/budworm has been simulated by manually removing terminal buds at various intensities.[74, 75] Rogers[76] manually excised buds to simulate flower-bud abortion of guar caused by larvae of the cecidomyiid *Contarinia texana.* Williams and Free[77] and Tatchell[78] manually removed flower buds and small pods to show that oilseed rape could compensate for flower bud and pod injury by adult pollen beetles, *Meligethes aeneus,* and cabbage seedpod weevil, *Ceutorhynchus assimilis.* Soybean seed destruction and destruction of the primary apical meristem caused by the seedcorn maggot has been simulated by manually removing the growing tip using forceps.[38, 79, 80]

However, damage simulation has been most extensively used to study crop yield loss caused by insect defoliation. Daily or incremental leaf-tissue consumption usually expressed as leaf area consumed is measured for a given pest in the laboratory. Defoliation is simulated by manually removing a specific amount or percentage of leaf mass at a growth stage when defoliation is likely to occur. Leaf mass removal can be done all at one time or be removed over time at a constant rate or according to a consumption model developed from laboratory development/feeding trials. Techniques for removing leaf tissue include picking entire leaves or leaflets,[81, 82] cutting a portion of a leaf with scissors,[60, 83] or punching holes in the leaf using a cork borer or paper punch.[84, 85] For example, Hammond and Pedigo[84] removed various percentage levels of soybean leaf tissue using a cork borer to simulate defoliation by the green cloverworm, *Plathypena scabra.* A larval consumption model developed from laboratory measurements of daily leaf consumption rates[86] was used to simulate foliage removal by a hypothetical cohort of larvae consuming a total of 54 cm^2 of leaf tissue over a 12-day period. More recent insect defoliation studies of soybean have picked whole leaves according to similar daily consumption models. Buntin and Pedigo[87] simulated complete defoliation of stubble regrowth in alfalfa by the variegated cutworm for several time periods up to ten days after cutting by manually picking all new shoots at two-day intervals.

Damage simulation probably is the most controversial method for damage yield loss assessment studies because of concerns about the ability of surrogate injury techniques to accurately simulate actual injury. The principal assumption about surrogate techniques is that the loss of tissue by pests does not cause systemic or chronic effects on plant growth and physiology that cannot be simulated by mechanical injury. Insect feeding activity or saliva secreted while chewing may induce plant chemical defenses[88] or possibly change plant physiological processes in ways that are not reproducible by simulated mechanical injury. Smith[89] cites a number of examples of wound-induced plant resistance caused by both insect and mechanical injury that induced plant resistance by insect feeding which can be simulated using mechanical injury techniques. Furthermore, simulating defoliation by mechanical removal of leaf tissue raises concerns about the effect of insect and mechanical injury affecting remaining leaf tissue. Welter[90] compared defoliation with scissors by *Manduca sexta* on tomato and found no significant changes in photosynthesis per unit leaf area of remaining leaf tissue of damaged and undamaged leaves. A number of other studies[91–95, 96] also have found that defoliation by insects does not affect photosynthesis of remaining leaf tissue and by remaining leaves and thus can be simulated by mechanical leaf removal techniques. However, Peterson et al.[97] found that out of seven defoliating insects on soybean, feeding injury by six did not affect photosynthetic rates of remaining leaf tissue; however, feeding injury by Mexican bean beetle, *Epilachna varivestris,* severely reduces photosynthetic rates of remaining leaf tissue. Unlike lepidopteran and other coleopteran defoliators which chew complete holes in soybean leaves, Mexican bean beetle adults and larvae skeletonize leaf tissue by scraping and crushing it which leaves most leaf veins unconsumed. Thus, mechanical removal of leaves or portions of leaves did not adequately simulate feeding damage by Mexican bean beetles.[97] Defoliation also can affect leaf transpiration and plant water balance. Ostlie and Pedigo[98] compared lepidopteran defoliation with picking and hole punching of leaves and found simulated and natural defoliation produced transient differences in whole-plant transpiration during the first 16 hours after defoliation with total loss not being different after 48 hours. In general, most studies comparing actual and simulated defoliation support the contention that the overriding cause of yield loss by insect defoliation is the loss of leaf area for light interception and photosynthesis and that mechanical leaf tissue removal techniques can adequately simulate this type of injury.

Another potential source of error in surrogate injury techniques includes failure to mimic the spatial distribution of injury within and between plants. In potato, Shields and Wyman[99] picked leaves from the canopy top down to simulate injury by the Colorado potato beetle and also picked leaves from the canopy bottom up to simulate variegated cutworm feeding injury. They found that potato was more sensitive to top-down than bottom-up defoliation. This pattern is probably true of most annual agronomic crops grown in a stand. The importance of spatial distribution of damage between plants has received little quantitative attention. Most simulated defoliation studies impose injury uniformly across a plant canopy. However, injury by some pests is distinctly clumped or aggregated. Hughes[100] reviewed the effect of spatial pattern of damage on crop loss models and found that, in general, crop loss was more severe when injury was aggregated than when injury was random or uniform, because aggregated damage reduces the potential for compensation to injury by adjacent uninjured plants when grown in a stand.

Likewise, the temporal pattern of injury also can affect damage loss relationships. Many damage simulation studies impose injury such as defoliation in a single day.[81, 99] However, except for an insect such as migratory locusts, insects typically feed and produce injury over a number of days or weeks, depending on environmental conditions. Ostlie[101] found that soybean was less sensitive to simulated defoliation imposed during a single day as compared with the same amount of injury imposed over 12 days. He concluded that the temporal pattern of injury should mimic as closely as possible the actual temporal pattern on injury by the insect in question. Ostlie and other authors[84, 102] have used temperature-driven leaf consumption models to impose injury in the same pattern as would occur as an insect develops, although the small amount of injury caused by early instars typically is pooled and imposed on the first day. Burkness et al.[83] also found that one-time simulated defoliation of seedlings with scissors had less impact than continuous simulated defoliation on cucumber yield. Simply imposing an equal portion of the total injury daily or every few days over a period approximating the duration of insect feeding has been done as an improvement over one time (i.e., single day) simulated injury without the added complexity resulting from the use of a leaf consumption model to dictate daily defoliation amounts.[59, 60, 85, 103]

Estimations of insect consumption rates are another potential source of error. Consumption rates can be affected by plant genotype, plant growing conditions and nutritional status, stage of plant development, plant tissue age, and environmental conditions. It should be mentioned that all these factors also can influence the results of other approaches used to quantify yield loss relationships. Consumption studies are conducted in the laboratory typically with greenhouse-grown leaves. Hammond et al.[86] found that because of differences in specific leaf weight of field-grown and greenhouse-grown leaves, soybean leaf area consumption by the green cloverworm was about 54% less on leaves from the field than the greenhouse. Insect population density also may affect individual insect consumption rates.[104, 105] If affected, consumption and injury per insect usually decline as population density increases.

These examples clearly demonstrate that it is important to understand the biology and pattern of injury by the insect in question. Indeed, preliminary studies may be necessary to validate the fidelity of surrogate techniques to adequately simulate insect injury. However, surrogate injury techniques can provide a valuable tool for the study of yield loss damage relationships by insect pests.

3.3   STATISTICAL DESCRIPTION OF THE DAMAGE/LOSS RELATIONSHIP

Accurate and biologically meaningful statistical description of the relationship between pest damage and yield loss is a critical step in modeling crop loss and developing pest management decision tools. Tammes[106] originally described the theoretical and empirical basis for a generalized damage loss curve. Pedigo et al.[2] named different segments of this generalized damage loss curve and argued that all pest loss curves could be described by a portion or all of the generalized curve. However, three generalized responses most often encountered, which represent part of the Tammes generalized curve, are[34] *tolerant response,* where some injury can occur before yield declines linearly with increasing injury; *susceptive response,* where little compensation occurs and yield declines linearly with increasing injury; and *hypersusceptive response,* where yield loss is greatest at low levels of injury and incremental losses become smaller as injury increases (Figure 3.1). Typically there is an approximately linear relationship between injury and pest numbers and crop yield over a midrange of injury for all three response curves. The main difference between curves is in the response of yield to low levels of injury. Because pest management decisions often occur at low levels of injury, accurate description of the damage loss curve at low levels of injury is crucially important.

In rare instances, low levels of injury may enhance yield somewhat in tolerant plants, thereby producing an over-compensatory response.[2,107] However, there are very few examples where a true over-compensatory response has been demonstrated[44] and is not an artifact of an improper statistical model describing the damage loss curve. Most importantly, statistical functions used to describe damage yield loss relationships should make biological sense. Indeed, it is all too easy to use a mathematical model to relate damage and yield loss with a good fit of the data, but this model may not reflect biological reality.[108] One reality is that zero yield loss should occur with zero injury. If the intercept of a simple linear regression is significantly different from zero, the relationship between yield loss and insect numbers may not actually be linear. Second, some sort of compensatory yield response may be expected for injury from planting until almost the final step of a yield determination, typically seed filling or fruit maturation. Before this stage of development, plants growing in a stand usually can compensate to a certain extent for indirect injury as long as the pest activity does not involve a severe phytotoxic response to insect injury or transmission of a plant pathogen. However, entomologists often use a straight line to describe a yield loss curve where a curvilinear response showing a tolerant yield response for low levels of injury would make more sense even if the curvilinear response does not provide a better fit than a linear regression.[1,109] The use of the linear regression may obscure the actual relationship by focusing solely on the linear portion of the response curve. In addition, many of the linear responses reported in

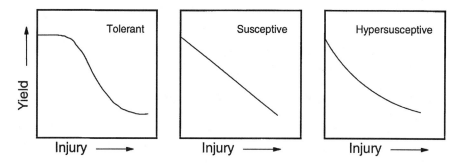

FIGURE 3.1   Three generalized crop-yield responses to insect injury (redrawn from Poston et al.[34]).

the literature most likely are actually tolerant or hypersusceptive, but the range of injury studied, amount of data collected, and variability in the data do not permit expression of the tolerant or hypersusceptive response.[34]

The quadratic regression has been used extensively to describe yield loss curves, but this model often does not describe well the yield loss relationship at low levels of injury, which is where pest management decisions often are made. Indeed, the quadratic typically produces an over-compensatory curve suggesting that yield actually increases at low levels of injury. I believe use of the quadratic equation to describe yield loss relationships accounts for some of the claims of overcompensation when in fact a tolerant response is more appropriate. The quadratic also may produce a distorted description at high levels of injury because it may curve more than the data indicate. Hopkins et al.[74] used a quadratic regression to relate cotton yield to terminal bud destruction which indicated an increase in yield at greater than 70% bud destruction. This most likely is an artifact of the quadratic regression and not a real phenomenon. Likewise, Jackai et al.[110] used a third-degree polynomial to describe the relationship between coreid bug injury and cowpea yield which showed declining yield until attaining 30 to 50 bugs per five plants where yield increased before deceasing again. This yield increase most likely is an artifact of the polynomial model rather than a true biological response of cowpea to bug damage.

Exponential, logistic, cumulative Weibull, and other functions have been used and probably better describe nonlinear response curves. Madden et al.[109] proposed a form of the Weibull function that effectively describes tolerant and hypersusceptive responses of crops to the injury by plant pathogens. The cumulative Weibull distribution function can fit most curves and is increasing in use in entomological literature. Exponential and logistic curves also may be useful in describing nonlinear response curves. A spline function or knotted regression divides a response curve into two sections and calculates a point of inflection in the curve where each section can be described by a linear regression.[46] This function is especially useful for describing tolerant response curves. Buntin[53] studied damage loss relationships of the Hessian fly in winter wheat using the approach of insecticide treated and untreated paired plots with resistant and susceptible varieties. Linear regression described the relationship between grain yield loss and percentage of infest tillers in autumn and spring; a Weibull function described the relationship between yield loss and number of immatures per stem in spring; and a knotted regression described the relationship between grain test weight and percentage of infested stems in autumn and spring. The knotted regression indicated that Hessian fly had little effect on grain test weight when infestations were below 20% infested tiller in autumn or 39% infested stems in spring. This verified that reductions in grain quality as measured by test weight occur at much higher levels of Hessian fly infestation than the levels needed to reduce grain yield.

Regardless of the mathematical model chosen, entomologists should be aware of the potential generalized response to injury when characterizing yield loss relationships. This way they may select functions that make sense biologically and therefore are useful in understanding plant crop yield loss relationships and in developing economic injury levels and decision tools for pest management.

REFERENCES

1. Madden, L. V., Measuring and modeling crop losses at the field level, *Phytopathology,* 73, 1591, 1983.
2. Pedigo, L. P., Hutchins, S. H., and Higley, L. G., Economic injury levels in theory and practice, *Annu. Rev. Entomol.,* 31, 341, 1986.
3. Weinzierl, R. A., Berry, R. E., and Fisher, G. C., Sweep-net sampling for western spotted cucumber beetle (Coleoptera: Chrysomelidae) in snap beans: spatial distribution, economic injury level, and sequential sampling plans, *J. Econ. Entomol.,* 80, 1278, 1987.
4. Dutcher, J. D., and All, J. N., Damage impact of larval feeding by the grape root borer in a commercial Concord grape vineyard, *J. Econ. Entomol.,* 72, 159, 1979.
5. Story, R. N., Keaster, A. J., Showers, W. B., Shaw, J. T., and Wright, V. L., Economic threshold dynamics on black and claybacked cutworm (Lepidoptera: Noctuidae) in field corn, *Environ. Entomol.,* 12, 1718, 1983.
6. Brown, M. W., Schmitt, J. J., Ranger, S., and Hogmire, H. W., Yield reduction in apple by edaphic woolly apple aphid (Homoptera: Aphididae) populations, *J. Econ. Entomol.,* 88, 127, 1995.
7. Hutchison, W. D., and Campbell, C. D., Economic impact of sugarbeet root aphid (Homoptera: Aphididae) on sugarbeet yield and quality in southern Minnesota, *J. Econ. Entomol.,* 87, 465, 1994.
8. Bode, W. M., and Calvin, D. D., Yield-loss relationships and economic injury levels for European corn borer (Lepidoptera: Pyralidae) populations infesting Pennsylvania field corn, *J. Econ. Entomol.,* 83, 1595, 1990.
9. Byers, R. A., and Calvin, D. D., Economic injury levels to field corn from slug (Stylommatophora: Agrolimacidae) feeding, *J. Econ. Entomol.,* 87, 1345, 1994.
10. Vail, K. M., Kok, L. T., and Lentner, M., Broccoli yield response to selected levels of cabbage looper (Lepidoptera: Noctuidae) larvae in southwestern Virginia, *J. Econ. Entomol.,* 82, 1437, 1989.
11. Breen, J. P., and Teetes, G. L., Economic injury levels for yellow sugarcane aphid (Homoptera: Aphididae) on seedling sorghum, *J. Econ. Entomol.,* 83, 1008, 1990.
12. Morrill, W. L., and Greene, G. L., Survival of fall armyworm larvae and yields of field corn after artificial infestations, *J. Econ. Entomol.,* 67, 119, 1974.
13. Jamjanya, T., and Quisenberry, S. S., Impact of fall armyworm (Lepidoptera: Noctuidae) feeding on quality and yield of coastal bermudagrass, *J. Econ. Entomol.,* 81, 922, 1988.
14. Mailloux, G., and Bostanian, N. J., Economic injury level model for tarnished plant bug, *Lygus lineolaris* (Palisot de Beauvois) (Hemiptera: Miridae), in strawberry fields, *Environ. Entomol.,* 17, 581, 1988.
15. Boote, K. J., Concepts for modeling crop response to pest damage, ASAE Paper 81-4007. St. Joseph, MI, American Society for Agricultural Engineering, 1983.
16. Hutchins, S. H., and Pedigo, L. P., Potato leafhopper-induced injury on growth and development on alfalfa, *Crop Sci.,* 29, 1005, 1989.
17. Hutchins, S. H., and Pedigo, L. P., Phenological disruption and economic consequences of injury to alfalfa induced by potato leafhopper (Homoptera: Cicadellidae), *J. Econ. Entomol.,* 83, 1587, 1990.
18. Todd, J. W., and Turnipseed, S. G., Effects of southern green stink bug damage on yield and quality of soybeans, *J. Econ. Entomol.,* 67, 412, 1974.
19. Hall, IV, D. G., and Teetes, G. L., Yield loss-density relationships of four species of panicle-feeding bugs in sorghum, *Environ. Entomol.,* 11, 738, 1982.

20. Ba-Angood, S. A., and Stewart, R. K., Economic thresholds and economic injury levels of cereal aphids on barley in southwestern Quebec, *Can. Entomol.,* 112, 759, 1980.

21. Michels, Jr., G. L., and Burkhardt, C. C., Economic threshold levels of the Mexican bean beetle on pinto beans in Wyoming, *J. Econ. Entomol.,* 74, 5, 1981.

22. Lerin, J., Assessment of yield loss in winter rape due to seed-pod weevil (*Ceutorhynchus assimilis* Payk.) II. Yield loss in a cage experiment, *Agronomie,* 4, 147, 1984.

23. Wise, I. L., and Lamb, R. J., Economic threshold for plant bugs, *Lygus* spp. (Heteroptera: Miridae), in canola, *Can. Entomol.,* 130, 825, 1998.

24. Huffman, F. R., and Mueller, A. J., Effects of beet armyworm (Lepidoptera: Noctuidae) infestation levels on soybean, *J. Econ. Entomol.,* 76, 744, 1983.

25. Showers, W. B., Kaster, L. V., and Mulder, P. G., Corn seedling growth stage and black cutworm (Lepidoptera: Noctuidae) damage, *Environ. Entomol.,* 12, 421, 1983.

26. Buntin, G. D., and Pedigo, L. P., Development of economic injury levels for last-stage variegated cutworm (Lepidoptera: Noctuidae) larvae in alfalfa stubble, *J. Econ. Entomol.,* 78, 1341, 1985.

27. Bechinski, E. J., and Hescock, R., Bioeconomics of the alfalfa snout beetle (Coleoptera: Curculionidae), *J. Econ. Entomol.,* 83, 1612, 1990.

28. Davis, P. M., and Pedigo, L. P., Yield response of corn stands to stalk borer (Lepidoptera: Noctuidae) injury imposed during early development, *J. Econ. Entomol.,* 83, 1582, 1990.

29. Parman, V. R., and Wilson, M. C., Alfalfa crop responses to feeding by the meadow spittlebug (Homoptera: Cercopidae), *J. Econ. Entomol.,* 75, 481, 1982.

30. Sutter, G. R., and Branson, T. F., A procedure for artificially infesting field plots with corn rootworm eggs, *J. Econ. Entomol.,* 73, 135, 1980.

31. Spike, B. P., and Tollefson, J. J., Relationship of plant phenology to corn yield loss resulting from western corn rootworm (Coleoptera: Chrysomelidae) larval injury, nitrogen deficiency, and high plant density, *J. Econ. Entomol.,* 82, 226, 1989.

32. Spike, B. P., and Tollefson, J. J., Yield response of corn subjected to western corn rootworm (Coleoptera: Chrysomelidae) infestation and lodging, *J. Econ. Entomol.,* 84, 1585, 1991.

33. Davis, P. M., Comparison of economic injury levels for western corn rootworm (Coleoptera: Chrysomelidae) infesting silage and grain corn, *J. Econ. Entomol.,* 87, 1086, 1994.

34. Poston, F. L., Pedigo, L. P., and Welch, S. M., Economic injury levels: Reality and practicality, *Bull. Entomol. Soc. Am.,* 29, 49, 1983.

35. Hartstack, A. W., Jr., Ridgeway, R. L., and Jones, S. L., Damage to cotton by the bollworm and tobacco budworm, *J. Econ. Entomol.,* 71, 239, 1978.

36. Shipp, J. L., Binns, M. R., Hao, X., and Wang, K., Economic injury levels for western flower thrips (Thysanoptera: Thripidae) on greenhouse sweet peppers, *J. Econ. Entomol.,* 91, 671, 1998.

37. Vea, E. V., and Eckenrode, C. J., Seed maggot injury on surviving bean seedlings influences yield, *J. Econ. Entomol.,* 69, 545, 1976.

38. Funderburk, J. E., and Pedigo, L. P., Effects of actual and simulated seedcorn maggot (Diptera: Anthomyiidae) damage on soybean growth and yield, *J. Econ. Entomol.,* 12, 323, 1983.

39. Cancelado, R. E., and Radcliffe, E. B., Action thresholds for potato leafhopper on potatoes in Minnesota, *J. Econ. Entomol.,* 72, 566, 1979.

40. Buntin, G. D., All, J. N., McCracken, D. V., and Hargrove, W. L., Cover crop and nitrogen fertility effects on southern corn rootworm (Coleoptera: Chrysomelidae) damage in corn, *J. Econ. Entomol.,,* 87, 1683, 1994.

41. Buntin, G. D., and Raymer, P. L., Hessian fly (Diptera: Cecidomyiidae) damage and forage production of winter wheat, *J. Econ. Entomol.*, 82, 301, 1989.

42. Buntin, G. D., and Raymer, P. L., Response of winter barley yield and yield components to spring infestations of the Hessian fly (Diptera: Cecidomyiidae), *J. Econ. Entomol.*, 85, 2447, 1992.

43. Wilson, M. C., Treece, R. E., and Shade, R. E., Impact of cereal leaf beetle larvae on yields of oats, *J. Econ. Entomol.*, 62, 699, 1969.

44. Hintz, T. R., Wilson, M. C., and Armbrust, E. J., Impact of alfalfa weevil larval feeding on the quality and yield of first cutting alfalfa, *J. Econ. Entomol.*, 69, 749, 1976.

45. Naranjo, S. E., Chang-Chi, C., and Henneberry, T. J., Economic injury levels for *Bemisia tabaci* (Homoptera: Aleyrodidae) in cotton: impact of crop price, control costs, and efficacy of control, *Crop Protect.*, 15, 779, 1996.

46. Wilson, L. T., Trichilo, P. J., and Gonzalez, D., Spider mite (Acari: Tetranychidae) infestation rate and initiation: effect on cotton yield, *J. Econ. Entomol.*, 84, 593, 1991.

47. Nault, B. A., and Kennedy, G. G., Evaluation of Colorado potato beetle (Coleoptera: Chrysomelidae) defoliation with concomitant European corn borer (Lepidoptera: Pyralidae) damage on potato yield, *J. Econ. Entomol.*, 89, 475, 1996.

48. Wyman, J. A., and Oatman, E. R., Yield response in broccoli plantings sprayed with *Bacillus thuringiensis* at various lepidopterous larval density treatment levels, *J. Econ. Entomol.*, 70, 821, 1977.

49. Ellis, S. A., Oakley, J. N., Parker, W. E., and Raw, K., The development of an action threshold for cabbage aphid (*Brevicoryne brassicae*) in oilseed rape in the UK, *Ann. Appl. Biol.*, 134, 153, 1999.

50. Bechinski, E. J., McNeal, C. D., and Gallian, J. J., Development of action thresholds for the sugarbeet root maggot (Diptera: Otitidae), *J. Econ. Entomol.*, 82, 608, 1989.

51. Butts, R. A., and Lamb, R. J., Pest status of lygus bugs (Hemiptera: Miridae) in oilseed *Brassica* crops, *J. Econ. Entomol.*, 84, 1591, 1991.

52. Butts, R. A., Thomas, J. B., Lukow, O., and Hill, B. D., Effect of fall infestations of Russian wheat aphid (Homoptera: Aphididae) in winter wheat yield and quality on the Canadian prairies, *J. Econ. Entomol.*, 90, 1005, 1997.

53. Buntin, G. D., Hessian fly (Diptera: Cecidomyiidae) injury and loss of winter wheat grain yield and quality, *J. Econ. Entomol.*, 92, 1190, 1999.

54. Walker, G. P., Voulgaropoulos, A. L., and Phillips, P. A., Effect of citrus bud mite (Acari: Eriophyidae) on lemon yields, *J. Econ. Entomol.*, 85, 1318, 1992.

55. Shelton, A. M., Andaloro, J. T., and Barnard, J., Effects of cabbage looper, imported cabbageworm, and diamondback moth on fresh market cabbage and processing cabbage, *J. Econ. Entomol.*, 75, 742, 1982.

56. Kirby, R. D., and Slosser, J. E., Composite economic threshold for three lepidopterous pests of cabbage, *J. Econ. Entomol.*, 77, 725, 1984.

57. Sears, M. K., Shelton, A. M., Quick, T. C., Wyman, J. A., and Webb, S. E., Evaluation of partial plant sampling procedures and corresponding action thresholds for management of Lepidoptera on cabbage, *J. Econ. Entomol.*, 78, 913, 1985.

58. Yencho, G. C., Getzin, L. W., and Long, G. E., Economic injury level, action threshold, and a yield–loss model for the pea aphid, *Acyrthosiphon pisum* (Homoptera: Aphididae), on green peas, *Pisum sativum, J. Econ. Entomol.*, 79, 1681, 1986.

59. Stewart, J. G., and Sears, M. K., Economic threshold for three species of lepidopterous larvae attacking cauliflower grown in southern Ontario, *J. Econ. Entomol.*, 81, 1726, 1988.

60. Stewart, J. G., McRae, K. B., and Sears, M. K., Response of two cultivars of cauliflower to simulated insect defoliation, *J. Econ. Entomol.,* 83, 1499, 1990.
61. Durant, J. A., Effect of treatment regime on control of *Heliothis virescens* and *Helicoverpa zea* (Lepidoptera: Noctuidae) on cotton, *J. Econ. Entomol.,* 84, 1577, 1991.
62. Archer, T. L., and Bynum, E. D., Jr., Economic injury level for Russian wheat aphid (Homoptera: Aphididae) on dryland wheat, *J. Econ. Entomol.,* 85, 987, 1992.
63. Maltais, P. M., Nuckle, J. R., and LeBlanc, P. V., Economic threshold for management of lepidopterous larvae on broccoli in southeastern New Brunswick, *J. Econ. Entomol.,* 87, 766, 1994.
64. Marini, R. P., Pfieffer, D. G., and Sowers, D. S., Influence of European red mite (Acari: Tetranychidae) and crop density on fruit size and quality and on crop value of 'Delicious' apples, *J. Econ. Entomol.,* 87, 1302, 1994.
65. Walgenbach, J. F., Effect of potato aphid (Homoptera: Aphididae) on yield, quality, and economics of staked-tomato production, *J. Econ. Entomol.,* 90, 996, 1997.
66. Fournier, F., Bovin, G., and Stewart, R. K., Effect of *Thrips tabaci* (Thysanoptera: Thripidae) on yellow onion yields and economic thresholds for its management, *J. Econ. Entomol.,* 88, 1401, 1995.
67. Welter, S. C., Freeman, R., and Farnham, D. S., Recovery of 'Zinfandel' grapevines from feeding damage by Willamette spider mite (Acari: Tetranychidae): implications for economic injury level studies in perennial crops, *Environ. Entomol.,* 20, 104, 1991.
68. Jones, V. P., Toscano, N. C., Johnson, M. W., Welter, S. C., and Youngman, R. R., Pesticide effects on plant physiology: integration into a pest management program, *Bull. Entomol. Soc. Am.,* 32, 103, 1986.
69. Ferree, D. C., Influence of pesticides on photosynthesis of crop plants, in *Photosynthesis and Plant Development,* Marcelle, R., Clijsters, H., and Van Poucke, M. Eds., W. Junk, The Hague, 1979, 331.
70. Toscano, N. C., Sances, F. V., Johnson, M. W., and LaPre, L. F., Effect of various pesticides on lettuce physiology and yield, *J. Econ. Entomol.,* 75, 738, 1982.
71. LaPre, L. F., Sances, F. V., Toscano, N. C., Oatman, E. R., Voth, V., and Johnson, M. W., The effects of acaricides on the physiology, growth, and yield of strawberries, *J. Econ. Entomol.,* 75, 616, 1982.
72. Trumble, J. T., Carson, W., Nakakihara, H., and Voth, V., Impact of pesticides for tomato fruitworm (Lepidoptera: Noctuidae) suppression on photosynthesis, yield and nontarget arthropods in strawberries, *J. Econ. Entomol.,* 81, 608, 1988.
73. Showers, W. B., Sechriest, R. E., Turpin, F. T., Mayo, Z. B., and Szatmari-Goodman, G., Simulated black cutworm damage to seedling corn, *J. Econ. Entomol.,* 72, 432, 1979.
74. Hopkins, A. R., Moore, R. F., and James, W., Economic injury levels for *Heliothis* spp. larvae on cotton plants in the four-true-leaf to pinhead-square stage, *J. Econ. Entomol.,* 75, 328, 1982.
75. Brook, K. D., Hearn, A. B., and Kelly, C. F., Response of cotton, *Gossypium hirsutum* L., to damage by insect pests in Australia: manual simulation of damage, *J. Econ. Entomol.,* 85, 1368, 1992.
76. Rogers, C. E., Economic injury level for *Contarinia texana* on Guar, *J. Econ. Entomol.* 69, 693, 1976.
77. Williams, I. H., and Free, J. B., Compensation of oil-seed rape (*Brassica napus* L.) plants after damage to their buds and pods, *J. Agric. Sci.,* 92, 53, 1978.
78. Tatchell, G. M., Compensation in spring-sown oil-seed rape (*Brassica napus* L.) plants in response to injury to their flower buds and pods, *J. Agric. Sci.,* 101, 565, 1983.

79. Higley, L. G., and Pedigo, L. P., Soybean growth responses and intraspecific competition from simulated seedcorn maggot injury, *Agron. J.,* 82, 1057, 1990.

80. Higley, L. G., and Pedigo, L. P., Soybean yield responses and intraspecific competition from simulated seedcorn maggot injury, *Agron. J.,* 83, 135, 1991.

81. Todd, J. W., and Morgan, L. W., Effects of defoliation on yield and seed weight of soybeans, *J. Econ. Entomol.,* 65, 567, 1972.

82. Peterson, R. K. D., Danielson, S. D., and Higley, L. G., Yield response of alfalfa to simulated alfalfa weevil injury and development of economic injury levels, *Agron. J.,* 85, 595, 1993.

83. Burkness, E. C., and Hutchison, W. D., Action thresholds for striped cucumber beetle (Coleoptera: Chrysomelidae) on 'Carolina' cucumber, *Crop Protect.,* 17, 331, 1998.

84. Hammond, R. B., and Pedigo, L. P., Determination of yield–loss relationships for two soybean defoliators by using simulated insect-defoliation techniques, *J. Econ. Entomol.,* 75, 102, 1982.

85. Shelton, A. M., Hoy, C. H., and Baker, P. B., Response of cabbage head weight to simulated Lepidoptera defoliation, *Entomol. Exp. Appl.,* 54, 181, 1990.

86. Hammond, R. B., Pedigo, L. P., and Poston, F. L., Green cloverworm leaf consumption on greenhouse and field soybean leaves and development of a leaf-consumption model, *J. Econ. Entomol.,* 72, 714, 1979.

87. Buntin, G. D., and Pedigo, L. P., Management of alfalfa stubble defoliators causing a complete suppression of regrowth, *J. Econ. Entomol.,* 79, 769, 1986.

88. Kogan, M., and Paxton, J., Natural inducers of plant resistance to insects, in *Plant Resistance to Insects,* Hedin, P. A., Ed., Am. Chem. Soc. Symp. Ser. 208. American Chemical Society, Washington, D.C., 1983, 153.

89. Smith, C. M., *Plant Resistance to Insects,* John Wiley & Sons, New York, 1989, chap. 7.

90. Welter, S. C., Responses of tomato to simulated and real herbivory by tobacco hornworm (Lepidoptera: Sphingidae), *Environ. Entomol.,* 20, 1537, 1991.

91. Poston, F. L., Pedigo, L. P., Pearce, R. B., and Hammond, R. B., Effects of artificial and insect defoliation on soybean net photosynthesis, *J. Econ. Entomol.,* 69, 109, 1976.

92. Welter, S. C., Arthropod impact on plant gas exchange, in *Insect–Plant Interactions, Vol. 1,* Bernays, E. A., Ed., CRC Press, Boca Raton, 1989, 135.

93. Higley, L. G., New understandings of soybean defoliation and their implications for pest management, in *Pest Management of Soybean,* Copping, L. G., Green, M. B., and Rees, R. T., Eds. Elsevier, London, 1992, 56.

94. Peterson, R. K. D., Danielson, S. D., and Higley, L. G., Photosynthetic responses of alfalfa to actual and simulated alfalfa weevil (Coleoptera: Curculionidae) injury, *Environ. Entomol.,* 21, 501, 1992.

95. Peterson, R. K. D., Higley, L. G., and Spomer, S. M., Injury by *Hyalophora cercropia* (Lepidoptera: Saturnidae) and photosynthetic responses of apple and crabapple, *Environ. Entomol.,* 25, 416, 1996.

96. Burkness, E. C., Hutchison, W. D., and Higley, L. G., Photosynthesis response of 'Carolina' cucumber to simulated and actual striped cucumber beetle (Coleoptera: Chrysomelidae) defoliation, *Entomologia Sinica,* 6, 29, 1999.

97. Peterson, R. K. D., Higley, L. G., Haile, F. J., and Barrigossi, J. A. F., Mexican bean beetle (Coleoptera: Coccinellidae) injury affects photosynthesis of *Glycine max* and *Phaseolus vulgaris, Environ. Entomol.,* 27, 373, 1998.

98. Ostlie, K. R., and Pedigo, L. P., Water loss from soybeans after simulated and actual defoliation, *Environ. Entomol.,* 13, 1675, 1984.

99. Shields, E. J., and Wyman, J. A., Effect of defoliation at specific growth stages on potato yield, *J. Econ. Entomol.,* 77, 1194, 1984.

100. Hughes, G., Incorporating spatial pattern of harmful organisms into crop loss models, *Crop Protect.,* 15, 407, 1996.

101. Ostlie, K. R., Soybean Transpiration, Vegetative Morphology, and Yield Components following Simulated and Actual Insect Injury. Ph.D. dissertation, Iowa State University, Ames, 1984.

102. Higgins, R. A., Pedigo, L. P., Staniforth, D. W., and Anderson, I. C., Partial growth analysis of soybeans stressed by simulated green cloverworm defoliation and velvetleaf competition, *Crop Sci.,* 24, 289, 1984.

103. Ramachandran, S., Buntin, G. D., and All, J. N., Response of canola to simulated insect defoliation at different growth stages, *Can. J. Plant Sci.,* 80, 2000.

104. Pedigo, L. P., Hammond, R. D., and Poston, F. L., Effects of green cloverworm larval intensity on consumption of soybean leaf tissue, *J. Econ. Entomol.,* 70, 159, 1977.

105. Bellows, T. S., Jr., Owens, J. C., and Huddleston, E. W., Model for simulating consumption and economic injury level for the range caterpillar (Lepidoptera: Saturniidae), *J. Econ. Entomol.,* 76, 1231, 1983.

106. Tammes, P. M. L., Studies of yield losses. II. Injury as a limiting factor of yield, *Tijdschr. Plantenziekten,* 67, 257, 1961.

107. Bardner, R., and Fletcher, K. E., Insect infestations and their effects on the growth and yield of field crops: a review, *Bull. Entomol. Res.,* 64, 141, 1974.

108. Higley, L. G., and Peterson, R. K. D., The biological basis of the EIL, in *Economic Thresholds for Integrated Pest Management,* Higley, L. G. and Pedigo, L. P., Eds., University of Nebraska Press, Lincoln, 1996, chap. 3.

109. Madden, L. V., Pennypacker, S-P., Antle, C. E., and Kingsolver, C. H., A loss model for crops, *Phytopathology,* 71, 685, 1981.

110. Jackai, L. E., Atropo, P. K., and Odebiyi, J. A., Use of the response of the coreid bug, *Clavigralla tomentosicollis* (Stål) to determine action threshold levels, *Crop Protect.,* 8, 422, 1989.

4 Yield Loss of Field Corn from Insects

Michael D. Culy

CONTENTS

0-8493-1145-4/01/$0.00+$.50

4.1 INTRODUCTION

Within any production system, actual crop yields are only a fraction of the maximum yields possible. Yield losses represented as the difference between maximum (potential) yields and actual yields are ascribed to various identified and unidentified environmental stresses.[1] These environmental stresses are represented by numerous abiotic and biotic factors. Such factors, and the accompanying stresses, are common occurrences within production systems, with essentially all crops being grown under some level, and complex, of environmental stresses. Natural environments are continuously suboptimal with respect to one or more environmental parameters, such as water or nutrient availability.[2] The impacts of such stresses on plant growth and crop yields are considerable. Indeed, stresses are estimated to limit overall productivity of U.S. agriculture to as little as 25% of its potential.[3]

Pest infestations, including insects and mites, nematodes, plant pathogens, and weeds, take a regular toll of crop yields, and the importance of quantitatively assessing losses associated with their presence has long been recognized.[1, 4] However, yield losses attributed to pests account for only a portion of total losses attributed to environmental stresses, and losses from insect and mite infestations specifically comprise an even smaller subset of total yield losses.

This does not mean that crop losses attributed to insects and mites are insignificant. Global corn production losses due to arthropod pests have been minimally estimated at 12%, with upside potential losses being much higher.[1] More precise estimates for corn, as well as other crops, have been difficult to establish because relationships between pest injury and host response are not well defined. This is largely because plant–arthropod relationships are strongly influenced by other environmental factors.[5, 6]

These "other environmental factors" refer to all other stressors acting upon both pest and crop biological systems. Environmental conditions will likely influence pest numbers (and/or behavior), the plant responses to pest attack, or both.[5] Because insect pests and crop plants are independent biological systems, the external stressors (abiotic or biotic) will have unique influences or effects on each—weakening or strengthening one biological system's position relative to the other. In spite of the inherent difficulties involved with quantifying individual insect stressors, much attention has been given to evaluation and characterization of crop yield losses due to insect injury.

4.2 YIELD DETERMINATION IN CORN: INFLUENCE OF ABIOTIC AND BIOTIC COMPONENTS

4.2.1 PRIMARY FACTORS INFLUENCING CORN YIELD POTENTIAL

It can be argued that three main factors (in addition to weather and latitude) determine the maximum yield potential for all crops. The hybrid or variety selected, the calendar date on which the crop is planted, and the soil type(s) present in the field all become fixed factors with a unique formula for determining maximum potential yields once the seeds are planted. With these key factors established, all other abiotic and biotic stressors work against this formula for potential yield and ultimately determine the actual yields observed at season's end. Because of the importance of these factors in overall corn production, their impact will be discussed briefly.

4.2.1.1 Maturity Group and Hybrid Type

According to the CERES–Maize corn growth model, hybrid selection is the most critical determinant of corn yields.[7] Genetically, a corn hybrid does not change, but its performance will vary in different environments as determined by seasonal and/or geographic alteration. This variability in relative performance is referred to as a *hybrid-x-environment interaction,* and points to the importance of proper hybrid selection.[8] Yields and harvest indices are greatly influenced by the corn genome.[1]

Physiological maturity of corn hybrids is a genetic characteristic generally defined as the period from germination to when the kernel ceases to increase in weight. Corn hybrids can vary widely in maturity (commonly 90 to 150 days) and have been adapted for production in a range of latitudes, thereby optimizing corn yield potentials with length of growing season. Maturity selection also can be used as a management tool to compensate for delayed planting dates, ensuring adequate grain-fill before freezing temperatures occur in colder climates. However, this flexibility in establishing maturity timelines also allows for wide variation in the "windows of susceptibility" relative to key crop pests, enhancing or diminishing their impacts relative to timing of events for the two biological systems.

Corn differs in its sensitivity to stresses at different growth stages. From germination to maturity, this sensitivity has the potential to modify partitioning of dry matter to the harvestable yield components.[1] Corn is a determinant crop and develops and flowers at a given time. Plant responses to pest injury can vary greatly depending on the developmental stage at the time of injury. In general, greater reductions in yield potential occur with injury inflicted during the reproductive stages of plant growth. A failed reproductive event due to stress cannot be corrected later in a determinant crop as it might be in indeterminant crops that continue to flower over extended periods.[1] After anthesis (pollination) in corn, most dry matter accumulation is diverted to the grain. Stresses during the time of grain fill obviously will cause reductions in corn yield.[1]

4.2.1.2 Planting Date

As mentioned above, planting date, once established, becomes a fixed factor which influences maximum potential yields in corn. This factor, combined with hybrid maturity characteristics, largely determines the plant growth stage present during pest attack. This is especially true with insect pests that migrate into growing areas or emerge and cause damage based on their physiological development as driven by temperature. For corn production in the midwestern U.S., this includes the majority of common arthropod pests.

Plant attractiveness and/or susceptibility to the pest based on planting date will vary widely. In areas where planting of a portion of the crop is delayed, due to weather or other circumstances, side-by-side plantings (with widely separated development) may experience dramatically different levels and degrees of pest attack and injury.

4.2.1.3 Soil Type

Although there are numerous agronomic factors associated with soil type that affect corn growth and development, these will not be discussed. In general, the physical attributes of soil that determine nutrient availability and suitability as a growth medium for corn are covered in a multitude of agronomy textbooks. Many of the deficiencies associated with soil nutrients and pH, which can influence corn development, can be modified artificially with proper fertilization and soil conditioning.

With regard to arthropod pest injury to corn, soil type probably plays two key roles. This is particularly true if it is assumed that proper plant nutrition and seed-bed preparation have been accomplished. These two roles include:

1. The condition of soil texture and its suitability for survival of soil pests. Nematode and soil arthropod survival varies widely with availability of soil pore space, abrasive nature of soil particles, available moisture, and other physical attributes associated with soil type. These characteristics may predispose certain soil types to aversion or habitation by one or more corn pests.
2. The water-holding capacity of soil and its impact on potential for water stress conditions in corn. The CERES–Maize corn growth model utilizes soil type information for the sole purpose establishing coefficients for water-holding capacity.[7] The model recognizes that corn yield potential is greatly influenced by the likelihood of water-stress conditions. Soil type is

one of the primary factors (rainfall being the other) which determines the likelihood of plant water stress. The effects of water stress will be discussed in more detail later in the chapter.

4.2.2 ABIOTIC AND BIOTIC STRESS-INDUCING FACTORS THAT INFLUENCE CORN GROWTH

There are numerous abiotic factors that influence plant growth. Each category holds the potential for a multitude of stress-producing scenarios that influence yield independently or interactively. These factors are discussed because of their importance for both singular and interactive influences on the primary determinants of corn yield—hybrid, planting date, and soil type. In addition, many of the plant stresses associated with these factors are markedly similar to stresses caused by insect injury to corn. A better understanding of the effects of insect injury on corn can be attained through a basic understanding of the stresses caused by these abiotic and biotic factors.

4.2.2.1 Weather and Other Edaphic Factors

Productivity of a corn crop is greatly influenced by the seasonal growing environment.[1] Weather is the most uncertain factor in farming. By itself, weather can make the difference between failure and a bumper crop.[9] In grass crops such as corn, environmental stresses during the time of rapid vegetative growth to achieve maximum photosynthetic rates may delay leaf expansion, reduce photosynthesis, reduce water availability through surface evaporation rather than transpiration, and influence flower initiation—all factors that can reduce corn yields.[1]

Adverse effects of weather come in a multitude of forms. Lack of water, too much water, temperatures too cool, temperatures too warm, impacts on nutrient uptake and utilization, and physical damage from strong winds, heavy rains, or hail, alone or in combination, can negatively (or positively) affect crop yields.

4.2.2.1.1 Moisture stress

Stress may result from too much or too little water being available for the corn plant. However, a shortage of plant water is by far the most frequently occurring and detrimental stressor. When the probabilities for water-stressed conditions are high, yield predictions (per the CERES–Maize corn growth model) decrease sharply.[7] Dry matter accumulation is somehow closely related to the amount of water transpired by the plant, with less dry matter assimilation (including grain-fill) observed when transpiration is reduced.[1] It is known that dry weather conditions result in water stress, restrict root growth, and reduce or prevent adequate nutrient uptake. Corn leaf phosphorus and potassium levels are often reduced, even with high fertility programs and/or fertilizer additions—likely contributing to lower yields in dry years.[9]

Plants under water stress may be less able to compensate for pest injury than plants that are fully hydrated. Additionally, plants suffering from water stress may become more (or less) attractive to arthropod pests. More specific impacts of water stress on pest damage will be discussed on a pest-by-pest basis later in the chapter.

4.2.2.1.2   Temperature stress

Plant stresses also can be induced by temperature extremes. Cold conditions generally slow both plant and arthropod growth and development. However, the degree to which activity is reduced in the plant and the pest is often not equal—providing the pest with a differential advantage relative to inflicting injury. Cold conditions may also alter the physiological or chemical make-up of plants—causing them to be less or more attractive to arthropod pests.

Stresses from temperatures that are too high may also adversely affect corn growth. Heat stress or heat shock (anoxia) has been shown to induce the appearance of otherwise unexpressed plant proteins, and extreme temperatures have been shown to modify the frequency of DNA transposition.[1] These chemical changes within the plant will also alter attractiveness to arthropod pests and plant susceptibility to injury.

High temperature conditions can have a significant negative impact on the success of corn pollination. Under extremely hot conditions, both pollen and silk viability is reduced, hindering complete fertilization of ears.

4.2.2.1.3   Fertility stress

Well-managed soil fertility programs do much to ease weather-induced stresses in corn production systems, reducing yield losses to some extent.[9] Likewise, yield losses from insect injury have been lessened under conditions of optimal (nitrogen) fertility.[10] Soil fertility levels and corresponding plant health can also serve to make corn more (or less) attractive to insect pests and subsequent pest injury. For example, improved soil fertility has been observed to favor oviposition by the stem borer, *Busseola fusca,* a noctuid moth, in corn.[11] Likewise, higher levels of organic matter were found to positively correlate with higher oviposition by *B. fusca* and other borers.[11]

Even with these findings, it is not the intention of this chapter to provide detailed insight and/or justification for maintaining effective fertility programs in corn. It is apparent that some arthropod pests are more likely to prefer corn plants with one or more nutrient deficiencies while others will likely prefer plants that are nutrient-rich. The point, however, is moot, when one considers that proper fertility management is likely to be prerequisite to most (if not all) corn pest management programs. Therefore, the assumption will be that proper fertility programs are in place within corn production systems, and that insect injuries and subsequent plant responses are assessed on nutritionally healthy plants.

4.2.2.1.4   Plant competition stress

Competition from neighboring plants, whether other corn plants or weeds, can create stresses in corn. Seeding rates and resulting plant populations that are excessive for a given hybrid, row spacing, and/or soil type can reduce plant vigor and productivity, forcing undue competition for moisture, light, and soil nutrients. These same factors also come into play for weed competition under less-than-effective herbicide programs.

As with soil fertility issues, it is not the intention of this chapter to espouse the benefits of optimal plant competition in corn production systems to any detail. It is assumed that planting/seeding rates utilized are within recommended populations for the hybrid, production practices utilized, and the soil types involved. It is likewise assumed that effective weed control is accomplished within the field to remove significant plant competition.

However, noncompetitive weed populations, such as infestations in field borders and waterways, have been known to provide some advantages relative to insect management. These benefits are similar to those observed for trap crops—reducing pest impacts within the field by diverting some of the pest population away from the targeted crop. This has been observed for stem borers, where wild grasses bordering fields were preferred by some moths for oviposition sites, resulting in inverse relationships in stem borer egg laying in fields where weeds were present peripherally.[11] This is an example of desired "plant competition" whereby plants outside of the field are competing as hosts for the insect pest.

4.2.2.1.5   Physical damage

Physical plant damage caused by adverse weather conditions can have deleterious effects on corn development similar to those observed for arthropod injury. Damage from strong winds, heavy rains, or hail has been known to significantly reduce corn yields, kernel weight, grain test weight, and shelling percentages.[1] As mentioned in previous discussions for hybrid selection and planting date issues, the corn growth stage present at the time of injury (whether pest injury or otherwise) plays a key role in the extent to which yield losses will occur.[1, 12]

With hail injury, grain yields have been observed to decrease with increasing severity of defoliation, and with later corn growth stages (nearer tassel stage).[12, 13] Defoliation of nearly 100% before the 7-leaf stage (V7) has not been observed to significantly reduce grain yields, while similar defoliation levels at tassel stage (V17–R1) were found to routinely result in total crop loss.[13] In addition, defoliation before tassel stage was found to slightly delay corn maturity, while defoliation after tassel stage appeared to hasten maturity.[12]

Wind damage to corn, which causes plant lodging, is known to have particularly severe consequences when plants are in the mid- to late-vegetative stages at time of injury when they have not yet developed sufficient brace roots to hold plants upright. As with hail injury, the greatest yield decreases (13 to 31%) resulting from plant lodging (due to wind) have been observed with injury occurring at the tassel stage (V17–R1). Slightly less injury has been observed with V13–V15 plants (5 to 15% yield reductions).[14] Unlike hail injury, plant lodging due to wind does not appear to affect subsequent timing of plant development.[14] It is assumed that developmental effects due to physical damage are driven by defoliation.

4.3   INSECT INJURY AND STRESS DEVELOPMENT IN CORN

Two important types of arthropod injury have been recognized based on the duration and magnitude of damage to the plant (immediacy of impact): acute injury and chronic injury. Acute injury results in immediate, noticeable damage (tissue removal, entrance holes, etc.), while chronic injury produces noticeable damage over longer periods of time (leaf chlorosis, plant malformation, stunting, etc.).[5]

Additionally, arthropod injury has been placed into several groupings based on physiological impact (response) to the plant (host). Pests producing similar physiological responses have been grouped into like injury types.[5] Higley and Peterson[5] categorized pests into ten groupings based on physiological impact on plants: population and stand reduction, leaf mass reduction, leaf photosynthetic rate reduction, leaf senescence alteration, light reduction, assimilate removal, water-balance disruption, seed or fruit destruction, architecture modification, and phenological disruption.

Arthropods injuring corn can be placed into the categories mentioned above based on physiological responses produced in the corn plant. However, I believe that a further simplified list of categories is appropriate for a discussion of key corn pests. The five categories to be used for purposes of this chapter will include: foliar feeding injury (tissue removal), vascular feeding injury (sap/water/nutrient removal), vascular disruption injury (internal tunneling/tissue removal), root feeding injury (root pruning/tissue removal), and reproductive disruption injury (pollination disruption, seed damage/removal).

4.3.1 FOLIAR FEEDING INJURY

Physical injury that removes foliar tissue from above-ground portions of the corn plant can have serious impacts on plant development and subsequent yields. Vigor and yield reductions result from reduced photosynthetic capacities within injured plants. With extensive tissue removal, injured plants may develop abnormally, producing a barren stalk, or die, reducing overall plant stands.

Foliar feeding injury typically occurs in two ways: (a) a partial removal of leaf tissues from leaf margins and whorl areas of the plant, appearing as ragged edges or holes in exposed leaf surfaces; and (b) a total removal of all leaf tissues above the soil line that requires complete regrowth of the photosynthetic portions of the plant.

In many ways, this type of arthropod feeding injury is similar to some physical injuries caused by weather events such as hail. With either source of injury, arthropod or hail, moderate injury levels incurred before V7 stage would generally have little or no impact on yield. Likewise, with injuries incurred in older plants, nearer tassel stage (V17–R1), yield reductions would be more likely, regardless of how injuries were sustained. Also, injuries incurred before tassel stage might result in slight delays in corn maturity, while injuries incurred after tassel stage would likely hasten maturity (as observed with hail injury).

4.3.1.1 Cutworms

Cutworms constitute a large group of serious, yet sporadic foliage-feeding lepidopteran pests in North American corn production. These insects attack corn in early growth stages when stand establishment is critical and plants are inherently more vulnerable. Cutworms can be grouped into three types based on the injuries they produce: the cutting species (such as black cutworm, *Agrotis ipsilon,* and claybacked cutworm, *Agrotis gladiaria*), the surface feeding species (such as dingy cutworm, *Feltia ducens,* bristly cutworm, *Lacinipolia renigera,* and sandhill cutworm, *Euxoa detersa*), and the climbing species (such as variegated cutworm, *Peridroma saucia* and spotted cutworm, *Xestia* spp.).[15]

Although all cutworms can be injurious to corn, those causing simple leaf injury are normally considered to be less injurious than those with feeding habits that include plant cutting. This is evident from established economic injury levels for cutworms. Thresholds for surface-feeding cutworms are typically higher than thresholds for cutworms with cutting behaviors.[15] In general, corn leaf feeding from cutworms is considered to be insignificant, with no evident effects demonstrated on yield.[16]

Cutworms that consume large amounts of leaf foliage (e.g., late stage surface feeders and climbing species) and cutworms with cutting behaviors are of significant economic importance in corn production.[15] These species have the potential to remove a majority, or all, of the above-ground foliage from young corn plants, dramatically affecting plant growth and development. The cutting behavior removes all leaf tissue from smaller corn plants by larval chewing injury that produces notches or complete cuts through plant stems near or below the soil surface. Large surface feeders and climbing species can produce a similar result through the exaggerated effects of grazing. Regardless of how the described injury occurs, impacts on plant health and yield are the same. On smaller plants (V1 to V5) damage can be extensive enough to cause wilting and death, particularly when injury occurs to the plant growing point. This leads to plant population reductions and reduced yield.

Researchers have studied the effects of cutworm injury on corn grain yields under Midwest corn-growing conditions. Collectively, these results have indicated that yield reductions are dependent upon the stage of plant development at the time of cutting injury and the location of the cutting injury on the plant.[17–20] Plants in the V2 to V5 stage are generally most vulnerable to cutworm injury, with the window for potentially serious damage ending at about the V7 stage, when corn stalks become too large to be cut. Cutting injury that occurs below the soil surface, a behavior of larger larvae, is usually more serious than cutting injury that occurs above the soil line.[15] Researchers have determined that relationships between cutworm injury (cut plants) and yield losses are linear, with greater yield losses occurring when older plants are damaged (V5 vs. V3 plants). Yield suppression is variable, with measured losses observed to range from 0 to 24% and 0 to 81% in V3 and V5 stage plants, respectively.[21]

4.3.1.2 Armyworm and Fall Armyworm

The armyworm, *Pseudaletia unipuncta,* and fall armyworm, *Spodoptera frugiperda,* are two lepidopteran pests which can occasionally inflict severe injury to above-ground foliar portions of the corn plant. In the Midwest, these pests are typically more injurious to corn in the mid- to late-vegetative or early reproductive stages, due to seasonal temperatures, migratory behaviors, and insect life cycles.[22] In the southern U.S., these insects may also attack younger whorl-stage corn.

Armyworm infestations typically develop in nearby grass pastures, fence rows, roadsides, or small grain fields with insects migrating into corn fields in late spring as food supplies dwindle or other host plants begin to mature. Because of this migratory behavior, armyworm injury often appears first at field edges. Exceptions can occur in reduced-till corn fields where cover crops or weedy conditions may provide attractive food sources for armyworm egg laying, resulting in a more even distribution of armyworm damage throughout the field. Armyworm injury gives the corn plant a ragged appearance, because leaf margins and whorl areas are consumed. With severe armyworm injury, most of the leaf area may be consumed.

Fall armyworm infestations typically develop from midsummer through harvest. Moths are attracted to late-maturing, late-planted corn fields for egg laying, distributing their eggs throughout the field. Injury from fall armyworm feeding appears as ragged holes in leaves, rather than ragged edges on leaf margins. In addition, fall armyworms may feed on tassels and corn ears later in the season.

Foliar injury to corn from both armyworm and fall armyworm consists of simple leaf tissue removal. The effects of foliar injury from fall armyworm infestations in corn are cumulative for vegetative growth stages with longer periods of injury having greater impacts on yield.[23] Corn plants are able to compensate for foliar injury incurred over short periods of time. Plant responses to feeding injury appear to be linear, with no one time period during vegetative growth being more critical in causing yield reductions than any other.[23, 24] Corn yield reductions of 7 to 45% have been observed as a result of 100% fall armyworm infestations during periods of vegetative growth.[23–25] Yield reductions from fall armyworm leaf feeding injury incurred nearer tassel stage were not observed to be significant.[25]

4.3.1.3 Grasshoppers and Other General Foliage Feeders

Grasshoppers, *Melanoplus* spp., and many other general foliage feeders that attack corn produce injury similar to that described for armyworm—removing leaf tissue from above-ground portions of plants. Injury from grasshoppers is often more severe in years when adverse conditions reduce natural vegetation and force them into cultivated crops. In addition, dry conditions can reduce the natural mortality of grasshopper eggs, increasing resident populations.[22]

Grasshopper injury can be devastating. They will feed on corn in many growth stages, but prefer the reproductive stages. They are known to attack all above-ground parts of the corn plant, including leaves, silks, and ear tips, potentially reducing yield through both indirect (reductions in plant vigor, pollination disruption) and direct (ear and seed damage) losses. When grasshopper populations are high, they may consume all above-ground plant parts except leaf mid-ribs, pruned ears, and stalks, giving the field the appearance of having suffered severe hail injury.[22]

4.3.2 Vascular Feeding Injury

Vascular feeding injury to corn is initiated by arthropod and nematode pests with piercing or sucking mouthparts. They remove sugars and other nutrients from the phloem and other vascular elements within leaves, stems, and roots, often resulting in significant damage to the crop.[26] Corn plant development and subsequent grain yields can be negatively impacted through general reductions in plant vigor by these pest-induced stresses. With higher pest populations, water and nutrient deficiencies within the plants may become evident. Slowed growth, wilting, and plant discoloration may become obvious, while external injuries go relatively unnoticed. Symptoms may mimic those observed for some plant diseases, herbicide injuries or negligent fertility programs.

In addition, vascular feeding injury may include the effects of arthropod or nematode toxins deposited into plants during feeding. While actively feeding, many sucking pests inject enzymes into the host plant to assist in digestion of plant tissues. These enzymes are toxins and when injected into plants, often result in partial destruction of plant tissues or plant deformities, such as irregular twisting or growth of stems and leaves. These secondary plant responses to the arthropod feeding often have more devastating impacts on development and yield than does the removal of resources.

4.3.2.1 Corn Leaf Aphid

The corn leaf aphid, *Rhopalosiphum maidis,* is a common pest of corn and numerous grass weeds in corn-growing regions. The insect removes plant fluids from the phloem tissues of whorl corn leaves. Corn leaf aphid rarely occurs in densities large enough to result in physiological yield losses. However, when heavy infestations occur, corn leaves may wilt, curl, and/or develop patches of yellow discoloration (chlorosis).[22]

Corn leaf aphids also deposit a sticky substance called "honeydew" on host plants (as a part of feeding). This honeydew often fosters the growth of molds, giving the top leaves and tassels a black sooty appearance. With excessive honeydew, tassels can become covered to the extent that anthesis is impeded, resulting in varying degrees of plant barrenness.[26] Although aphid outbreaks and excessive honeydew can hinder pollen-shed in commercial fields, the phenomenon is primarily of concern in seed production fields where pollen levels are already reduced by detasseling practices.

Aphid numbers necessary to cause yield reductions in corn will vary based on the interval between pest infestation and plant development to the tassel stage. The degree of environmental stress (water stress) under which the corn crop is being grown will also play a role in the amount of aphid injury that can be tolerated. Generally, fewer aphids are required to justify control treatments when the interval between infestation and tassel stage is lengthened (15 to 30 aphids per plant). Likewise, fewer aphids are required to justify control measures when plants are experiencing stressed conditions (10 to 15 aphids per plant).[22]

4.3.2.2 Chinch Bugs and Stink Bugs

Hemipteran pests of corn can have devastating effects on yield when sufficiently high numbers are present. Stink bugs, *Euschistus* spp., *Nezara viridula,* and chinch bugs,

Blissus leucopterus leucopterus, are known to remove plant fluids from vascular tissues at the base of corn plants. Injury from these insects can range from temporary reductions in plant vigor to stunted corn growth and serious plant malformation.[22, 27–29]

Chinch bugs feed on a wide range of grass species including wheat, barley, rye, oats, corn, sorghum, and numerous weeds. Among preferred hosts, corn often figures prominently in pest population build-ups. Chinch bug levels are correlated with meteorological conditions, with economically important populations being associated with above-normal temperatures and below-normal rainfall.[26] Chinch bug injury to corn can occur from spring brood infestations on early corn plantings, but more commonly occurs when small grains (which serve as host for a majority of spring populations) begin to mature and chinch bug nymphs move into neighboring corn fields. A second generation of chinch bugs will then develop within the corn crop.

Chinch bugs feed in large numbers at the base of corn plants, obtaining food and water from the phloem elements. As a result of chinch bug feeding, vascular bundles become clogged, thereby restricting water and nutrient transport within the plant.[26] Injury symptoms associated with chinch bug feeding range from reductions in plant height (stunting) to the presence of curved or twisted stems, then severe leaf malformation and dead leaves.[29] Younger plants (V2 stage) are more susceptible to chinch bug injury than are older plants (V5 stage), which require greater chinch bug numbers to sustain serious injury.[29, 30] Following chinch bug feeding at the V2 and V5 growth stages, ear weight and ear length decreases with increasing insect numbers. Infestations ranging from 2 to 20 chinch bugs per plant elicited yield responses.[29]

Several stink bug species are also known to cause injury to corn. The brown stink bugs, *Euschistus servus* and *E. servus euschistoides,* the onespotted stink bug, *Euschistus variolarius,* and the southern green stink bug, *Nezara viridula,* have been observed feeding on most corn growth stages.[22, 27, 28] Like chinch bugs, many adult and nymphal stink bugs move into corn fields from small grains or other host plants upon harvest, removal by herbicide use, mowing, or other means.

Stink bugs feed in the vascular elements of the plant by puncturing plant tissues and removing water and nutrients. The volumes of plant sap removed by stink bug feeding are generally of little consequence, even with heavy infestations. During the feeding process, these insects also inject enzymes into the plant that assist in digestion of tissue.[22] It is the presence of these enzyme toxins (and direct injury from puncturing mouthparts) that produces the majority of corn injury. Plant tissues at or near the feeding puncture often undergo partial or complete destruction, producing holes that are ringed with yellow or brown tissue. Puncture injury also may appear as elongate holes in expanding leaves that have unrolled after injury is incurred.[22] In addition, plant deformities including irregular twisting or growth of stalks and leaves, and excessive tillering, may result from enzyme toxins introduced through stink bug feeding. Plant death also may occur from heavy feeding on smaller plants.[28]

Injury to corn from brown or onespotted stink bug feeding is most significant in younger (V2) plants.[22, 28] Severe wilting of corn seedlings can occur when eight stink bugs per plant are confined to plants for two days.[27] Four or more stink bugs per plant for a period of 13 days resulted in high plant mortality.[27] On smaller plants, aside from mortality, excessive plant tillering is the most apparent response from stink bug feeding, with 39 to 52% of plants exposed to stink bugs (one stink bug per plant) producing tillers.[28] Due to excessive tillering, silk development was delayed and mean extended leaf height and grain weights per ear were reduced in stink bug injured plants.[28] Significant plant injury also has been observed for corn injured in late vegetative and early reproductive stages (V15 to R2).[27] In older corn plants (V15), two adult southern green stink bugs per plant were sufficient to cause reductions in ear weight and length.[27] Overall yield decreases resulting from southern green stink bug feeding on larger corn were attributed to production of fewer ears rather than reduction in kernel weight.[27] Ear deformity from stink bug feeding was also observed, but only when insects were intentionally confined to ears for extended periods.[27]

4.3.2.3   Nematodes

A variety of species of plant parasitic nematodes can be found in every acre of farmland soil. Nematodes are microscopic roundworms that feed on plant roots by withdrawing the liquid contents from individual plant cells. In sufficient numbers, feeding by plant parasitic nematodes can cause severe injury to corn. Symptoms of nematode feeding injury can include yellowing or distortion of foliage, stunting, and/or wilting of plants, resembling symptoms brought on by low fertility, drought conditions, or soil compaction.[22]

Plant parasitic nematodes are capable of injecting enzymes and other substances into the plant cells. These substances may be toxic, killing the cells, altering their growth, or predisposing the plants to injury by other agents. In corn, common plant responses to toxic enzymes introduced by nematodes include: (1) suppressed or clubbed roots from which fine feeder roots arise, giving the appearance of a witches broom (dagger nematodes, *Xiphinema americanum,* and needle nematodes, *Longidorus breviannalatus*), and (2) brown streaks of discoloration on roots (lesion nematodes, *Pratylenchus hexincisus*).[22] In addition, feeding wounds left by nematodes may provide portals for bacterial and fungal pathogens not otherwise able to enter corn roots.

Attack from plant parasitic nematodes can occur throughout the growing season, and rescue treatments currently do not exist for nematode pests. Combinations of effective crop rotation (to a non-host crop) and effective utilization of soil nematicides at planting are most appropriate under heavy infestations. Once plant parasitic nematodes are confirmed as the causal agent for symptoms observed in the field, management tactics are likely warranted. Nematode populations at numbers too low to cause visual responses (but possibly causing subtle yield depressions) in the corn crop are normally neither detected nor treated.

4.3.2.4   Spider Mites

Spider mite feeding can have serious effects on corn. This is particularly true when environmental conditions such as hot, dry weather or soil compaction enhance plant stress. The twospotted spider mite, *Tetranychus urticae,* and Banks grass mite, *Oligonychus pratensis,* feed on corn by piercing cell walls and sucking out the contents of the cells, rendering cells nonfunctional.[22] Injury to corn produced by both species is similar.[31]

Mites typically overwinter on native grasses in field borders or waterways, or in plantings of winter wheat. Infestations in corn generally appear first near overwintering sites, with populations establishing on the undersides of lower corn leaves. Highest colonization occurs along leaf midribs, near the natural bend of the leaves.[22] Injury first appears as whitish or yellowish stippling, which is visible on upper leaf surfaces. As mite injury progresses, leaves may become brown and a general decline in plant growth may be observed. Banks grass mite usually infests corn earlier in the season (May and June), while twospotted spider mite arrives later in the year (July and August).[31–33]

Corn yield losses from spider mite injury have been observed to range from 0 to 47%, depending on plant growth stage and level of environmental stress. Whole plant injury ratings have been found to be highly correlated with corn yields, with higher injury ratings predicting lower grain production.[31] Spider mite infestations of 90 to 120 mites per plant, incurred before the dent stage of corn (R5), have been found to significantly reduce corn yields. Mite infestations that occur after the dent stage have not been observed to reduce corn yields.[31]

4.3.3   VASCULAR DISRUPTION INJURY

Insects that tunnel and feed within vascular tissues of corn disrupt water and nutrient transport within plants. Plant stunting, deformed growth, wilting of entire plants or whorl leaves (dead-heart), and even plant death are all symptoms that can be associated with this type of internal injury. In addition, mechanical tissue injury, including insect entry holes and consumption of internal vascular structures, can lead to stalk breakage, or lodging, and possible ear droppage. Both of these effects may result in yield reductions.[22]

The physiological impacts of vascular disruption in corn from insect injury are likely similar to those observed for environmental conditions that produce water stresses in the plant. Reduced water and nutrient transport to expanding leaves and plant energy sinks will have detrimental effects on plant development whether produced by mechanical injury to vascular bundles (insect injury) or by simple lack of adequate water (drought conditions). In either case, plant development would be slowed and dry matter assimilation would be reduced.[1] This has been confirmed to some degree through observations indicating greater effects on corn from stressed conditions (drought) when plants were not injured by vascular disruptive insect injury.[34] Water-deficient conditions were of less consequence to the plant when the capabilities to transport water and nutrients were already diminished.

Severity of damage resulting from vascular disruption injury is dependent upon stage of corn growth at the time of injury, the extent of injury incurred, and distribution of injury within the plant.[35] Earlier infestations, relative to plant development, generally result in greater impacts on yield than do later infestations.[36–40] Likewise, a significant negative correlation has been observed between grain weight (yield) and the amount of insect tunneling injury.[40–42]

Vascular disruption injury to corn from insect feeding can also increase the potential for plant diseases. Even with minimal stalk injury, European corn borer injury was found to significantly predispose plants to anthracnose stalk rot development.[43] Although the phenomenon is likely disease specific, the link between insect tunneling injury and stalk rot disease is fairly consistent.[44, 45]

4.3.3.1 European Corn Borer

The European corn borer (ECB), *Ostrinia nubilalis,* is a major pest of corn in the U.S. The insect typically develops through two (midwest regions) or three (southern regions) generations per year, with the first generation infesting corn in the whorl stage (approximately V6 to V16). Plant injury from all ECB generations may result in serious corn yield losses.[22, 40–42, 46] As would be expected, plants injured from more than one generation of ECB will experience yield losses greater than plants injured by a single pest generation.[40] Detrimental effects of ECB tunneling in corn stalks are similar to those described in the general discussion of vascular disruption injury, producing physiological stress in plants, and with sufficient injury, stalk breakage and ear droppage.

4.3.3.1.1 First generation European corn borer (ECB1)
Newly hatched larvae of ECB1 begin feeding on leaves of whorl-stage corn (approximately V6 to V16), giving infested plants a "shot-hole" appearance.[22] Even with heavy infestations, this leaf feeding injury is of little consequence because insects are small (2 mm in length) and leaf area consumed is minimal. However, as ECB1 larvae mature, they move deeper into corn whorls and into leaf sheaths and midribs, with later instars eventually boring into stalks. This injury results in adverse physiological effects that can ultimately translate to yield reductions. These yield losses are dependent on the extent of tunneling injury, timing of the injury (relative to plant growth stage), and distribution of injury within the plant.[35]

4.3.3.1.2 Second generation European corn borer (ECB2)
Eggs that produce ECB2 are deposited near the ear zone on corn plants in the late vegetative and early reproductive stages of growth (approximately V17 to R6 growth stages). Newly hatched larvae quickly move to leaf axils and sheaths to feed on pollen and leaf collar tissue. As these larvae mature, they may be found feeding on developing kernels and tunneling into ear shanks and stalks.[22] Injury from ECB2 may result in reduced corn yields through both physiological disruption (as previously described) and stalk breakage or ear droppage.

4.3.3.1.3 Third generation European corn borer (ECB3)
A third generation may occur during years with particularly early and warm growing seasons, and in southern corn growing regions. When ECB3 injury occurs in corn (approximately VT to R6 growth stages), the effects are similar to those described for ECB2.[22]

4.3.3.1.4   The impacts of European corn borer feeding

Although the literature contains a few discrepancies, it can generally be stated that yield losses resulting from ECB1 infestations are typically higher than those observed for ECB2. Lynch[39] reported that physiological yield losses were lower with ECB infestations at blister stage (R2) than with infestations at earlier growth stages including whorl, pre-tassel, and pollen-shed. Similar reports were made by Berry and Campbell.[40] This would indicate that early disruption of vascular systems within the plant has far-reaching effects on subsequent plant development. A significant relationship has been observed between grain weight and the number of ECB tunnels per plant, with a greater number of tunnels resulting in lower grain yields.[41] However, this response appears to be non-linear, with yield losses per cavity (tunnel) decreasing as the number of cavities per plant increases.[40, 47] Regardless of response function, corn yields will typically decrease as severity of ECB infestations increase. This is supported by pest management scouting and sampling research, which identified a significant relationship between the percentage of plants showing ECB leaf-feeding injury and subsequent corn yields.[42, 48] Yield losses of a "substantial nature" were observed when 50% of the plants sustained ECB1 leaf feeding injury.[42]

Although physiological impacts of ECB1 injury may outweigh those observed for ECB2, actual yield reductions associated with ECB2 infestations can be significant, depending on timing, severity, and distribution of injury incurred. Maximum potential yield decreases from ECB2 tunneling injury occur when plants are attacked during initiation of the blister stage (R2).[38, 46, 49] Percentage yield decreases per borer decline as tunneling is initiated nearer plant physiological maturity (R6), or when initiated before pollination (VT to R2 stages), as compared to injury incurred at blister stage.[38] In addition, losses due to stalk breakage (lodging) or ear droppage are significant under heavy infestations. Actual losses due to insect-induced breakdowns in plant structure are largely dependent upon environmental conditions. Excessive winds that accentuate stalk breakage or lodging, and temperature or moisture conditions that encourage stalk rot diseases, can contribute significantly to mechanical losses brought on by ECB2 injury.

A significant negative linear relationship between grain yield and level of ECB2 infestation has been observed for insects attacking the bottom and middle strata of corn plants. This has not been observed for infestations found to occur in the upper strata (above corn ears).[38, 49] This indicates that vascular disruption that occurs above the ear zone will have little impact on water and nutrient delivery to the seed.

The effect of corn hybrid or variety on impact of ECB injury has been investigated extensively. Plant breeders and seed producers continue to search for sources of ECB tolerance or resistance. However, it is not the intention of this chapter to discuss the effects of innate plant toxins, such as cyclic hydroximates (DIMBOA), or transgenic gene insertions, such as *Bacillus thuringiensis* (Bt), on ECB survival and subsequent plant injury. It should suffice to say that hybrid selection can and will have significant impacts on potential ECB injury that is observed in corn plantings. However, once ECB infestations have been established in corn, the effects on plant physiology and productivity should be similar, regardless of resistance levels.[50] Aside from the effects of plant toxins, there are other general factors associated with hybrid selection that may int ıence ECB injury to corn. One such factor is corn maturity group. Although research results are mixed, there is evidence to suggest that yield losses due to ECB2 injury are more significant in shorter-season hybrids, especially when planted late, as compared with longer-season hybrids.[51] This would suggest that the best strategy for avoiding ECB2 injury is attained by planting long-season hybrids early, a strategy that is consistent with optimization of other agronomic factors.[51]

4.3.3.2   Southwestern Corn Borer

The southwestern corn borer (SWCB), *Diatraea grandiosella,* is a common pest in south-central and southwestern corn growing regions of the U.S. In most of these corn growing areas, the insect completes two generations per year.[36] Substantial yield losses from SWCB injury can result independently from physiological stresses associated with stalk tunneling and from plant lodging as a result of larval stalk girdling behavior.[36] Yield impacts arising from physiological stresses associated with SWCB stalk tunneling are likely similar to those observed for European corn borer stalk tunneling, with the severity of yield depressions being dependent upon extent of tunneling, timing of injury, and distribution of damage within the plant.

As with European corn borer, early SWCB infestations (relative to plant development) generally result in greater yield reductions than those observed with later infestations.[36] First and second instar feeding occurs primarily on leaves and sheaths of corn, producing damage of little or no economic significance. However, as larvae develop to the third stage and beyond, stalk tunneling injury becomes more and more evident, with total tunneling per larva observed to be about 20 cm. Although infestations in smaller plants typically result in greater yield reductions, injury at all corn growth stages has been observed to cause yield reductions.[36, 53] The rate of SWCB tunneling injury has been found to increase with advancement of corn phenology, with older (larger) plants sustaining more tunneling damage.[36] As SWCB larvae mature, they locate near the base of corn plants, where their tunneling activities have a girdling effect on the plant. This plant girdling can lead to significant stalk breakage and yield decreases.

Yield losses associated with SWCB infestations in whorl-stage corn have been observed to range from 20 to 30%.[53–56] Under severe infestations to whorl-stage corn (30 larvae per plant), yield reductions of 57% were observed along with reductions in ear height of 50%.[57] Yield losses associated with second generation SWCB in older plants have been observed to range from 12 to 28%.[36] Harvest losses, resulting from stalk girdling by second generation SWCB, have been effectively minimized through the planting of shorter-season hybrids or harvesting grain at higher moisture contents before larval girdling is completed.[36]

4.3.3.3   Stalk Borers and Stem Borers

In addition to European and southwestern corn borers, several other lepidopteran pests are known to cause vascular disruption injury in corn. In the U.S., the common stalk borer, *Papaipema nebris,* is a sporadic, yet potentially serious pest of maize. Although similarities in corn injury exist for ECB, SWCB, and common stalk borer, some feeding habits for common stalk borer are known to be different from those observed for the corn borers. Common stalk borer is a "general feeder" that tunnels into stems of various grass and broadleaf species. After smaller common stalk borer larvae outgrow the stems in which they are feeding, they leave the host and migrate to plants with larger stems, such as corn. Larvae then bore into the new host and normally feed within it until development is completed.[58] Larvae may enter the corn plant at the base of the stem and burrow upward, causing the heart of the plant to die (dead-heart). They may also enter small corn plants from the whorl and burrow downward, causing part of the plant to wilt or die. Injury may also be exhibited as unnatural growth, including twisted or deformed leaves, plant stunting, or a general wilting or death of the plant.[22]

Globally, several noctuid and pyralid pests are known to cause injury similar to that observed for common stalk borer. A substantial amount of information is known for six lepidopteran species common to Africa and India. Referred to collectively as maize stem borers, research information on *Busseola fusca, Eldana saccharina, Sesamia calamistis, Chilo partellus, Chilo aleniellus,* and *Sesamia botanephaga,* will be utilized to strengthen a discussion of yield impacts from injury synonymous to that of common stalk borer.

Stalk borer and stem borer injury to corn is similar to that observed for European and southwestern corn borer in several respects:

a) Plants which sustain only leaf feeding injury, exhibit little or no corn yield loss.[59]
b) In general, earlier infestations (relative to plant development) cause greater yield reductions than those occurring later in the season.[37, 60, 61] Maximum yield losses and higher plant damage are observed when infestations occur before stem elongation.[61–63]
c) Yield reductions from stalk and stem borer injury are more pronounced under conditions of drought stress, with drought stress conditions alone being more detrimental (from a relative perspective) to uninjured plants.[34, 61]

The most significant difference between stalk and stem borer injury to corn, as compared with injury from European and southwestern corn borers, is the frequency of dead-heart symptoms. Occurrence of dead-heart is the main cause of crop losses from stalk borer and stem borer injury.[60] Corn stalks with dead-heart injury have been found to produce significantly less dry matter, often resulting in barren plants or plants with partially-filled ears (scattered kernels).[59] Grain yield reductions of approximately 70% have been observed with stalk and stem borer injury of a magnitude sufficient enough to induce plant tillers.[59] In addition, with the presence of dead-heart injury, occurrence of barren stalks has been observed to range from 25 to 63%.[59]

4.3.4 ROOT FEEDING INJURY

Arthropods that feed on root systems of corn can place severe physiological stresses on plants. Much like the stresses caused by injury to above-ground vascular tissues, root feeding injuries can reduce water and nutrient uptake, negatively influencing yield,[64, 65] reducing vegetative and reproductive biomass,[20, 66, 67] causing asynchrony between tassel and silk development (increasing plant barrenness),[68] increasing incidence of stalk lodging,[69] and altering corn nutrient content[68, 70] and gas exchange parameters.[71, 72]

In addition to the physiological stresses associated with arthropod-induced root injury, mechanical removal of root tissues can weaken the plant's foothold in the soil. With significantly reduced root systems, plant lodging can occur, complicating harvest and contributing to mechanical yield losses.[22] Plant lodging can also result from root rot and stalk rot diseases introduced through arthropod feeding sites on corn roots. These opportunistic plant pathogens can enhance the incidence of stalk lodging beyond that observed for arthropod injury alone.[22]

The negative effects of root feeding injury to corn are normally more pronounced under conditions of moisture and fertility stress.[22, 68] It is probable that the effects of both stressors (biotic and abiotic) are additive, with both conditions contributing to reduced water and nutrient transport within the plant. The importance of soil moisture in determining the actual impact of arthropod-induced root injury cannot be overemphasized. Indeed, it largely explains the lack of correlation between root injury and yield with lower pest pressures. When one factor (root injury or dry soil conditions) becomes critically limiting, the impact of the other factor is difficult to quantify, often appearing less damaging than similar scenarios observed without competing stresses.

4.3.4.1   Corn Rootworms

The western corn rootworm, *Diabrotica virgifera virgifera,* the northern corn rootworm, *Diabrotica barberi,* and the southern corn rootworm, *Diabrotica undecimpunctata howardi,* are all considered major pests of corn in North America. All three species (collectively referred to as "rootworms") can inflict serious injury through larval feeding on or in corn roots. Rootworm larvae pass through three stages before pupating. Newly hatched larvae feed primarily on root hairs and outer root tissue, causing damage of little or no significance. However, as larvae grow and food requirements increase, they consume greater amounts of root tissue, often causing extensive injury. Older larvae may be found tunneling into larger roots and sometimes feeding in plant crowns.

Rootworm injury to corn may include some or all of the physiological impacts outlined for root feeding injury in the general discussion above. Injury is normally most severe when the secondary root system is well established and brace root development is underway. Root tips will be brown and often show signs of tunneling. Growing points at root tips can also be killed with excessive rootworm feeding, leading to varying degrees of fibrous secondary root growth.[26]

Rootworm injury to corn roots reduces plant turgor pressure, altering phenological events and influencing plant height.[73] Tasseling can be delayed and silking period prolonged with sufficient corn rootworm injury. This reduced synchrony of male and female flowering can reduce pollination success.[68] Feeding injury from rootworms also can interfere with nutrient uptake.[68] This is evidenced by the partial compensation of yield losses that have been observed with the addition of some fertilizers following rootworm injury.[74]

The effects of rootworm injury seem to be complicated by moisture-stress.[68] Dry soil conditions have been found to accentuate the impacts of moderate rootworm feeding injury. Conversely, dry soil conditions have been found to decrease larval densities, lessening the potential for rootworm injury.[74] Root compensatory regrowth is also affected by moisture level. Such root regrowth, following rootworm injury, has been found to positively affect yields when soil moisture levels are inadequate, and to negatively affect yields when soil moisture is not a limiting factor.[75] The severity of root injury combined with the level of root compensatory growth plays an important role in mediating shoot growth and carbon-dioxide assimilation responses imposed by rootworm larval feeding.[76]

Although corn plants can tolerate a certain amount of rootworm injury,[68] negative yield impacts can be directly attributed to this insect injury.[64] Researchers have estimated yield losses of 0.8 to 2.5% per larva per plant, with higher losses per larva associated with lower rootworm infestations.[64, 74] A precise correlation of root injury with yield has been difficult to establish because hybrid root characteristics (root volume), seasonal moisture levels, and potential for root regrowth work in concert to either enhance or negate the effects of rootworm injury. [68, 75]

Corn yield losses from stalk lodging also are commonly associated with severe rootworm pressures (more than one node of adventitious root axes destroyed). Weakened brace roots and reduced root masses negatively impact plant stability and anchorage in the soil, reducing the likelihood of plants remaining upright when subjected to turbulent weather. With stalk lodging, a characteristic "sled-runner" or "gooseneck" shape appears in plants that have fallen over due to weakened brace roots. In a cropping situation, these lodged and misshaped plants make mechanical harvest difficult or nearly impossible.[22] Corn yield decreases that result from stalk lodging can often be greater than those observed from physiological stresses related to injury.

The larger the root system, the more tolerant the corn plant is to rootworm larval injury.[75] The compensatory abilities of the plant to generate new roots following injury have also been associated with tolerance.[75] These characteristics are closely tied to hybrid or variety selection and the influences (positive or negative) of edaphic factors.

4.3.4.2 Other Root Feeding Insects

Several other arthropod pests cause root feeding injuries to corn. In general, the impacts of their injuries are similar to those described for corn rootworms, because physiological effects of root pruning are not largely species-specific. Exceptions to this rule may exist for those root-feeding pests that attack corn early in stand establishment (VE to V4 stages). Root injury during these critical stages of plant development may affect plant productivity and yield in manners different from injury incurred later in plant development. Root-feeding pests of importance during the early season include several white grubs, *Phyllophaga* spp., *Cyclocephala* spp., and *Popilla japonica,* and wireworms, *Melanotus* spp., *Agriotes mancus,* and *Limonius dubitans.*[22]

White grub and wireworm infestations can have similar impacts on corn physiology and yield. Although the intensity and degree of injury produced per individual may differ for the two pests, the effects of severe infestations on corn growth are essentially the same. The most obvious injury symptoms associated with early-season feeding by both pests are reduced plant stands and poor seedling vigor.[22] Stunted or wilted plants, and discolored, dead, or dying seedlings are commonly associated with heavier infestations.[22]

White grubs congregate near the bases of young corn plants, where they feed on and sever young roots.[26] When corn is planted into fields infested with sufficient white grub populations, the impacts can be devastating. Wireworms are likely more damaging (on an individual basis) than are white grubs, with half as many individuals (per cubic foot of soil) required to cause economic losses.[22] In addition to general root pruning, wireworms are notorious for attacking plant growing points below the soil surface. It is this feeding behavior that raises the probability of death of whorl leaves (dead-heart) or entire plants as compared to feeding targeted at fibrous roots. Regardless, sufficient levels of either pest can result in plant death or vigor reduction, negatively affecting corn yields.

4.3.5 REPRODUCTIVE DISRUPTION INJURY

One of the most important concerns with hybrid corn production is complete ear pollination. The entire corn crop is dependent upon the degree to which ear fertilization and kernel-set are achieved.[77] During the sensitive and critical flowering period, plant stresses due to adverse environmental conditions and pest injury can negatively influence pollination, reducing kernel-set and subsequent corn yields.[78, 79]

Pollen-shed and silking of corn plants usually takes place during the hottest days of the growing season. All major growth of the plant has taken place by the time of flowering, with metabolic activity at peak levels. When fertilization disruptions occur, reductions in ear weight and kernel numbers are evident.[80] Plants compensate for reduced kernel-set with increased kernel weight, but the response is weak when compared to the trend for reduced ear weight.[80]

Arthropods that interfere with pollen availability and/or silk receptivity can dramatically impact end-of-season corn yields. In open-pollinated corn, the limiting factor that assures complete ear pollination is most often silk receptivity and not availability of pollen. Arthropod injuries that remove corn silks, or sufficiently reduce silk lengths, will decrease the probability of complete ear fertilization. Several insect pests are known to feed on green silks of corn. Silk clipping that reduces (and maintains) average silk length to 0.5 to 0.75 inches on 20% of plants can significantly reduce corn yields.[78, 80]

Reproductive disruptions in corn can also occur after pollination is completed. Arthropod injuries to developing ears and kernels can negatively impact grain yields and quality. These direct losses to grain production can also be significant, with severity of injury related to species-specific feeding behaviors and the magnitude of pest populations.

4.3.5.1 Corn Earworm

The corn earworm, *Helicoverpa zea,* attacks fresh silks and ears of corn. Corn plants with loose husks and exposed ear tips, and/or full-season maturities are usually the most likely candidates for infestation by corn earworms.[22] Upon egg hatch, first stage larvae migrate from silks to the ear, following a silk channel to the ear tip. The larvae are cannibalistic, resulting in only one insect establishing per ear. Throughout the larval stages, corn earworms, under the protection of the husks, continue to feed on developing kernels in the ear tips.

Corn earworm larvae can destroy numerous kernels as they tunnel along the sides of corn ears.[26] Higher larval densities, on loose-husked inbreds and hybrids, have been observed to reduce kernel production by 30 to 40%.[22]

4.3.5.2 Fall Armyworm

The fall armyworm, *Spodoptera frugiperda,* has been discussed previously as a foliage-feeding insect. However, its impact as a reproductive disrupter is equally important, contributing to yield losses through direct feeding on corn ears. With the onset of anthesis and ear development, fall armyworm larvae have been observed to feed on tassels, husks, silks, kernels, and in the ear shanks.[22, 26] Fall armyworm injury in the ear tip can be easily confused with that caused by the corn earworm. Negative impacts on corn yield due to fall armyworm are therefore similar to those reported for corn earworm.

4.3.5.3 Western Bean Cutworm

The western bean cutworm, *Loxagrotis albicosta,* is a native pest of corn in the western U.S. Larvae actively feed on whorl foliage and developing tassels (florets and pollen) before ear development, and move to leaf sheaths, husks and developing ears as plants mature. Larvae enter the ears through the silk channel, or by boring directly through the husks.[22] Injury to the corn silks during initial ear infestation can result in poor pollination. As larvae continue to feed and mature, they routinely damage developing kernels, often causing severe injury and kernel loss.[26]

The western bean cutworm is a univoltine pest that is not cannibalistic in the larval stage. Therefore, several western bean cutworms may establish in a single ear of corn, further enhancing the potential for significant yield losses. Actual yield losses attributed to infestations of western bean cutworm in corn have been estimated to average 3.7 bushels per acre for each larva per plant as infested at dent stage (R5).[81]

4.3.5.4 Adult Corn Rootworms

The corn rootworms, *Diabrotica* spp., have been discussed previously as root-feeding insects. However, the impacts of adult feeding on green corn silks during plant flowering can be equally important from a yield perspective. High adult populations have been observed to trim silks back to the ear tips, resulting in partial kernel set and poorly filled ears.[77] With scatter-grained ears, kernel size is increased, while yields are reduced.[22]

Little has been written concerning the impact of adult corn rootworm silk feeding on corn yields. Leva[80] reported that with open pollination, three beetles per ear on inbred corn and five beetles per ear on hybrid corn could significantly reduce ear weight and kernel number per ear. Research to date indicates that silk length plays a significant role in determining severity of corn rootworm feeding damage, with injury tied more closely to the "mechanics of pollination" than to physiological changes within the plant.[77] Silk lengths of 0.75 to 1 inch (inbred corn) or 0.5 inch (hybrid corn) have been determined as bordering on minimum lengths required to assure optimal pollination.[77]

4.3.5.5 Japanese Beetles

Adult Japanese beetles, *Popilla japonica,* can disrupt corn reproduction through silk clipping and silk feeding behaviors similar to those observed for adult corn rootworms. Although Japanese beetles will feed gregariously on numerous plant species, corn in the silking stage (R1) is a highly preferred host. Adult beetle populations therefore will concentrate in corn fields during silk emergence, hindering pollination and seed set.[22, 26]

Yield losses associated with Japanese beetle are tantamount to those observed for other silk clipping insect pests. Average silk length during pollination (as regulated by beetle feeding) is likely a reliable indicator of potential yield impacts. The impacts associated with reduced silk lengths are most probably similar to those observed for corn rootworm.

4.3.5.6 Grasshoppers

Grasshoppers, *Melanoplus* spp., were also discussed earlier as foliage feeding pests. Although it is known that grasshoppers will readily feed on corn foliage, a distinct preference for corn silks is regularly observed. Heavy grasshopper populations will regularly eat corn silks down to the cob, a practice that interferes with pollination through mechanical injury to kernels.[26]

Injury first appears at field margins, as grasshoppers rarely originate from within fields.[82] As with corn rootworm and Japanese beetles, silk length during pollination and yield reductions are likely associated with injury that results in silks being trimmed to critically short lengths.[78]

4.4 SUMMARY AND CONCLUSIONS

Welter[6] noted that what is perhaps most interesting about the pattern of plant responses to herbivory is not the uniqueness of these responses, but rather the similarity between plant responses to a diversity of biotic and abiotic stresses. Parallel responses to herbivory have been observed for plant pathogens, drought, nutrient stress, and/or light stress.[6] Although not all corn responses to arthropod injury are aligned with those observed for other sources of stress, several analogies can be drawn. In this chapter, I suggest that corn plant responses to foliar insect feeding are similar to those observed for hail and other physical leaf injuries. Likewise, plant responses to vascular disruption and root pruning injuries were observed to be arguably similar to those observed for water and fertility stresses.

Of equal importance is the degree of commonality between plant responses to numerous arthropod pests inflicting injury to plants. In this chapter, arthropods causing injury to corn were grouped into five categories for general discussions of biotic stress and plant response. Within each injury grouping (foliar feeding, vascular feeding, root feeding, and vascular and reproductive disruption), similar impacts to corn growth and development were often observed for more than one key pest. This trend is apparent for both acute and chronic injury.

Although generalizations can often be drawn relative to injury and plant responses associated with arthropod pests having similar feeding behaviors, many arthropod pests are known to be quite species-specific in this regard. Unique feeding behaviors that result in plant injuries to critical plant parts (growing points, flowers, ear shanks, etc.) during critical stages of plant development can result in physiological responses that are equally unique. Arthropods that introduce toxins into plants (stink bugs, aphids, etc.) also are generally associated with injury symptoms and physiological responses that are more specific (plant stunting, abnormal growth, or destruction of tissues). Arthropod injuries and plant responses that are unique and closely aligned also are evident in corn production systems.

Regardless of uniqueness or specificity of arthropod injury, when imposed at sufficiently high levels, all will stimulate physiological responses in host plants. The exact impact of arthropod injury is critically dependent upon a host of physical, environmental, and genetic factors. These abiotic influences have been discussed in general terms in the previous pages.

The combinatorial nature of arthropod injury and other plant stressors will result in varied impacts on plant productivity. The range of these impacts is likely broad and diverse; possible scenarios affecting plant function are almost limitless. It should be noted that while the discussions included have focused on corn and corn growth, the philosophies presented are readily transferable to other crops and related pests. Yield losses associated with induced stresses in plants are as real as those observed for stresses imposed on animal systems. Losses due to both acute and chronic effects of arthropod injury in plants are to be expected, and therefore should be fundamentally understood. Through a founded and yet general knowledge of the impacts of insect stressors, scientists and practitioners alike can manage arthropod pests as part of overall production systems rather than as independent events, thereby "integrating" pest management.

REFERENCES

1. Vaadia, Y., The impact of plant stresses on crop yields, in *Cellular and Molecular Biology of Plant Stress,* Key, J. L., and Kosuge, T., Eds., Alan R. Liss, New York, 1985, 13.
2. Chapin, F. S., III, Integrated responses of plants to stress, *BioScience,* 41, 29, 1991.

3. Boyer, J. S., Plant productivity and environment, *Science,* 218, 443, 1982.

4. Judenko, E., Analytical method for assessing yield losses caused by pests on cereal crops with and without pesticides, *Trop. Pest Bull. No. 2,* Center for Overseas Pest Research, Oak House, Kent, U.K., 31, 1973.

5. Higley, L. G., and Peterson, R. K. D., The biological basis of the EIL, in *Economic Thresholds for Integrated Pest Management,* Higley, L. G., and Pedigo, L. P., Eds., University of Nebraska Press, Lincoln, 1996, 22.

6. Welter, S.C., Responses of plants to insects: eco-physiological insights, in *International Crop Science I,* Buxton, D. R., Shibles, R., Forsberg, R. A., Blad, B. L., Asay, K. H., Paulson, G. M., and Wilson, R. F., Eds., Crop Science Society of America, Madison, WI, 1993, 773.

7. Jones, C. A., and Kiniry, J. R., Eds., CERES–Maize: A simulation model of maize growth and development, Texas A&M University Press, College Station, 1986.

8. Nielsen, R. L., *National Corn Handbook,* CES, Purdue University, Department of Agronomy, West Lafayette, IN, 1987.

9. Johnson, J. W., and Wallingford, W., Weather-stress yield loss, proper fertilization reduces the risk, *Crops & Soils Mag.,* 35, 15, 1983.

10. Setamou, M., Schulthess, F., Bosque-Perez, N. A., and Thomas-Odjo, A., The effect of stem and cob borers on maize subjected to different nitrogen treatments, *Entomologia Experimentalis et Applicata,* 77, 205, 1995.

11. Cardwell, K. F., Schulthess, F., Ndemah, R., and Ngoko, Z., A systems approach to assess crop health and maize yield losses due to pests and diseases in Cameroon, *Agric. Ecosys. Environ.* 65, 33, 1997.

12. Hicks, D. R., Nelson, W. W., and Ford, J. H., Defoliation effects on corn hybrids adapted to the northern corn belt, *Agron. J.,* 69, 387, 1977.

13. Shapiro, C. A., Peterson, T. A., and Flowerday, A. D., Yield loss due to simulated hail damage on corn, a comparison of actual and predicted values, *Agron. J.,* 78, 585, 1986.

14. Carter, P. R., and Hudelson, K. D., Influence of simulated wind lodging on corn growth and grain yield, *J. Prod. Agric.,* 1, 295, 1988.

15. Story, R. N., Keaster, A. J., Showers, W. B., Shaw, J. T., and Wright, V. L., Economic-threshold dynamics of black and claybacked cutworms (Lepidoptera: Noctuidae) in field corn, *Environ. Entomol.,* 12, 1718, 1983.

16. Levine, E., Clement, S. L., Rubink, W. L., and McCartney, D. A., Regrowth of corn seedlings following injury at different growth stages by black cutworm larvae, *J. Econ. Entomol.,* 76, 389, 1983.

17. Crookston, R. K., and Hicks, D. R., Early defoliation affects corn grain yields, *Crop Sci.,* 18, 485, 1978.

18. Johnson, R. R., Growth and yield of maize as affected by early-season defoliation, *Agron. J.,* 70, 996, 1978.

19. Showers, W. B., Sechriest, R. E., Turpin, F. T., Mayo, Z. B., and Szamtari-Goodman, G., Simulated black cutworm damage to seedling corn, *J. Econ. Entomol.,* 72, 432, 1979.

20. Gibb, T. J., and Higgins, R. A., Aboveground dry weight and yield responses of irrigated field corn to defoliation and root pruning stresses, *J. Econ. Entomol.,* 84, 1562, 1991.

21. Santos, L., and Shields, E. J., Yield responses of corn to simulated black cutworm (Lepidoptera: Noctuidae) damage, *J. Econ. Entomol.,* 91, 748, 1998.

22. Edwards, C. R., Obermeyer, J. L., Jordan, T. N., Childs, D. J., Scott, D. H., Ferris, J. M., Corrigan, R. M., and Bergman, M. K., *Seed Corn Pest Management Manual for the Midwest,* Purdue University CES and Department of Entomology, West Lafayette, IN, 36–124, 183–186, 1992.

23. Hruska, A., and Gladstone, S. M., Effect of period and level of infestation of the fall armyworm, Spodoptera frugiperda, on irrigated maize, *Florida Entomol.,* 71, 249, 1988.

24. Harrison, F. P., The development of an economic injury level for low populations of fall armyworm (Lepidoptera: Noctuidae) in grain corn, *Florida Entomol.,* 67, 335, 1984.

25. Evans, D. C., and Stansly, P. A., Weekly economic injury levels for fall armyworm (Lepidoptera: Noctuidae) infestations of corn in lowland Ecuador, *J. Econ. Entomol.,* 83, 2452, 1990.

26. Dicke, F. F., and Guthrie, W. D., The most important corn insects, in *Corn and Corn Improvement,* Sprague, G. F., and Dudley, J. W., Eds., American Society of Agronomy, Madison, WI, 1988, 767–867.

27. Negron, J. F., and Riley, T. J., Southern green stink bug, *Nezara viridula* (Heteroptera: Pentatomidae), feeding in corn, *J. Econ. Entomol.,* 80, 666, 1987.

28. Apriyanto, D., Townsend, L. H., and Sedlacek, J. D., Yield reduction from feeding by *Euschistus servus* and *E. variolarius* (Heteroptera: Pentatomidae) on stage V2 field corn, *J. Econ. Entomol.,* 82, 445, 1989.

29. Negron, J. F., and Riley, T. J., Long-term effects of chinch bug (Hemiptera: Heteroptera: Lygaeidae) feeding in corn, *J. Econ. Entomol.,* 83, 618, 1990.

30. Negron, J. F., and Riley, T. J., Effect of chinch bug (Heteroptera: Lygaeidae) feeding in seedling field corn, *J. Econ. Entomol.,* 78, 1370, 1985.

31. Archer, T. L., and Bynum, E. D., Jr., Yield loss to corn from feeding by Banks grass mite and two-spotted spider mite (Acari: Tetranychidae), *Exp. Appl. Acarol.,* 17, 895, 1993.

32. Logan, J. A., Congdon, D. B., and Allredge, J. K., Ecology and control of spider mites on corn in northeastern Colorado, *Colorado State Univ. Exp. Stn. Bull.,* 585S, 1, 1983.

33. Sloderbeck, P. E., Morrison, W. P., Patrick, C. D., and Buschman, L. L., Seasonal shift in species composition of spider mites on corn in the western Great Plains, *South. Entomol.,* 13, 63, 1988.

34. van Rensburg, J. B. J., Plant resistance x environment interaction: perspectives on yield losses caused by the maize stalk borer, *Busseola fusca* (Fuller), *S. Afr. J. Plant Soil,* 13, 61, 1996.

35. Labatte, J. M., and Got, B., Modelling damage on maize by the European corn borer, *Ostrinia nubilalis, Ann. Appl. Biol.,* 119, 401, 1991.

36. Whitworth, R. J., Poston, F. L., Welch, S. M., and Calvin, D., Quantification of southwestern corn borer feeding and its impact on corn yield, *Southwestern Entomol.,* 9, 308, 1984.

37. van Rensburg, J. B. J., Walters, M. C., and Giliomee, J. H., Response of maize to levels and times of infestation by *Busseola fusca* (Fuller) (Lepidoptera: Noctuidae), *S. J. Ent. Soc. S. Afr.,* 51, 283, 1988.

38. Calvin, D. D., Knapp, M. C., Xingquan, K., Poston, F. L., and Welch, S. M., Influence of European corn borer (Lepidoptera: Pyralidae) feeding on various stages of field corn in Kansas, *J. Econ. Entomol.,* 81, 1203, 1988.

39. Lynch, R. E., European corn borer: yield losses in relation to hybrid and stage of corn development, *J. Econ. Entomol.,* 73, 159, 1980.

40. Berry, E. C., and Campbell, J. E., European corn borer: relationship between stalk damage and yield losses in inbred and single-cross seed corn, *Iowa State J. Res.,* 53, 49, 1978.

41. Umeozor, O. C., Van Duyn, J. W., Kennedy, G. G., and Bradley, J. R., Jr., European corn borer (Lepidoptera: Pyralidae) damage in maize in eastern North Carolina, *J. Econ. Entomol.,* 78, 1488, 1985.

42. Berry, E. C., Guthrie, W. D., and Campbell, J. E., European corn borer: relationship between leaf-feeding damage and yield losses in inbred and single-cross seed corn, *Iowa State J. Res.,* 53, 137, 1978.

43. Keller, N. P., Bergstrom, G. C., and Carruthers, R. I., Potential yield reductions in maize associated with an anthracnose/European corn borer pest complex in New York, *Phytopathology,* 76, 586, 1986.

44. Hudon, M., Bourgeois, G., Boivin, G., and Chez, D., Yield reductions in grain maize associated with the presence of European corn borer and *Gibberella* stalk rot in Quebec, *Phytoprotection,* 73, 101, 1992.

45. Bosque-Perez, N. A., and Mareck, J. H., Effect of the stem borer *Eldana saccharina* (Lepidoptera: Pyralidae) on the yield of maize, *Bull. Entomol. Res.,* 81, 243, 1991.

46. Lynch, R. E., Robinson, J. F., and Berry, E. C., European corn borer: yield loss and damage resulting from a simulated natural infestation, *J. Econ. Entomol.,* 73, 141, 1980.

47. Jarvis, J. L., and Guthrie, W. D., Effect of first-generation European corn borer on yield and plant height of popcorn, *J. Agric. Entomol.,* 5, 179, 1988.

48. Sayers, A. C., Johnson, R. H., Arndt, D. J., and Bergman, M. K., Development of economic injury levels for European corn borer (Lepidoptera: Pyralidae) on corn grown for seed, *J. Econ. Entomol.,* 87, 458, 1994.

49. Calvin, D. D., Knapp, M. C., Xingquan, K., Poston, F. L., and Welch, S. M., Influence of European corn borer (Lepidoptera: Pyralidae) feeding on various stages of field corn in Kansas, *J. Econ. Entomol.,* 81, 1203, 1988.

50. Jarvis, J. L., Guthrie, W. D., and Berry, E. C., Time and level of infestation by second-generation European corn borers on a resistant and susceptible maize hybrid in relation to yield losses, *Maydica XXVIII,* 391, 1983.

51. Jarvis, J. L., Guthrie, W. D., and Robbins, J. C., Yield losses from second-generation European corn borers (Lepidoptera: Pyralidae) in long-season maize hybrids planted early compared to short-season hybrids planted late, *J. Econ. Entomol.,* 79, 243, 1986.

52. Bode, W. M., and Calvin, D. D., Yield-loss relationships and economic injury levels for European corn borer (Lepidoptera: Pyralidae) populations infesting Pennsylvania field corn, *J. Econ. Entomol.,* 83, 1595, 1990.

53. Maredia, K. M., and Mihm, J. A., Damage by southwestern corn borer (*Diatraea grandiosella* Dyar) on resistant and susceptible maize at three plant growth stages in Mexico, *Trop. Pest Manage.,* 36, 141, 1990.

54. Scott, G. E., and Davis, F. M., Effects of southwestern corn borer feeding on maize, *Agron. J.,* 66, 773, 1974.

55. Davis, F. M., Scott, G. E., and Williams, W. P., Southwestern corn borer: effects of levels of first brood on maize, *J. Econ. Entomol.,* 71, 244, 1978.

56. Hinderliter, D. G., Host Plant Resistance in Two Tropical Maize, *Zea mays* L., Populations to the Southwestern Corn Borer, *Diatraea grandiosella* Dyar, and the Sugarcane Borer, *D. saccharalis* F., Ph.D. dissertation, University of Wisconsin, Madison, 1983, 113.

57. Williams, W. P., and Davis, F. M., Response of corn to artificial infestation with fall armyworm and southwestern corn borer larvae, *Southwest. Entomol.,* 15, 163, 1990.

58. Meyer, S. J., and Peterson, R. K. D., Predicting movement of stalk borer (Lepidoptera: Noctuidae) larvae in corn, *Crop Protect.,* 17, 609, 1998.

59. Bailey, W. C., and Pedigo, L. P., Damage and yield loss induced by stalk borer (Lepidoptera: Noctuidae) in field corn, *J. Econ. Entomol.* 79, 233, 1986.

60. Moyal, P., Crop losses caused by maize stem borers (Lepidoptera: Noctuidae, Pyralidae) in Cote d'Ivoire, Africa: statistical model based on damage assessment during the production cycle, *J. Econ. Entomol.*, 91, 512, 1998.

61. Davis, P. M., and Pedigo, L. P., Economic injury levels for management of stalk borer (Lepidoptera: Noctuidae) in corn, *J. Econ. Entomol.*, 84, 290, 1991.

62. Seshu Reddy, K. V., and Sum, K. O. S., Determination of economic injury level of the stem borer, *Chilo partellus* (Swinhoe) in maize, *Zea mays* L. Insect, *Sci. Applic.*, 12, 269, 1991.

63. Sarup, P., Sharma, V. K., Panwar, V. P. S., Siddiqui, K. H., Marwaha, K. K., and Agarwal, K. N., Economic threshold for *Chilo partellus* (Swinhoe) infesting maize crop, *J. Ent. Res.*, 1, 92, 1977.

64. Petty, H. B., Kuhlman, D. E., and Sechriest, R. E., Corn yield losses correlated with rootworm larval populations, *Entomol. Soc. Amer. N. Cent. Br. Proc.*, 24, 141, 1969.

65. Turpin, F. T., Dumenil, L. C., and Peters, D. C., Edaphic and agronomic characters that affect potential for rootworm damage to corn in Iowa, *J. Econ. Entomol.*, 65, 1615, 1972.

66. Spike, B. P., and Tollefson, J. J., Response of western corn rootworm-infested corn to nitrogen fertilization and plant density, *Crop Sci.*, 31, 776, 1991.

67. Godfrey, L. D., Meinke, L. J., and Wright, R. J., Vegetative and reproductive biomass accumulation in field corn: response to root injury by western corn rootworm (Coleoptera: Chrysomelidae), *J. Econ. Entomol.*, 86, 1557, 1993.

68. Spike, B. P., and Tollefson, J. J., Relationship of plant phenology to corn yield loss resulting from western corn rootworm (Coleoptera: Chrysomelidae) larval injury, nitrogen deficiency, and high plant density, *J. Econ. Entomol.*, 82, 226, 1989.

69. Sutter, G. R., Fisher, J. R., Elliott, N. C., and Branson, T. F., Effect of insecticide treatments on root lodging and yields of maize in controlled infestations of western corn rootworms (Coleoptera: Chrysomelidae), *J. Econ. Entomol.*, 83, 2414, 1990.

70. Kahler, A. L., Olness, A. E., Sutter, G. R., Dybing, C. D., and Devine, O. J., Root damage by western corn rootworm and nutrient content in maize, *Agron. J.*, 77, 769, 1985.

71. Godfrey, L. D., Meinke, L. J., and Wright, R. J., Effects of larval injury by western corn rootworm (Coleoptera: Chrysomelidae) on gas exchange parameters of field corn, *J. Econ. Entomol.*, 86, 1546, 1993.

72. Riedell, W. E., Rootworm and mechanical damage effects on root morphology and water relations in maize, *Crop Sci.*, 30, 628, 1990.

73. Davis, P. M., and Coleman, S., Managing corn rootworms: (Coleoptera: Chrysomelidae) on dairy farms: the need for a soil insecticide, *J. Econ. Entomol.*, 90, 205, 1997.

74. Smith, B. C., Population changes of the northern corn rootworm (Coleoptera: Chrysomelidae) and corn yield losses in southwestern Ontario, *Proc. Entomol. Soc. Ont.*, 1979, 85.

75. Gray, M. E., and Steffey, K. L., Corn rootworm (Coleoptera: Chrysomelidae) larval injury and root compensation of 12 maize hybrids: an assessment of the economic injury index, *J. Econ. Entomol.*, 91, 723, 1998.

76. Reidell, W. E., and Reese, R. N., Maize morphology and shoot CO_2 assimilation after root damage by western corn rootworm larvae, *Crop Sci.*, 39, 1332, 1999.

77. Culy, M. D., The Impact of Western Corn Rootworm (Coleoptera: Chrysomelidae) on Pollination Success in Seed Corn Production Fields, Ph.D. dissertation, Purdue University, West Lafayette, IN, 113, 1987.

78. Culy, M. D., Edwards, C. R., and Cornelius, J. R., Effect of silk feeding by western corn rootworm (Coleoptera: Chrysomelidae) on yield and quality of inbred corn in seed corn production fields, *J. Econ. Entomol.*, 85, 2440, 1992.

79. Culy, M. D., Edwards, C. R., and Cornelius, J. R., Minimum silk length for optimum pollination in seed corn production fields, *J. Prod. Agric.*, 5, 387, 1992.

80. Leva, D. M., Population Suppression of Adult Corn Rootworms (*Diabrotica* spp.) and the Effect of Rootworm Adults on Pollination in Indiana Cornfields, Ph.D. dissertation, Purdue University, West Lafayette, IN, (Diss. Abstr. 80-27299), 168, 1980.

81. Appel, L. L., Wright, R. J., and Campbell, J. B., Economic injury levels for western bean cutworm, *Loxagrotis albicosta* (Smith) (Lepidoptera: Noctuidae), eggs and larvae in field corn, *J. Kan. Ent. Soc.*, 66, 434, 1993.

82. Metcalf, R. L., and Metcalf, R. A., *Destructive and Useful Insects: Their Habits and Control*, 5th ed., McGraw-Hill, New York, 1993.

5   Phenological Disruption and Yield Loss from Insects

Scott H. Hutchins

CONTENTS

5.1   INTRODUCTION

Injury has been characterized as the effect of pest activities on host physiology that is usually deleterious.[1] The actual characterization and quantification of specific forms of injury, however, are very complex and dynamic, affected by both the pest and the host. Injury can be a discrete event such as a larva cutting a young seedling where the resultant damage is immediate, acute, and easily measured. In contrast, injury may also be a slow and continuous process, such as soil insects affecting root tissue, which reduces transpiration over a season-long host growth cycle, but resultant damage is highly dependent upon environmental circumstances. This latter case, where injury is a continuous process to reduce host vigor, frequently is difficult to assess and represents one of many components to the "black box" relationship with regard to host yield (see Chapter 1). To establish a cause-and-effect relationship, some rational segregation and quantification of component injury must be achieved.

Insect-induced injury to plants has been classified within six host physiology-based categories: stand reducers, leaf-mass consumers, assimilate sappers, turgor reducers, fruit feeders, and architecture modifiers.[1, 2] The premise for segregation is that all (or nearly all) forms of insect-induced injury act upon one of these distinct physiological processes, thereby enabling a relational map to be constructed for growth "efficiency" as it relates to intensity of insect-induced injury. With this concept, insects producing similar forms of injury conceivably can be managed as injury guilds.[1, 3, 4] In addition to the practical management value of injury guilds, the concept places a strong research emphasis on understanding the true context of insect-induced injury, from the plant perspective. This emphasis has not only resulted in novel new techniques to assess plant stress, but it also has allowed for successful modeling of plant development based upon imposed pest-induced reductions in physiological efficiency.

5.2 PHYSIOLOGICAL MATURITY AS A RELATIVE MEASURE OF PLANT STRESS

Modern agriculture and horticulture require prediction of crop growth rates and the associated milestones (e.g., first bloom, pod fill, harvest) to manage the mechanical, labor, and marketing components of the enterprise. Although not originally considered a category of physiological injury, phenological disruption of plant growth is a critical form of pest-induced injury, and one that clearly affects crop quality. The delay in maturity, in particular, can have a significant impact on both the biomass and quality of harvested crops.[5] This impact will be magnified if delays in maturity provide a wider window for additional pest problems (insect, weed, nematode, disease), extend maturity (and therefore harvest) past the optimal date, or shorten the development cycle for subsequent plantings.

As with other forms of chronic injury, it may be very difficult to detect delays in phenological development because of confounding rate-limiting factors such as temperature and soil moisture. Or, conversely, favorable environmental conditions may "mask" the deleterious effects of pest-induced delays. Nonetheless, the impact of insects on host phenological development has been clearly established[5, 6] and must be factored into a successful pest management strategy. Although all cropping systems are susceptible to phenological delay as a key form of injury (as defined previously), forage crops are particularly susceptible because of the special focus on quality (for animal feed) and multiple harvest management schemes for the crop.

5.3 MANAGEMENT OF PEST-INDUCED PHENOLOGICAL DELAY IN FEED-BASED CROPPING SYSTEMS

Plants and herbivores have coevolved. The evolutionary balance between allocation of metabolic resources for establishment of structural or competitive/defensive mechanisms vs. reproductive mechanisms is always in flux.[7] The fundamental approach by plants, however, has been one of reducing herbivore intake directly (e.g., spines or toxins) or indirectly (e.g., poor nutrition, digestibility, or availability of energy). Given that direct methods are not compatible with production of a feed, agronomic practice designed to improve the proportion of plant intake and digestibility by animal consumers is the agronomic objective. Similarly, the pest management objective is to reduce any detrimental influence of pest-induced injury to crop yield and quality.

5.3.1 Relationship of Feed Quality to Plant Maturity

Managing feed quality of crops to maximize animal growth and production is a complex venture that requires knowledge of the physiology of plant biomass development. In addition, details about the development and relative value of qualitative attributes in plants, their relationship to feed utilization by animal consumers, and the temporal development of both biomass and qualitative attributes are central to the design of management systems. Indeed, prerequisite knowledge of nutrient development in plants and subsequent utilization in animals is necessary for intensive production of forages. Because of this complexity, alfalfa production and pest management provide a model system from which to illuminate the influence of pests on crop phenological development as well as crop management principles to minimize economic losses.

Forages should be managed based on the principle of maximizing total animal growth. In the alfalfa model system, certain qualitative attributes are widely recognized as indications of ultimate utility to the animal consumer. The character and nutritive value of forages are determined by two essential factors: the proportion of plant cell wall and its corresponding degree of lignification.[7] The feeding value to animals is limited by the daily intake of digestible nutrients and the efficiency with which these digested nutrients can be used for metabolic processes.[8] Available energy is particularly important because forage rations frequently are limited by energy content rather than nutritional content (e.g., protein) per se.[9] From a pest management perspective, therefore, plant characteristics that promote the rate of digestion or the proportion of total intake that is successfully digested are favorable and represent feed of higher nutritive value.

The percent of feed that is digestible by ruminants varies among plant species, but, more important to intraseasonal pest management decisions, also varies among plant parts and at different phenological stages of development.[10] The underlying factor regulating digestibility of a feed relates to the level of lignification at the time of consumption. Lignin limits the extent of digestion, but has minimal influence on the rate of digestion.[11] Alfalfa stems, dominated by a dense lignified cell-wall, are not susceptible to the enzymatic action of gastric chemicals or symbiotic organisms of the rumen designed to "release" the plant energy for use by the animal. By contrast, alfalfa leaves contain a smaller proportion of cell wall fractions (and lignin), which makes them relatively high in digestibility. As plants continue to mature toward the reproductive phases, even leaf tissues become less available through lignification as the carbohydrate sink shifts to reproductive organs and eventually seed. In addition to digestibility, feed intake also is a critical aspect of forage quality. The species of animal, its physiological status, energy demand, and its individual preference all affect intake.[7] Intake is in proportion to the degree of structural volume (cell-wall content).

The final assessment of high-quality alfalfa goes beyond the digestive nutrient content and is compounded by a potential for being consumed at greater levels, a faster rate of digestibility, and perhaps a more efficient conversion of digested energy to productive energy.[8] The ultimate expression of animal desire for a particular feed is greater at lower cell-wall concentrations of immature plants. Indeed, voluntary intake clearly is a key quality attribute, as it may account for two thirds of the variability of growth-rate performance of animals.[12]

Although forage quality frequently is assessed on a total-herbage basis, digestibility does indeed differ by plant part as suggested previously. Buxton et al.[10] demonstrated this clearly and further demonstrated that the lower stem component was less digestible than the apical portions. Subsequently, Buxton and Hornstein[13] determined that cell-wall concentration was low in leaves and greatest in basal stem segments. All this leads to the practical management conclusion that the leaf-to-stem biomass ratio is important to forage quality along with the age or maturity of the crop.

5.3.2 RELATIONSHIP OF HOST MATURITY TO INSECT-INDUCED INJURY

Although insects affect forages in all the injury categories described previously, assimilate removal and leaf-mass consumption are considered to be the two primary forms of injury.[14] Host response in the form of reduced carbohydrate synthesis, canopy development, or feeding value reflects the specific mechanism of insect-induced injury to the plant.

A primary indication of final host response is the measured impact on leaf-to-stem (L/S) ratio. Assimilate removal by insects tends to increase the L/S ratio. Although the specific sites and mechanisms of injury vary among pest species, the final result frequently is exhibited as reduced stem length. From an absolute quality standpoint, the reduced stem component may slightly increase feeding value, but the associated loss in biomass and nutrient yield may offset any marginal gains in quality. Another view of the effect on the crop relates to assessment of the crop by physiological age, or maturity. Indeed, insect-induced injury in the form of slower crop development may be the primary form of damage. Accordingly, days of regrowth delay may serve as the estimate of injury when making management decisions for assimilate removing or leaf-mass consuming insects.[5]

5.3.3 RELATIONSHIP OF INSECT-INDUCED INJURY TO HOST DAMAGE

Damage is considered to be the measurable reduction in plant growth, development, or reproduction resulting from injury. In the case of quality, damage per unit of injury may include a measurable loss in the feed value as a result of insect-induced physiological injury. Assessment of damage when using feed value as a management focus requires knowledge of how pests simultaneously affect both plant biomass and plant quality components under a continuum of injury intensities. As noted previously, structural components (e.g., stems) that contribute to plant establishment, competition for nutrients, and biomass must be considered as trade-offs with non-structural components (e.g., leaves) that contribute a disproportionately high level of digestible energy and protein per unit of biomass. To manage both biomass and quality, a nutrient-yield determination for the feed-value approach is recommended. Energy yield (Mcal/ha), for example, is calculated as the product of biomass (kg/ha) and digestible energy (Mcal/kg[15]). For protein yield (kg/ha) determinations, biomass production (kg/ha) yield can be multiplied by percentage of crude protein.

When using the feed-value approach for managing quality, the integrated rate of nutrient yield loss per pest should be determined and used as a basis for modeling the relationship of injury and damage (i.e., the damage curve). Assimilate removal by insects tends to increase the L/S ratio.[14] Although the specific sites and mechanisms of injury vary among piercing–sucking pest species, the final result frequently is exhibited as reduced stem length. The feeding value per unit of dry matter herbage, therefore, is maintained or slightly increased. The reduced stem component, however, depresses dry matter and overall nutrient yields.

Leaf-mass consumption from insects disrupts resource partitioning, reflecting a basic alteration in carbohydrate source/sink relationships for severely defoliated plants. Regrowth of defoliated plants relies on currently produced photosynthates, which are adequate for leaf growth but only minimal stem growth. Defoliated plants tend to compensate for increased carbohydrate demand by producing thinner leaves, which compensates for the reduced total leaf area but does not significantly increase leaf biomass.

5.4 THE AGRONOMIC VALUE OF TIME IN CROP PRODUCTION

Reduced or delayed phenological development most likely is the most fundamental form of injury associated with leaf-mass consuming or assimilate-removing insects. Many of the physical observations for plant injury are believed to be symptomatic of immature plants when compared with uninfested plants.[16] In the case of forages, delayed canopy development affects the carbohydrate reserves available for present and future regrowth and may adversely affect subsequent harvest schedules.[5] Indeed, excessive delay may require the loss of a full or partial harvest (Figure 5.1). In this stylized illustration, leaf-mass consumption at the earliest phase of regrowth has a very significant deleterious effect on nutrient yield development.[14] Assimilate removal also has a deleterious effect on nutrient yield development. Insect-induced injury later in the regrowth cycle continues to have a negative effect, but is less severe. The principal interpretation is that insect-induced injury can be compensated for by the crop, but not in the time frame normally associated with the production system. If the harvest cycle is on a calendar-date basis, then the phenological delay results in the harvesting of an immature crop, which severely limits biomass production and may impact stored carbohydrates for future growth and vigor. If the harvest cycle is based on a phenological indicator (e.g., first bloom), then the delay results in the

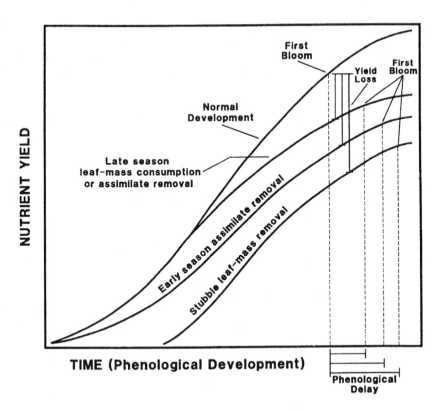

FIGURE 5.1   Stylized representation of disrupted harvest schedule resulting from crop development delays associated with insect-induced leaf-mass removal and assimilate removal (from Hutchins et al.[14]).

inefficiency of regrowth within a limited seasonal time period. In this case, the actual loss is related to the severity of the delay and the total herbage production capability within the growing season (Figure 5.2).

5.5   INTEGRATING PHENOLOGICAL DELAY WITHIN CONVENTIONAL DECISION INDICES

The use of economic injury levels (EILs) is amenable to pest management scenarios where the crop is injured via phenological delay. The biologic variable of injury should represent the relationship of insect density to crop delay.[5] The biologic variable of damage relates to the influence of injury (days delay) to the production of nutrient yield (quality × quantity). The economic variable of market (feed) value should be based upon the value of substitute feeds. In cases where forages are produced on-farm, the final utility of the forage should be measured within the context of specific needs of the animal consumer.[17] Specific examples for establishing EILs, including details on each of the variables, have been reviewed[1] and recently

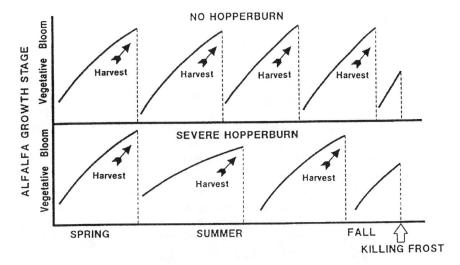

FIGURE 5.2 Generalized alfalfa growth response from leaf-mass consumers and assimilate removers. Reduced phenological rates affect nutrient yields harvested on a calendar date basis (from Hutchins et al.[14]).

updated.[18] In addition, specific examples and considerations for pests affecting quality and on-farm EILs are available for review.[16, 17]

Although much of this review has focused on the special case of addressing phenological delay in forage crops because of the many complex factors affected, phenological delay induced by pest injury also is a significant issue for annual crops with a single harvest event. Specifically, delayed development opens the window of host susceptibility and eventual yield loss from pests because they may be able to extend their life cycles longer or continue through additional generations. Indeed, much of current theory on avoiding pest-induced injury is focused on preventive agronomic tactics to "de-synchronize" the host:pest phenologies. The general management principles include establishing crops as early as possible, promoting their rapid growth and maturation, minimizing loss of metabolic energy focused on grain production, and harvesting as soon as possible.

In addition, crop development that is unnaturally delayed may extend the harvest event into a seasonal period that makes harvest risky from a weather standpoint, especially in more northern latitudes. In situations where double or continuous cropping is practiced, final harvests may be placed at risk as a result of the growth delays (similar to the multiple harvest of forages example).

Delayed development and excessive stress on crops may lead to the invasion and biotic release of secondary or tertiary pests from all classes, further affecting the rate of development and yield potential. Early delays in canopy development, for example, have been documented as leading to increased weed pressure[19] and eventual loss of yield. Extended growth periods may also magnify the subtle influence of generally subeconomic pests, reducing the overall host tolerance and compensatory ability.

5.6 FUTURE FOCUS FOR DEFINING THE INFLUENCE OF PHENOLOGICAL DELAY

Pest-induced injury to crops requires a clear understanding of the physiological disruption to the host. Although the use of injury guilds represents a large step forward, care must be taken to ensure that injury is related to and adjusted for host maturity and growth rate. Indeed, specific forms of injury may be undetected if not directly related to host growth and development without pest-induced stress. Research focused on assessing the relationship of pest-injury to host development must consider the possibility of delayed maturity. As a general practice, treatments attempting to produce ideal growth conditions should be included within the experimental design as a means to (1) compare the influence of pest-induced injury on host response and (2) measure and segregate the comparative host response as it relates to physiological state. If these procedures are not included, the researcher runs the significant risk of making erroneous conclusions using chronological age comparisons when physiological age comparisons would be more appropriate.

The above recommendations assume "conventional" techniques for assessing plant stress and development are employed. The hope and clear need is that future technological advances enable researchers (and eventually practitioners) to assess plant stress instantaneously through continuous monitoring of key physiological processes. This would allow for real-time diagnosis (vs. delayed use of symptoms), integration of phenological delay considerations, and a clear illumination of the black box relationship between pest-induced injury and crop damage.

REFERENCES

1. Pedigo, L. P., Hutchins, S. H., and Higley, L. G., Economic-injury levels in theory and practice, *Ann. Rev. Entomol.,* 31, 341, 1986.
2. Boote, K. J., Concepts for modeling crop response to pest damage, ASAE Pap. 81-4007. American Society of Agricultural Engineers, St. Joseph, MI, 1981.
3. Hutchins, S. H., Higley, L. G., and Pedigo, L. P., Injury equivalency as a basis for developing multiple-species economic injury levels, *J. Econ. Entomol.,* 81, 1, 1988.
4. Hutchins, S. H., and Funderburk, J. E., Injury guilds: a practical approach for managing pest losses to soybean, *Agric. Zool. Rev.,* 4, 1, 1991.
5. Hutchins, S. H., and Pedigo, L. P., Phenological disruption and economic consequence of injury to alfalfa induced by potato leafhopper (Homoptera: Cicadellidae), *J. Econ. Entomol.,* 83, 1587, 1990.
6. Santos, L., and Shields, E. J., Yield responses of corn to simulated cutworm (Lepidoptera: Noctuidae) damage, *J. Econ. Entomol.,* 91, 748, 1998.
7. Van Soest, P.J., *Nutritional Ecology of the Ruminant,* O and B Books, Corvallis, OR, 1982.
8. Barnes, R. F., and Gordon, C. H., Feeding value and on-farm feeding, in *Alfalfa Science and Technology,* Hanson, C. H., Ed., American Society Agronomy, Madison, WI, 601, 1972.
9. Gordon, C. H., Derbyshire, J. C., Wiseman, H. G., Kane, E. A., and Melin, C. G., Preservation and feeding value of alfalfa stored as hay, haylage and direct-cut silage. *J. Dairy Sci.,* 44, 1299, 1961.

10. Buxton, D. R., Hornstein, J. S., Wedin, W. F., and Marten, G. C., Forage quality in stratified canopies of alfalfa, birdsfoot trefoil, and red clover, *Crop Sci.,* 25, 273, 1985.
11. Smith, O. E., Goering, H. K., Waldo, D. R., and Gordon, C. H., In vitro digestion rate of forage cell wall components, *J. Dairy Sci.,* 54, 71, 1971.
12. Byers, J. H., and Ormiston, L. E., Nutrient value of forages. II. The influence of two stages of development of an alfalfa-borne grass hay on consumption and milk production, *J. Dairy Sci.,* 45, 693, 1962.
13. Buxton, D. R., and Hornstein, J. S., Cell-wall concentration and components in stratified canopies of alfalfa, birdsfoot trefoil, and red clover, *Crop Sci.,* 26, 180, 1986.
14. Hutchins, S. H., Buntin, G. D., and Pedigo, L. P., Impact of insect feeding on alfalfa regrowth: a review of physiological responses and economic consequences. *Agron. J.,* 82, 1035, 1990.
15. National Research Council, *Nutrient Requirements of Dairy Cattle,* 5th edition, National Academy of Sciences, Washington, D. C., 1978.
16. Hutchins, S. H., Thresholds involving plant quality and phenological disruption, in *Economic Thresholds for Integrated Pest Management,* Higley, L. G. and Pedigo, L. P., Eds., University of Nebraska Press, Lincoln, 275, 1996.
17. Hutchins, S. H., and Pedigo, L. P., Feed-value approach for establishing economic-injury levels, *J. Econ. Entomol.,* 91, 347, 1998.
18. Higley, L. G., and Pedigo, L. P., Eds., *Economic Thresholds for Integrated Pest Management,* University of Nebraska Press, Lincoln, 1996.
19. Buntin, G. D., and Pedigo, L. P., Enhancement of annual weed populations in alfalfa after stubble defoliation by variegated cutworm (Lepidoptera: Noctuidae), *J. Econ. Entomol.,* 79, 1507, 1986.

6 Photosynthesis, Yield Loss, and Injury Guilds

Robert K. D. Peterson

CONTENTS

6.1 INTRODUCTION

The title and subject of this book indicate that biotic stressors do indeed impact plant yield. The impact of insects, weeds, and plant pathogens on crop yields within agroecosystems has long been recognized and quantified. Further, the impact of biotic stressors on plant yield and fitness within natural ecosystems also is well known. Biotic stressors influence plant evolution through their impact on plant population dynamics, life history strategies, community structure, and ecosystem structure. But what do we know about how biotic stressors impact yield? What are the physiological mechanisms underlying yield loss or fitness changes in response to biotic stressors? Why are they important to understand?

Despite the importance of biotic stress on crop yield and plant fitness, the physiological mechanisms by which plants respond to these stresses continue to be poorly understood. In Chapter 1, Peterson and Higley argue that a focus on plant physiology provides a common language for characterizing plant stress and is essential for integrating understandings of stress. In this chapter, I build on that theme and discuss why plant gas exchange processes provide a critical foundation for that common language.

6.2   PLANT GAS EXCHANGE AND YIELD LOSS

6.2.1   Insect Injury and Plant Gas Exchange

Plant gas exchange processes represent a subset of a plant's physiological processes. Understanding how insect injury influences photosynthesis, water-vapor transfer, and respiration is important because these are the primary processes determining plant growth, development, and, ultimately, fitness. Consequently, understanding how insect injury impacts these primary processes is crucial to developing mechanistic explanations of yield loss.

Welter[1] and Peterson and Higley[2] reviewed the literature relatively recently with respect to current understandings of arthropod injury and plant gas exchange, focusing on insect injury guilds. Therefore, the literature will not be reviewed in this chapter. In this section, I highlight two examples of insect injury and their impact on plant gas exchange processes.

6.2.1.1   Leaf-mass Consumption

Most insect–plant gas exchange studies have used leaf-mass consumers (defoliators) as the stressor. This is understandable given that this type of injury is readily visible and quantifiable. Further, the insects, as a group, are relatively easy to rear and manipulate. Despite the relatively large number of studies that have been conducted during the past 40 years, general models of response have only been developed relatively recently.[1] This was partly due to observations of highly variable photosynthetic responses in response to defoliation injury. Indeed, studies have revealed a continuum of responses, from decreases to increases in photosynthetic rates of remaining leaves, plants, and plant canopies.[1] This has made generalizations difficult.

Even though a continuum of responses has been observed, most studies indicate that removal of either partial or entire leaves by insect herbivores increases photosynthetic rates of remaining leaf tissue. Most of these studies did not involve elucidation of the mechanisms underlying the photosynthetic increases. However, Welter[1] proposed several physiological (intrinsic) and environmental (extrinsic) causal factors for the increases in photosynthesis that have been observed. Physiological factors include increased assimilate demand after defoliation,[1] reduced competition between leaves for mineral nutrients necessary for cytokinin production,[1] and delayed leaf senescence.[3–7]

Reductions in photosynthetic rates have been observed, but the reductions typically were temporary.[8-10] Reductions generally were observed at the canopy level and were caused by decreased leaf-area indices, smaller leaf size, and decreased light interception.[11-13]

Several researchers also have observed no changes in photosynthetic rates of remaining leaf tissue in response to insect defoliation.[9, 14-22] These responses suggest that the photosynthetic apparatus of many plant species is not affected directly by leaf-mass consumption injury. Therefore, the principal effect of this injury type seems to be the reduction of photosynthesizing leaf area, not reduction or enhancement of photosynthetic capacity of remaining tissue of injured leaves.[21]

The most extensive work on insect defoliation injury and gas exchange responses has been on soybean, *Glycine max*.[13, 15, 18, 21, 23, 24] These studies have demonstrated that both simulated and actual insect defoliation do not perturb photosynthetic rates of remaining tissue of individual remaining leaves.

Poston et al.[15] used green cloverworm larvae, *Plathypena scabra*, painted lady larvae, *Vanessa cardui*, and hole punches from a paper punch and cork-borer punch to injure soybean leaflets. They observed small reductions in photosynthetic rates per unit area after 50% defoliation from painted lady larvae, paper punches, and cork-borer punches. These reductions were observed 12 hours after injury, but no reductions were observed 24, 48, 72, or 96 hours after injury.

Hammond and Pedigo[23] characterized water loss of individual soybean leaflets after defoliation injury by green cloverworm larvae, hole punches, and excision of whole leaflets. Leaves injured by green cloverworm larvae or by using a hole punch had significantly greater water loss than remaining leaflets after excision of one or two whole leaflets. Ostlie and Pedigo[24] observed water loss of soybean leaves after defoliation by cabbage looper larvae, *Trichoplusia ni*, green cloverworm larvae, hole punches, and picking whole leaflets. Water-loss differences among all treatments were transient, occurring for approximately 16 hours after injury.

Even though the soybean studies discussed above demonstrated leaf-mass consumption injury did not perturb gas exchange rates of remaining tissue of individual remaining leaves, this injury type does affect soybean gas exchange of the whole plant and canopy by reducing leaf area and delaying leaf senescence.[18, 25] Additionally, Ostlie and Pedigo[24] observed reductions in transpiration and water loss of the soybean canopy after defoliation.

6.2.1.2 Mexican Bean Beetle Injury

As discussed above, studies on soybean and many other plant species have shown that both simulated and actual insect defoliation do not perturb photosynthetic rates of remaining tissue of individual injured leaves.[12, 17-19] However, there are exceptions to this generalized photosynthetic response. One such exception is Mexican bean beetle, *Epilachna varivestis*, injury and photosynthetic responses of soybean and dry bean, *Phaseolus vulgaris*. Visually, injury by adult and larval Mexican bean beetles is physically different from injury by other lepidopteran and coleopteran soybean defoliators. Adults and larvae scrape, crush, and then consume leaf tissue, leaving both large and small leaf veins unconsumed, but often injured. The injured leaflet appears "laced" or "skeletonized."

The visual differences in leaflet injury by Mexican bean beetles also reflect physiological differences. Injury to plant leaves by both adults and larvae reduced photosynthetic rates of the remaining tissue of the injured leaflet.[26] Further, an inverse relationship was observed between photosynthetic rate and percentage injury. Additionally, there was no recovery of photosynthetic rates after injury to an individual leaflet. In other words, injury reduced the gas exchange capacity of the leaflet until it underwent normal, progressive photosynthetic and physical senescence.

The ultimate question is, what are the physiological mechanisms responsible for reductions in photosynthetic activity after Mexican bean beetle injury? Fortunately, we are now able to answer this question because of advances in *in vitro* and *in vivo* experimental methodologies. Sharkey[27] suggested three categories for all limitations to photosynthesis: the supply or utilization of CO_2, the supply or utilization of light, and the supply or utilization of phosphate. Using ecophysiological instrumentation and biochemically based models,[27–29] researchers have determined the role of stomatal and nonstomatal limitations to photosynthesis in several plant systems.

Specific biochemical limitations, such as ribulose 1,5-bisphosphate carboxylase/ oxygenase (rubisco) activity, ribulose 1,5-bisphosphate regeneration, and triose phosphate utilization can be determined for C_3 species using a combination of assimilation-intercellular CO_2 response curves, quantum efficiency determinations, fluorescence measurements, and metabolite assays.[27]

Many studies have characterized photosynthetic limitations during drought stress. However, only a few studies have examined photosynthetic limitations in response to biotic stress. Bowden et al.[30] observed that photosynthesis rates of leaves of potato, *Solanum tuberosum,* were reduced by *Verticillium dahliae,* a vascular fungal pathogen. Using gas exchange measurements, light response curves, and CO_2 response curves, they concluded that photosynthetic reductions primarily were caused by stomatal closure. However, Pennypacker et al.[31] observed that photosynthetic reductions of alfalfa, *Medicago sativa,* infected with *Verticillium albo-atrum* were caused by a reduction in the total activity and amount of rubisco, and not by stomatal closure.

Using some of the physiological measurement techniques described above, Peterson et al.[26] determined the likely biochemical and physiological mechanisms resulting in reductions in photosynthesis after Mexican bean beetle injury. By eliminating the possibility of reduced CO_2 availability (stomatal conductance limitations) and light-reaction limitations, they determined that the limitations to photosynthesis most likely were attributable to the utilization of CO_2 or the supply or utilization of phosphate. Therefore, the limitations were associated with rubisco activity, ribulose bisphosphate regeneration, or phosphate utilization. Ribulose bisphosphate regeneration can be affected by the photosynthetic electron transport chain which ATP and NADPH, or by insufficient capacity of the carbon reactions of the photosynthetic carbon reduction cycle. Because experimental results suggested that photosynthetic electron transport was not limiting, ribulose bisphosphate seemed to be limited by alterations in metabolite pools associated with the photosynthetic carbon reduction cycle (Figure 6.1).

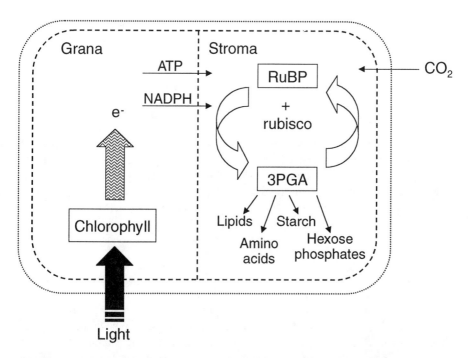

FIGURE 6.1 The light and dark reactions of photosynthesis in the chloroplast. Mexican bean beetle injury induces a mesophyll limitation to photosynthesis by reducing phosphate utilization, such as RuBP. (RuBP = ribulose bisphosphate; rubisco = ribulose bisphosphate carboxylase/oxygenase; 3PGA = 3 phosphoglycerate.)

Even though gas exchange responses and mechanisms to Mexican bean beetle injury arguably have been better characterized than any other type of insect injury, we still do not understand precisely where and how the limitations are occurring. Determining where the limitations are occurring would require, at the very least, quantitative assays for rubisco, ribulose bisphosphate, as well as other carbon reduction cycle metabolites. Determining how limitations are occurring is more problematic. Because the entire soybean and dry bean leaflet is affected by injury, endogenous signals such as phytohormones may be involved, given that phytohormones have been implicated in rate limitations of photosynthesis.[27] Alternatively, plant cell wall fragments (specifically oligosaccharides) are known to act as wound signals in response to some pathogens.[32] A similar signal transduction system may occur with Mexican bean beetle feeding, resulting in altered leaflet gas exchange.

Determining the potential mechanisms for photosynthetic rate limitations under any environmental stress, whether biotic or abiotic, is not simple, given the lack of understanding of many basic plant physiological processes associated with photosynthesis in the absence of stress.[27] However, detailed knowledge of the biochemical and cellular mechanisms underlying gas exchange responses to Mexican bean beetle injury is critical if we are to develop more encompassing understandings of the physiology of this type of biotic stress. Physiological responses need to be determined for this injury type at all levels of plant organization, from cellular to population. Whole plants may respond differently than individual leaflets to this injury. Whole plants and plant populations (canopies) may compensate for injury through various interactions with intrinsic and extrinsic factors (discussed below). This is likely, given that soybean and dry bean compensate for another type of injury (leaf-mass consumption) through delayed leaf senescence.

6.2.2 APPROACHES FOR SYNTHESIS

As stated above, identifying the physiological mechanisms underlying plant responses to arthropod injury is critical if we are to explain adequately yield loss and develop general models of response. To integrate and synthesize explanations of plant response to insect injury, we must consider five key areas: (1) injury guilds, (2) plant organization, (3) extrinsic factors, (4) experimental limitations, and (5) research objectives. Injury guilds will be discussed at length in a subsequent section.

6.2.2.1 Plant Organization

Plants (indeed all living things) can be considered to be organized at many levels, from biochemicals and their symphony of reactions, to cells, tissue, organs, organisms, and finally populations. Insect injury may impact plants at several or all of these levels (Figure 6.2). Moreover, responses to injury may be dramatically different at these different organizational levels. For example, in alfalfa and soybean, defoliation does not alter photosynthetic rates of the remaining tissue of individual leaflets (plant organs).[15, 18, 19] However, defoliation does alter the pattern of normal progressive leaf senescence of plants (organisms).[18, 19]

Differences in responses to insect injury among organization levels undoubtedly have led to the variability of photosynthetic responses observed in studies. Therefore, it is critical that researchers identify their research questions carefully and interpret their results accurately. Welter[1] put it succinctly, "... if the question is to examine the effects of herbivory on the productivity or "fitness" of a plant then total canopy measurements would provide a better indicator. If the authors are interested in specific

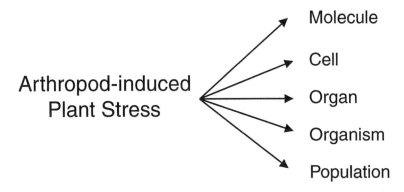

FIGURE 6.2 Arthropod-induced stress impacts plants at several levels of organization. Gas exchange responses to stressors may differ among different levels.

plant buffering mechanisms, then use of individual leaves should provide a more detailed understanding."

Describing how individual leaves respond to injury is important given that these are the organs most immediately affected by herbivory. This is reflected in the literature; the research area on plant gas exchange and herbivory is replete with papers on individual leaf responses to herbivory.[1] However, describing leaf gas exchange responses to injury clearly is not sufficient. Developing a more complete understanding of injury requires that greater attention be given to responses at organizational levels above and below that of individual leaves. Mechanistic understanding of gas exchange responses necessitates work at lower levels of organization (e.g., molecular and cellular levels). The Mexican bean beetle example discussed above is an example of research at these levels of biological organization.

In a similar vein, evolutionary, ecological, and agricultural understandings of the impact of herbivory depend upon characterizing how plant populations are affected by insect injury. Unfortunately, responses of individual leaves have not been related to responses at higher levels of organization. Indeed, I am aware of only four studies directly examining canopy responses to herbivory.[12, 13, 18, 33]

6.2.2.2   Extrinsic Factors

By determining the direct responses of insect injury on plant physiology, interactive effects can be evaluated more accurately (Figure 6.3). This is especially salient in natural systems because extrinsic factors constraining optimal plant growth, both

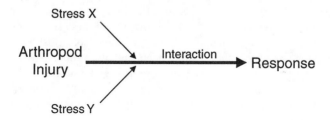

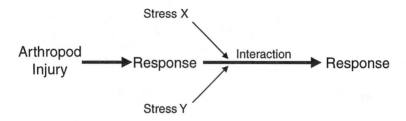

FIGURE 6.3   Arthropod injury interacts with other biotic and abiotic stresses to produce plant responses (top). Substantial progress will be made by first characterizing the direct responses of plants to arthropod injury, and then determining how extrinsic factors interact with arthropod stress (bottom).

temporally and spatially, are more common than in agricultural systems. Numerous extrinsic factors can interact with the direct effects of injury. Extrinsic factors may include light penetration, water availability, and nutrient availability.[1, 19] The variable responses reported in the literature may be attributable to the interaction of these extrinsic factors with insect injury.

If physiological responses to injury among plant species are similar, then general models can be constructed and indirect factors that influence plant response can be evaluated more effectively.[34] For example, in an optimal growing environment defoliation injury to a leaf may not alter its photosynthetic rate per unit area of remaining tissue. If water and nitrogen are limiting, defoliation injury may result in increases in photosynthetic rates of remaining leaves because the remaining leaves would no longer be water and nitrogen deficient. This mechanism may have occurred in results from grassland system studies by McNaughton[35] and Detling et al.[11]

6.2.2.3 Experimental Limitations

There is no doubt that progress in understanding herbivory and plant responses has been constrained by experimental limitations. These limitations involve both experimental objectives and experimental procedures. Peterson and Higley[2] stated, "Experimental objectives focused research onto specific questions; are we asking questions about herbivory and plant response that lead to broad understandings or are we posing narrow questions that preclude such understandings? Much research on herbivory and plant gas exchange has been observational, in the sense that we observe plant responses to different types of insect injury. This work is important in establishing the general nature of plant response to herbivory, but of itself it does not provide explanations for the observed effects." In support of these statements, Higley et al.[34] argue for an emphasis on mechanisms of stress, specifically physiological responses to injury, as a basis for broader understandings of plant stress.

The second experimental issue is that of experimental procedures. Difficulties in conducting experiments on plant responses to herbivory have constrained, and likely will continue to constrain, our understanding of herbivory. Procedural issues can be separated into questions of quantification (or measurement) and of control (maintaining treatment integrity).[2]

To study plant gas exchange, the ability to measure gas exchange accurately and conveniently has been a daunting obstacle. However, the development of portable infrared gas analyzers for measuring carbon exchange rates solved this problem, at least for single leaf or leaflet measurements. Measurements of canopy gas exchange parameters are more difficult, and although canopy measurement systems can be constructed,[36] their complexity has limited their use.

Another important measurement issue involves quantifying insect injury. Physiological, growth, and yield responses to herbivory are a function of the amount of injury. However, many studies on herbivory and plant response have not adequately assessed injury. This problem is particularly relevant for some types of insect injury that are difficult to quantify, such as leaf mining or those producing chlorosis. Recent advances in digital imaging hardware and software provide a means for discriminating between even subtle differences between most injured and uninjured tissues.[2]

Another experimental quantification issue relates to biochemical measurements. Peterson and Higley[2] stated, "For example, assessing differences in ribulose bisphosphate carboxylase/oxygenase (rubisco) levels between injured and uninjured plants can be important in determining direct effects of injury on the photosynthetic apparatus. However, rubisco assays are expensive, time consuming, and cannot be conducted for all plant species. Similar arguments apply to other molecules of interest in stress response. These limitations tend to restrict our work to plant species whose biochemistry and physiology are relatively well known."

We discussed above the influence of extrinsic factors that can confound understandings of the direct effects of insect injury on photosynthesis. Because plant gas exchange processes are highly sensitive, considerations and control of confounding factors are crucial during experimentation. Plant gas exchange processes vary across individual plants, plant ages, and plant tissues, and are highly sensitive to many factors, including light, temperature, relative humidity, and plant water status.[2] These factors are difficult to control across treatments, but lack of control may lead to unacceptable variation among treatments, masking the effects of herbivory. Unfortunately, much of the literature on plant gas exchange and insect injury does not include thorough discussions of potential confounding or of experimental procedures used to maintain treatment uniformity.[2]

Another problem in experimentation is that injury treatments are imposed. Techniques such as caging or insecticides used to manipulate insect numbers (and thereby injury) have the potential to interact with injury. For example, cages reduce the light environment of the plant, and insecticides can alter plant physiology and photosynthetic responses. These potential problems need to be discussed by the study authors, but this is seldom the case. (See Chapter 3 on techniques for evaluating yield loss from insects for more information.)

The issue of control and confounding in experiments makes research in most natural systems extremely difficult. The finding that natural and cultivated species do not have qualitative differences in response to insect injury[37] suggests it may be possible to develop models of plant response in cultivated species, where external factors are more easily controlled. These models then can be used to characterize plant responses in more variable systems.[2]

6.2.2.4 Research Objectives

Most of the studies on plant gas exchange and insect injury to date address physiological responses in that they document those responses. However, very few of those studies incorporate mechanistic understandings as part of their research objectives. This is reasonable given current understandings. Indeed, considerable documentation research is needed, especially for injury types such as leaf mining, assimilate removal, root feeding, stem boring, and leaf skeletonizing.

Current descriptive understandings of gas exchange responses to herbivory in some systems have allowed for a transition to more explanatory research.

Comprehensive mechanistic explanations of leaf-mass consumption injury have been determined for soybean. Further, Peterson et al.[26] determined likely cellular and sub-cellular mechanisms underlying reductions in photosynthetic rates in response to Mexican bean beetle injury. The desired goal would be to have general models that provide mechanistic physiological explanations of impacts on yield loss and fitness. Therefore, research objectives should focus on explanations of gas exchange responses, not merely cataloging responses. This would lead to broader understandings of biotic stressors, yield loss, and plant-stressor evolution. Additionally, from a practical perspective, mechanistic physiological explanations will lead to more comprehensive pest management strategies that focus on management of a group of stressors that cause homogeneities of injury. As I will discuss in the next section, this would be a considerable improvement over managing individual pest species.

6.3   PHOTOSYNTHESIS AND INJURY GUILDS

6.3.1   THE EVOLUTION OF THE INJURY GUILD CONCEPT

Insects historically were placed into feeding guilds according to their taxonomic status. However, studies revealed that injury mechanisms are not unique to each insect species. Consequently, insects herbivores can be grouped into injury guilds based on the general physical appearance of the injury.[38, 39] Examples include leaf mining, leaf skeletonizing, stem boring, fruit scarring, and seed feeding. Boote[25] further refined the classification of injury guilds by emphasizing the physiological responses of the plant to unique injury types. According to Boote's scheme, there are five injury guilds: stand reducers, leaf-mass consumers, assimilate sappers, turgor reducers, and fruit feeders. Pedigo et al.[40] proposed an additional injury guild — plant architectural modifier. Higley et al.[34] incorporated the six injury types, and suggested several more, into categories of physiological impact. These include: population or stand reduction, leaf-mass reduction, leaf photosynthetic-rate reduction, leaf senescence alteration, light reduction, assimilate removal, water-balance disruption, seed or fruit destruction, architecture modification, and phenological disruption. "Using this scheme, insects can be grouped into categories that better describe their differential impact on host physiology and yield."[2] Further refinements of injury types are possible with better knowledge of plant physiological and biochemical processes. However, evolving from categories of pest taxonomy to plant physiological impact is a tremendous improvement. The focus now clearly is on the host and injury vs. the biotic stressor. This allows for more comprehensive understandings of yield loss and fitness impacts.

6.3.2   INJURY EQUIVALENCY AND GUILDS

Homogeneities in physiological response have been identified for different pest species.[1, 2, 17, 19, 34, 41] This is especially well known for leaf-mass consumers.[19, 42–44] If pests produce similarities in plant response, they then can be placed into injury guilds. A tremendous practical advantage to placing those pests into injury guilds is that pest management programs can be developed for the entire injury guild, as opposed to managing individual pest species. Most of the research in this area has focused on the identification of injury types, the construction of injury guilds,[34, 42] and the development of multiple-species economic injury levels (EILs).[43] To date, injury guilds have been developed for weeds,[45] defoliating caterpillars in soybean,[42, 43] and stubble regrowth defoliators in alfalfa.[44]

To use injury guilds for pest management, several requirements have been promulgated.[34, 42, 43] The pest species must: (1) produce a similar type of injury, e.g., individuals cannot be leaf-mass consumers and fruit feeders at the same point in time; (2) produce injury within the same phenological time frame of the host, e.g., insect species consuming leaf tissue in the early vegetative stages of soybean growth cannot be compared to late-season defoliators; (3) produce injury of a similar intensity, e.g., the slope of the damage curves for each member species should be functionally similar; and (4) affect the same plant part, e.g., leaf-mass consumption in the lower canopy stratum may produce a different host response than consumption of terminal leaves.

First, *injury equivalents* need to be determined. An *injury equivalent* is the amount of injury that could be produced by one pest through its complete life cycle.[42] *Injury equivalency* is the total number of injury equivalents for a population. In other words, injury equivalency is injury by one species expressed in terms of another species or by different life stages within a species.[42] Hutchins and Funderburk[43] stated, "For example, a small larva (instar 1 or 2), which can consume only about 1% of its total lifetime potential, would be considered the equivalent of 1/100th of a large larva. Hence, leaf tissue loss from 100 small larvae is required to equal the foliage consumption of one large larva." When considering different species, "a medium velvetbean caterpillar larva consumes the equivalent of 5% of the leaf tissue that a large soybean looper larva consumes."[43]

Once injury guilds and injury equivalents have been determined, multiple-species EILs can be calculated. Hutchins et al.[42] developed multiple-species EILs for insect defoliators of soybean. Because soybean defoliators produce similar responses, a single damage function was determined for all species (damage functions were based on injury equivalencies). A matrix of injury equivalency coefficients by larval size and species was used to assess the collective impact of several insect pest species on yield.

6.3.3 PHOTOSYNTHESIS AND THE DEVELOPMENT OF INJURY GUILDS

As discussed above, the injury guild concept has evolved from grouping pests into taxonomic categories to grouping them into categories of plant physiological impact. Therefore, each injury guild species must produce an injury that elicits a similar physiological response. Improved knowledge of the impact of biotic stressors on plant gas exchange processes may improve the injury guild, injury equivalency, and multiple-species EIL concepts. More emphasis on photosynthetic responses and injury guilds would center on the first injury guild requirement as promulgated by

Hutchins et al.[42] Injury guild species must produce a similar type of injury. Consequently, similar types of injury can be determined by assessing homogeneities of photosynthetic responses. This approach has been used to support injury guild status in soybean and alfalfa.

6.3.3.1 Soybean Leaf-mass Consumer Injury Guild

Because several soybean leaf-mass consumers elicit similar physiological impacts on soybean, they can be placed into the same injury guild. As discussed above, soybean leaf-mass consumption does not alter photosynthesis in remaining leaf tissue. Canopy gas exchange responses are altered by leaf-mass consumption injury, but they most likely are similar among pest species. The other requirements for injury guild membership (discussed above) also are met for soybean leaf-mass consumers, so several species can be placed into the guild, such as soybean looper, *Pseudoplusia includens,* velvetbean caterpillar, *Anticarsia gemmatalis,* green cloverworm, beet armyworm, *Spodoptera exigua,* and bean leaf beetle, *Ceratoma trifurcata.*[42]

However, not all species that consume soybean leaves can be members of the leaf-mass consumption injury guild. For example, Mexican bean beetle larvae and adults produce a physiological response (reduced photosynthetic rates) different than that produced by members of the injury guild.[26] Therefore, the exclusion of the Mexican bean beetle from the injury guild is based on physiological response, not physical appearance of the injury.

6.3.3.2 Alfalfa Stubble Defoliator Injury Guild

Alfalfa stubble defoliation occurs after the hay crop is cut and the stubble is regrowing. Several insects may defoliate alfalfa stubble, including alfalfa weevil larvae and adults, *Hypera postica,* clover leaf weevil larvae and adults, *Hypera punctata,* and variegated cutworm larvae, *Peridroma saucia.* These insects consume dry matter, delay regrowth initiation and subsequent plant maturity, and reduce growth rates after defoliation.[46]

Peterson et al.[44] concluded that alfalfa responses to clover leaf weevils, alfalfa weevils, and variegated cutworms are similar. The three species are leaf-mass consumers. Research on leaf-mass consumption in alfalfa showed no alterations in photosynthetic rates of remaining tissue.[19] Similarities in gas exchange responses in addition to similarities in consumption patterns and timing of injury meet the requirements for placing the three species into a common injury guild in alfalfa. Peterson et al.[44] developed injury equivalencies for the three species, which could then be used for multiple-species management guidelines.

6.4 FUTURE DIRECTIONS

Future research must emphasize how and why changes in gas exchange rates occur in response to biotic stress. This type of research will do more to advance our understanding of plant stress from biotic stress than simply characterizing gas exchange responses. Using this approach, insect injury and other factors, such as plant competitive interactions, plant diseases, mineral stress, and moisture stress can be integrated into a more encompassing view of plant stress at all levels of plant organization.[34, 47-49]

Advances in plant physiology instrumentation and biotechnology will help determine both how and why changes in plant gas exchange occur after the initiation of biotic stress. In recent years, portable photosynthesis systems (infrared CO_2 gas analyzers) have been developed that allow light response and CO_2 assimilation curves to be determined more easily. Determining the genetic basis underlying plant physiology mechanisms to biotic stress clearly would have far-reaching consequences. Plant transgenic approaches could be employed to better understand and manage biotic stress. For example, whole plants and canopies of some species, such as soybean, respond to leaf-mass consumption injury by delaying normal progressive photosynthetic senescence. If there is a genetic basis for this phenomenon, transgenic techniques could be utilized that would result in greater delays of photosynthetic senescence after injury, thus tolerating injury better than standard cultivars.

Chapin[50] promulgated a conceptual integration for abiotic stresses. With the appropriate research objectives, a conceptual integration for biotic stresses also is possible. Finally, integration of abiotic and biotic stress will lead to a synthesis for all types of plant stress. This will have tremendous value for our understanding of natural and agricultural ecosystems.

REFERENCES

1. Welter, S. C., Arthropod impact and plant gas exchange, in *Insect-plant Interactions, Vol. 1,* Bernays, E. A., Ed., CRC Press, Boca Raton, 135, 1989.
2. Peterson, R. K. D., and Higley, L. G., Arthropod injury and plant gas exchange: current understanding and approaches for synthesis, *Entomol. (Trends Agric. Sci.),* 1, 93, 1993.
3. Neales, T. F., and Incoll, L. D., The control of leaf photosynthesis rate by the level of assimilate concentration in the leaf: a review of the hypothesis, *Bot. Rev.,* 34, 107, 1968.
4. Wareing, P. F., Khalifa, M. M., and Treharne, K. J., Rate-limiting processes in photosynthesis at saturating light intensities, *Nature,* 220, 453, 1968.
5. Gifford, R. M., and Marshall, C., Photosynthesis and assimilate distribution in *Lolium multiflorum* Lam. following differential tiller defoliation, *Aust. J. Biol. Sci.,* 26, 517, 1973.
6. Caldwell, M. M., Richards, J. H., Johnson, D. A., and Dzurec, R. S., Coping with herbivory: photosynthetic capacity and resource allocation in two semiarid *Agropyron* bunchgrass, *Oecologia,* 50, 14, 1981.
7. Nowak, R. S., and Caldwell, M. M., A test of compensatory photosynthesis in the field: implications for herbivore tolerance, *Oecologia,* 61, 311, 1984.
8. Alderfelder, R. G., and Eagles, C. F., The effect of partial defoliation on the growth and photosynthetic efficiency of bean leaves, *Bot. Gaz.,* 137, 351, 1976.
9. Hall, F. R., and Ferree, D. C., Effects of insect injury simulation on photosynthesis of apple leaves, *J. Econ. Entomol.,* 69, 245, 1976.
10. Li, J., and Proctor, T. A., Simulated pest effects [sic] photosynthesis and transpiration of apple leaves, *HortScience,* 19, 815, 1984.

11. Detling, J. K., Dyer, M. I., and Winn, D. T., Effect of simulated grasshopper grazing on carbon dioxide exchange rates of western wheatgrass leaves, *J. Econ. Entomol,* 72, 403, 1979.

12. Boote, K. J., Jones, J. W., Smerage, G. H., Barfield, C. S., and Berger, R. D., Photosynthesis of peanut canopies as affected by leafspot and artificial defoliation, *Agron. J.,* 72, 247, 1980.

13. Ingram, K. T., Herzog, D. C., Boote, K. J., Jones, J. W., and Barfield, C. S., Effects of defoliating pests on soybean CO_2 exchange and reproductive growth, *Crop Sci.,* 21, 961, 1981.

14. Davidson, J. L., and Milthorpe, F. L., The effect of defoliation on the carbon balance in *Dactylis glomerata Ann. Bot.,* 30, 185, 1966.

15. Poston, F. L., Pedigo, L. P., Pearce, R. B., and Hammond, R. B., Effects of artificial and insect defoliation on soybean net photosynthesis, *J. Econ. Entomol.,* 69, 109, 1976.

16. Syvertsen, J. P., and McCoy, C. W., Leaf feeding injury to citrus by root weevil adults: leaf area, photosynthesis, and water use efficiency, *Florida Entomol.,* 68, 386, 1985.

17. Welter, S. C., Responses of tomato to simulated and real herbivory by tobacco hornworm, *Environ. Entomol.,* 20, 1537, 1991.

18. Higley, L. G., New understandings of soybean defoliation and their implications for pest management, in *Pest Management in Soybean,* Copping, L. G., Green, M. B., and Rees, R. T., Eds., Elsevier, London, 1992, 56.

19. Peterson, R. K. D., Danielson, S. D., and Higley, L. G., Photosynthetic responses of alfalfa to actual and simulated alfalfa weevil (Coleoptera: Curculionidae) injury, *Environ. Entomol.,* 21, 501, 1992.

20. Peterson, R. K. D., Higley, L. G., and Spomer, S. M., Injury by *Hyalophora cecropia* (Lepidoptera: Saturniidae) and photosynthetic responses of apple and crabapple, *Environ. Entomol.,* 25, 416, 1996.

21. Peterson, R. K. D., and Higley, L. G., Temporal changes in soybean gas exchange following simulated insect defoliation, *Agron. J.,* 88, 550, 1996.

22. Burkness, E. C., Hutchinson, W. D., and Higley, L. G., Photosynthesis response of 'Carolina' cucumber to simulated and actual striped cucumber beetle (Coleoptera: Chrysomelidae) defoliation. *Entomologia Sinica,* 6(1), 29, 1999.

23. Hammond, R. B., and Pedigo, L. P., Effects of artificial and insect defoliation on water loss from excised soybean leaves, *J. Kansas Entomol. Soc.,* 54, 331, 1981.

24. Ostlie, K. R., and Pedigo, L. P., Water loss from soybeans after simulated and actual insect defoliation, *Environ. Entomol.,* 31, 341, 1984.

25. Boote, K. J., Concepts for modeling crop response to pest damage, ASAE Pap. 81-4007, American Society of Agricultural Engineers, St. Joseph, MI, 1981.

26. Peterson, R. K. D., Higley, L. G., Haile, F. J., and Barrigossi, J. A. F., Mexican bean beetle (Coleoptera: Coccinellidae) injury affects photosynthesis of *Glycine max* and *Phaseolus vulgaris, Environ. Entomol.,* 27, 373, 1998.

27. Sharkey, T. D., Photosynthesis in intact leaves of C_3 plants: physics, physiology, and rate limitations, *Bot. Rev.,* 51, 53, 1985.

28. Farquhar, G. D., and von Caemmerer, S., Modelling of photosynthetic response to environmental conditions, in *Encyclopedia of Plant Physiology, New Series,* Lange, O. L., Nobel, P. S., Osmond, C. B., and Zeigler, H., Eds., Springer, New York, 549, 1982.

29. Farquhar, G. D., and Sharkey, T. D., Stomatal conductance and photosynthesis, *Annu. Rev. Plant Physiol.,* 33, 317, 1982.

30. Bowden, R. L., Rouse, D. I., and Sharkey, T. D., Mechanism of photosynthesis decrease by *Verticillium dahliae* in potato, *Plant Physiol.,* 94, 1048, 1990.

31. Pennypacker, B. W., Knievel, D. P., Leath, K. T., Pell, E. J., and Hill, Jr., R. R., Analysis of photosynthesis in resistant and susceptible alfalfa clones infected with *Verticillium albo-atrum*, *Phytopathology*, 80, 1300, 1990.

32. Fry, S. C., Aldington, S., Hetherington, P. R., and Aitken, J., Oligosaccharides as signals and substrates in the plant cell wall, *Plant Physiol.*, 10, 1, 1993.

33. Daley, P. F., and McNeill, J. N., Canopy photosynthesis and dry matter partitioning of alfalfa infested by the alfalfa blotch leafminer (*Agromyza frontella* (Rondani)), *Can. J. Plant Sci.*, 67, 433, 1987.

34. Higley, L. G., Browde, J. A., and Higley, P. M., Moving towards new understandings of biotic stress and stress interactions, in *International Crop Science I*, Buxton, D. R., Shibles, R., Forsberg, R. A., Blad, B. L., Asay, K. H., Paulson, G. M., and Wilson, R. F., Eds., Crop Science Society of America, Madison, WI, 1993, 749.

35. McNaughton, S. J., Physiological and ecological implications of herbivory, in *Encyclopedia of Plant Physiology, New Series. Vol. 15.*, Functional Responses to the Chemical and Biological Environment, Lange, O. L., Osmond, C. B., Nobel, P. S., and Ziegler, H., Eds., Springer, New York, 1983, 12.

36. Garcia, R. L., Norman, J. H., and McDermitt, D. K., Measurements of canopy gas exchange using an open chamber system, *Remote Sensing Rev.*, 5, 141, 1990.

37. Welter, S. C., and Steggall, J. W., Contrasting the tolerance of wild and domesticated tomatoes to herbivory: agro-ecological implications, *Ecol. Appl.*, 3, 271, 1993.

38. Metcalf, C. L., Flint, W. P., and Metcalf, R. L., *Destructive and Useful Insects*, McGraw-Hill, New York, 1962.

39. Bardner, R., and Fletcher, K. E., Insect infestations and their effects on the growth and yield of field crops: a review. *Bull. Entomol. Res.*, 64, 141, 1974.

40. Pedigo, L. P., Hutchins, S. H., and Higley, L. G., Economic injury levels in theory and practice, *Annu. Rev. Entomol.*, 31, 341, 1986.

41. Welter, S. C., Responses of plants to insects: eco-physiological insights, in *International Crop Science I*, Buxton, D. R., Shibles, R., Forsberg, R. A., Blad, B. L., Asay, K. H., Paulson, G. M., and Wilson, R. F., Eds., Crop Science Society of America, Madison, WI, 1993, 773.

42. Hutchins, S. H., Higley, L. G., and Pedigo, L. P., Injury equivalency as a basis for developing multiple-species economic injury levels, *J. Econ. Entomol.*, 81, 1, 1988.

43. Hutchins, S. H., and Funderburk, J. E., Injury guilds: a practical approach for managing pest losses to soybean, *Agric. Zool. Rev.*, 4, 1, 1991.

44. Peterson, R. K. D., Higley, L. G., and Danielson, S. D., Alfalfa consumption by adult clover leaf weevil (Coleoptera: Curculionidae) and development of injury equivalents for stubble defoliators, *J. Econ. Entomol.*, 88, 1441, 1995.

45. Wilkerson, G. G., Modena, S. A., and Coble, H. D., HERB: Decision model for postemergence weed control in soybean, *Agron. J.*, 83, 413, 1991.

46. Hutchins, S. H., Buntin, G. D., and Pedigo, L. P., Impact of insect feeding on alfalfa regrowth: a review of physiological responses and economic consequences, *Agron. J.*, 82, 1035, 1990.

47. Louda, S. M., Keeler, K. H., and Holt, R. D., Herbivore influences on plant performance and competitive interactions, in *Perspectives in Plant Competition*, Grace, J. B., and Tilman, D., Eds., Academic Press, New York, 1990.

48. Tilman, D., Constraints and tradeoffs: toward a predictive theory of competition and succession, *Oikos*, 58, 3, 1990.

49. Louda, S. M., and Collinge, S. K., Plant-resistance to insect herbivores: a field test of the environmental-stress hypothesis, *Ecology*, 73, 153, 1992.

50. Chapin, F. S., Integrated responses of plants to stress, *BioScience*, 41, 29, 1991.

7 The Influence of Cultivar and Plant Architecture on Yield Loss

Fikru J. Haile

CONTENTS

7.1 INTRODUCTION

Solar energy plays an indispensable role in ecosystem processes. Plants convert solar energy into chemical energy, via photosynthesis. The carbohydrates and other macromolecules that are synthesized form the basis for the food webs on which all heterotrophes depend. Because light energy supports life on earth, understanding how plants intercept light and convert it into macromolecules has been, and still is, a critical area of investigation.

Light interception is partly modified by plant architecture. In both natural and agricultural systems, therefore, plant architecture affects plant fitness and yield. Plants have evolved different adaptive traits to maximize canopy light interception. One major adaptive trait is modification of plant architecture. Plant architecture used to maximize light interception involves modification of leaf size, leaf shape, leaf angle, plant height, branches, and tillers. In addition, some plants can modify their canopy architecture transiently to maximize light interception by leaf solar tracking. These adaptations suggest that plants may be limited by light to carry out photosynthesis at a full efficiency.

Plant architecture is a genetically controlled trait and therefore it is heritable. However, environmental factors, both abiotic and biotic, can modify canopy architecture and alter light interception. Abiotic factors that affect canopy architecture include soil moisture content, nutrient availability, temperature, and light. Biotic factors capable of altering plant architecture are herbivores, pathogens, and competition with other plants.

Interactions among environmental factors also occur, modifying plant canopy architecture. These environmental factors interact to affect plant architecture and subsequently plant fitness. Conversely, plant architecture can affect the distribution and survival of arthropods and their natural enemies and the competitive ability of neighboring plants. In agricultural ecosystems, understanding how plant canopies affect the fitness of arthropod pests, weeds, and diseases is useful for devising effective pest management strategies.

Plant architecture modification following arthropod injury may contribute to plant tolerance. Consequently, traits such as leaf size and shape, canopy size, tillering, and branching need to be considered when selecting cultivars tolerant to arthropod injury. In this chapter, we examine the ecological significance of plant architecture modification by environmental stresses and how yield loss to arthropods can be minimized by using cultivars with improved architectures.

7.2 UNDERSTANDING RADIATION INTERCEPTION AND CANOPY PROCESSES

Canopy structures generally refer to the volume and distribution of above-ground plant parts. Norman[1] defines canopy structure as the shape, size, orientation, and distribution of leaves, stems, branches, flowers, and fruits. Detailed information about canopy structure can be obtained by measuring canopy volume and light interception. Information about canopy structures can be generated either by direct or indirect methods. The direct method is destructive and involves measuring leaf angles, leaf areas, and leaf positions in the canopy. It also is time consuming. The indirect method is nondestructive and relies on instruments such as plant canopy analyzers and light sensors.[2] These techniques are reliable and can also be used to study canopies defoliated by insect pests.[3]

7.2.1 LIGHT INTERCEPTION AND PRODUCTIVITY

Canopy light interception, which determines plant biomass production, depends on leaf area index (LAI) and canopy architecture. LAI is the ratio of total leaf area per unit ground area, and it indicates canopy volume. Usually, larger canopy volume or greater LAI values may indicate greater light interception. However, LAI does not accurately indicate light interception. In addition to canopy volume (LAI), canopy architecture modifies canopy light penetration and also affects total canopy light interception. Canopy light penetration is measured by the light extinction coefficient (k). Consequently, canopy photosynthesis and productivity can be modified by canopy architecture because of the depth of light penetration and changes in k.

Canopy photosynthesis and productivity is a function of canopy volume and longevity.[4] Leaf longevity can be modified by plant canopy architecture because of changes in the amount of light penetrating canopies. In plant canopies with a greater k, the contribution of leaves on the lower nodes to canopy productivity is greater than canopies with a smaller k. This is because canopies with a smaller k have horizontal leaf orientation and once the canopy is closed, leaves on the lower nodes are shaded and senesce faster than canopies with a greater k. In canopies with a greater k value, light reaches leaves on the lower nodes that contribute to plant productivity without senescing as quickly as in canopies with a smaller k.

Canopy longevity depends on the longevity of individual leaves, which can determine plant productivity. Plant physiologists agree that improvement in seed yield has been achieved primarily by improving leaf longevity rather than photosynthetic rate.[4] The contribution of leaf longevity or leaf duration to yield is derived from the following relationship:

$$\text{Yield} = \text{Rate} \times \text{Duration} \qquad [7.1]$$

This relationship can be applied to individual leaves or canopies. Canopy architecture, which also impacts leaf longevity, determines plant productivity. Environmental factors, such as nutrients, soil moisture, and herbivory can greatly impact the longevity and photosynthetic rates of plant canopies.[5]

7.2.2   LIGHT INTERCEPTION AND PLANT TOLERANCE

Leaf area index indicates the relative tolerance of cultivars to insect injury. Cultivars that can maintain excess leaf tissues, which may not contribute to seed yield, can tolerate insect injury without a significant yield loss. Consequently, high LAI values may be important in selecting cultivars tolerant to insect injury.

Light extinction coefficient, which affects light reaching leaves on the lower nodes, is another important aspect of plant architecture that can affect plant tolerance to insect injury. Mathematically, k is derived from the fraction of photosynthetically active radiation intercepted by canopy and LAI. According to Beer's Law:

$$\text{IPAR} = \text{PAR} \times (1 - \exp)(-k \times \text{LAI}) \qquad [7.2]$$

where IPAR is the fraction of photosynthetically active radiation intercepted by the canopy, and PAR is photosynthetically active radiation.[6]

The values for k may be variable depending on cultivar[7, 8] and row spacing.[9] Like LAI, k plays an important role in the tolerance response of plants to insect injury. Yield depends on the amount of canopy light interception after insect injury, and k

can greatly modify light interception because of changes in canopy light penetration. Consequently, depth of light penetration and light extinction coefficients can be important in selecting cultivars that are tolerant to insect injury.[10]

7.3   ADAPTIVE SIGNIFICANCE OF PLANT ARCHITECTURE

Light is an important environmental factor that can reduce plant fitness if it is not available in adequate supply for normal plant physiological processes. The evolution of plant architecture to maximize light interception can be regarded as the major adaptation to enhance plant fitness. Plant architecture not only maximizes light interception but it also increases plant ecological resistance to pests. Because changes in microenvironments of arthropods and pathogens are greatly mediated by plant architecture, the evolution of a modified plant architecture could also be an adaptation to minimize plant damage from arthropods and pathogens.

Ecological studies have examined the evolution and adaptive significance of plant architectures. These studies suggest that competition for light is the primary selection pressure that led to the evolution of plant architectures. Even within the same plant species, there are genetic differences in radiation-use-efficiencies under different light intensities.[11] This suggests light alone can be an important environmental variable to influence plant genetic diversity. The selection pressure for light is so great that some plants evolved leaves with more than one morphology and orientation type at later plant growth stages to maximize light interception.[12] The evolution of such a trait is considered an adaptation for arboreal life.

Plant features that contribute to plant architecture include plant height, leaf size, leaf shape, leaf angle, leaf area index, and canopy light interception. Models have been developed to examine how these plant features alter plant architecture and light interception. King[13] studied the ecological importance and the contribution of heights to plant fitness. These studies demonstrate that plant features, such as plant height, are adaptations for arboreal life to increase light interception. This is particularly true in trees, where plant height is considered the major adaptation. In addition, changes in leaf shape, size, and arrangement patterns on the nodes are known to play significant roles in light interception and can impact plant fitness.[14] Even within a given community structure, light is one of the environmental requisites that affect the evolution and distribution of plants adapted to exploit a specific niche in a community.[15]

In agricultural ecosystems, the yield potential of crop plants can be increased by improving plant architecture. Dwarf cultivars have been developed with modified canopy architectures capable of better light interception in different crops.[16] Cultivar differences in plant architecture, resulting from differences in LAI, light interception, and light extinction coefficients contribute to differences in cultivar yield. Although the major factor that determines the evolution of such plant features is light, cultivars may exhibit differences in light interception also as an adaptation to different growing conditions. Therefore, yield potential of cultivars cannot be accurately estimated without considering plant architecture. Consequently, models used to predict cultivar yield potential must incorporate variations in plant architectures.

7.4   PLANT ARCHITECTURE AND INSECT FITNESS

Plant architecture greatly mediates insect–plant interactions. Plant architecture also impacts the fitness of natural enemies of insect herbivores. Plant architecture influences the abundance and distribution of insects and their natural enemies primarily by modifying canopy microenvironments. Conversely, plant architectures can be modified by environmental factors such as temperature and moisture stress. See Chapter 8 for a discussion on the impact of moisture stress and changes in canopy microenvironments and insect fitness.

Insect herbivores inhabit plant canopies, if host plants are nutritionally acceptable and canopy microenvironments are suitable, for survival and reproduction. In addition, insect herbivores need hiding places from their natural enemies. These hiding places, which are known as enemy-free zones, also determine insect distribution in plant canopies. Further, oviposition sites, shelter, and places to overwinter also can influence insect distribution in plant canopies.

The diversity and abundance of arthropods in plant canopies may depend on canopy size. Tree canopies, which are larger and provide diverse microenvironments, have richer insect diversity compared to herbs and annual plants.[17] The diversity of insect fauna in tree canopies has been explained by the resource diversity hypothesis.[17] This hypothesis assumes that, because trees provide more diverse resources, there is more diverse insect fauna on tree canopies, including natural enemies, than on other canopies. The same hypothesis can be applied to crop canopies. Larger crop canopies may provide more resources for insects and, therefore, may have more diverse insect fauna than smaller canopies. However, the diversity of insect fauna in monoculture farming may be limited compared to mixed farming, regardless of canopy sizes.

The distribution of arthropods within crop canopies may be variable, depending on the microenvironments at different heights in a canopy. Some arthropods prefer lower canopies to seek a more suitable temperature and relative humidity,[18] while others may prefer the upper canopies. Also, the microenvironment in plant canopies can affect the fitness of natural enemies. Consequently, the success of pest management programs, particularly biological control, may depend on plant canopy architecture. This is primarily because plant architecture impacts microenvironments for biological control agents affecting predator or parasite searching ability[19] or the incidence of fungal pathogens.[20]

7.5   CANOPY ARCHITECTURE MODIFICATION

Plant architecture is an inherited trait. However, the expression of plant architecture can be greatly modified depending on the growing conditions of plants. Plant architectural modifications, in response to changes in environmental factors, are adaptations to environmental stresses. These environmental stresses can be abiotic factors that are related to resource acquisition, or biotic factors including competition with neighboring plants and the impact of herbivores and plant pathogens. In agricultural ecosystems, cultural practices such as row spacing, plant density, and cultivar can also affect plant architecture.

7.5.1 ABIOTIC FACTORS

Plant architecture is greatly mediated by environmental resources under which plants are grown. Abiotic factors that can affect plant architecture include resources for plant growth such as soil moisture, temperature, and light. If these resources are sufficient, plants can attain a growth rate close to their genetic potential, maximize fitness, and express typical architectures. However, if these resources are not sufficient, plants undergo physiological and growth changes to modify their architectures in an attempt to increase fitness.

Moisture stress is the most important abiotic factor limiting plant productivity.[21] As an adaptation to moisture stress, plants undergo architectural modifications. Clearly, plants adapted to drought conditions, such as desert environments, display unique plant architectures to minimize water loss. When plants are moisture-stressed, they alter resource allocations to different plant parts. Bloom, Chapin, and Mooney[22] suggested that plants allocate more photosynthates to structures that are used to acquire resources limiting plant growth. Because water is the most limiting resource under moisture stress, more resources are allocated to roots to acquire more water. Consequently, moisture-stressed plants have a higher root:shoot ratio. Because allocation of resources to the above-ground plant parts is relatively smaller, moisture stressed plants are shorter and have smaller leaves and canopies.

In addition to plant architectural modification because of changes in resource allocation to roots and shoots, moisture stressed plants undergo changes in canopy architecture because of changes in leaf orientation toward the sun. In soybean, moisture stressed plants reduce heat load by reflecting the incident radiation. This is achieved by inverting the leaves and exposing the undersides of the leaves to the sun.[23, 24] The underside of the leaf is lighter than the upper surface and can reflect more light away from soybeans. This adaptation to moisture stress condition minimizes transpirational water loss. Changes in leaf orientation in response to moisture stress, resulting in canopy architectural modification, also occur in other crop plants.[25, 26] In cowpeas, Shackel and Hall[25] observed that moisture stressed plants showed a different diurnal leaf orientation towards the sun from plants provided with ample moisture. Moisture stressed plants tracked the sun in the morning (diaheliotropism) and avoided the sun (paraheliotropism) in the afternoon, when the heat load is greatest.

This change in leaf orientation, also called leaf solar tracking, is an adaptation that helps plants reduce heat load and minimize water loss when plants are subjected to moisture stress. However, leaf solar tracking is not limited to moisture stress. It can also be expressed in response to other stresses, particularly to limited light.[27] Leaf solar tracking enables plants to increase the canopy light interception for photosynthesis. Regardless of the nature of the stress, leaf solar tracking has important consequences because it impacts canopy architecture by altering canopy light interception transiently.

Leaf solar tracking is considered adaptive only for canopies with smaller LAIs. For larger canopies, the major limitation to productivity is the amount of light penetrating into a canopy, and the contribution of leaf solar tracking is considered insignificant.[27, 28] Based on canopy models, leaf solar tracking restricts light penetration into the lower canopy and may be counterproductive. Travis and Reed,[29] however, demonstrated that a closed alfalfa canopy increased productivity by leaf solar tracking. Whether productivity can be increased in all plant species by leaf solar tracking has not been established. However, leaf solar tracking can increase plant fitness by improving canopy architecture and light interception, at least in some systems.

Other abiotic factors that affect plant architectures include temperature and photoperiod. These factors affect canopy architecture because they impact plant growth rate and resource allocation to vegetative and reproductive structures. Temperature contributes significantly to plant growth because temperature influences enzymatic activity and other plant biochemical processes. In muskmelons, modification of canopy architecture by temperature is related to its impact on plant height, number of nodes, and flowering.[29]

7.5.2 CULTURAL PRACTICES

In agricultural ecosystems, crop architecture can be modified by cultural practices such as row spacings and planting densities.[30, 31, 32] In grain crops, narrow row spacings may increase light interception earlier in the season, before canopy closure, and can result in greater yields compared to the standard 76-cm spacings. In addition, narrow spacing modifies canopy size and improves the competitive ability of some crops against weeds.[33] Likewise, native plants also alter plant architecture depending on the density and competitive ability of neighboring plants.[34] Therefore, changes in plant architecture resulting from altered row spacings or plant density affect plant productivity because such changes alter canopy light interception and canopy apparent photosynthesis.

Another cultural practice that also can improve plant productivity by altering canopy architecture and light interception is selection of cultivars with different canopy architectures. These three cultural practices (row spacings, planting densities, and cultivars) may interact to affect crop productivity. Willcott et al.[35] demonstrated that there was a significant interaction among these three cultural practices, impacting light interception, canopy light penetration, and soybean yield. Consequently, there may be specific row spacing and planting density for a cultivar with a given plant architecture that can result in a maximum light interception and yield.

Although modifying row spacing and plant density can result in improved seed yield depending on the cultivar used, the wider adoption of such a practice may be limited by the technology used for planting, cultivating, and harvesting. For instance, the current U.S. farm machinery used for grain crop production is based on 76-cm row spacing. Altering such machinery for narrow row spacing may not be feasible, at least in the short term. In addition, increased crop yields by using a narrow spacing may vary depending on canopy architectures of cultivars used, and it may not apply to all crops. Westgate et al.[36] did not observe significant yield increases in maize using narrow row-spacings. However, their study showed that narrow spacing increased radiation-use-efficiency, and canopy closure was achieved earlier in the season compared to the standard row spacing.

7.5.3 BIOTIC FACTORS

Plants actively respond to biotic stressors such as arthropods and pathogens. Plant response to these stressors involves physiological processes aimed at healing and repairing injured cells and tissues, thus ensuring structural and physiological integrity. Plants respond to injuries by biotic stressors in different ways. The most common is a compensatory response to outgrow injury without reduction in plant fitness or yield. The other less common plant response is induced plant resistance, which involves an increased concentration of some compounds to deter further injury.

Insect injury can modify plant canopy architecture in different ways. Injury by defoliating insects involves removal of leaf mass tissue and results in reduced canopy light interception.[37, 38] Defoliating insects typically impose only mechanical leaf removal, without interfering with plant physiological processes such as photosynthesis. However, possibly because of improved water and light status, defoliated plants can delay leaf senescence and may compensate for defoliation. In addition, defoliated plants may produce new leaves to replace lost leaf tissues. Consequently, the major impact of defoliating insects is reduced canopy photosynthesis because of smaller canopy size,[39] without reductions in photosynthetic rates per unit leaf area of the remaining leaves.[37, 40–42]

Canopy architecture modification after injury by defoliating insects involves improved light penetration into canopies, new leaf tissue production, branching, and tillering.[10, 43, 44] This likely is an adaptation to offset the impact of defoliation injury. In addition to changes in canopy structure, defoliated plants also may exhibit delayed maturity compared to undefoliated plants. Although delayed maturity potentially contributes to yield compensation in defoliated plants because of an extended growing period, it may not be desirable in an agricultural ecosystem, where uniform field maturity is required for mechanized harvesting, or in areas where the growing season is too short to allow normal plant growth and maturity.

In addition to defoliation injury, plants also experience injury from piercing and sucking, skeletonizing, stem-boring, or root-feeding arthropods. Unlike defoliators, these arthropods typically remove plant sap or leaf epidermal tissues, or feed on roots. In addition to removing plant tissue, these arthropods can also interrupt normal plant physiological processes. Typically, injuries from these arthropods result in yield losses because of their impact on plant physiology, including reduced photosynthesis per unit leaf area, interruption of water and sap flow, or reduced water uptake from the soil. Consequently, these injury types can reduce plant canopy sizes, indirectly, by altering plant physiology.

Arthropods that remove plant sap, such as spider mites, can also reduce the radiation-use-efficiency of crop plants. These types of injuries also can cause leaf senescence in severe infestations, leading to reductions in canopy size and canopy light interception.[45] Skeletonizing insects, such as the Mexican bean beetle, *Epilachna varivestis,* can reduce photosynthetic rates per unit leaf area[46] and can potentially modify plant canopy architecture also by causing leaf senescence. Arthropods that feed in plant stems such as the European corn borer, *Ostrinia nubilalis,* or those that feed on plant roots such as the western corn rootworm, *Diabrotica virgifera virgifera,* can directly reduce canopy light interception when injured plants are dislodged.[47, 48] The primary mechanism by which root-feeding insects impact canopy architectures may be by altering sink-source relationships and limiting resource allocation to canopies. For example, injury to corn by western corn rootworm can reduce allocation of photosynthates to leaves as more resources are allocated to replace injured roots. Such injuries can also reduce canopy light interception, canopy photosynthesis, and seed yield.[49]

Plant canopy architecture also can be modified by plant pathogens and weeds. Parallels can be drawn between plant physiological responses to injuries from arthropods that alter plant physiology and injuries from plant pathogens. Plant pathogens reduce photosynthetic rates per unit leaf area and canopy radiation-use-efficiencies.[50] Under severe disease incidence, leaves senesce, reducing canopy light interception.[51–53] Consequently, the primary mechanism by which plant pathogens injure their host plants is by reducing canopy light interception and canopy photosynthesis, similar to arthropods.

Unlike arthropods and plant pathogen injuries, weeds or competing neighboring plants impose their stress differently. Stress from neighboring plants involves competition for light, water, and nutrients. Such competition can affect resource allocation patterns in different plant parts and can also modify plant architecture.[34] Light can be a major resource for which plants compete and may have slightly more impact on plant architecture than other resources. In agricultural ecosystems, the ability of crop plants to compete successfully with weeds partly depends on crop growth rate. Some crops, such as soybean, are capable of closing their canopy earlier in a season and can suppress weed population better than other crops. Consequently, canopy traits such as leaf area index, rate of canopy closure, and plant height[54] are known to modify plant competitive ability against and tolerance to weeds.

7.6 YIELD LOSS: INTERACTION OF INSECT INJURY WITH ABIOTIC FACTORS

Yield loss can occur when plants are subjected to stresses. Environmental resources, such as soil moisture content, soil nutrient status, and temperature are rarely optimal for normal plant growth. Stresses from these environmental factors, particularly from moisture stress, represent a major impediment to agricultural productivity.[21]

The magnitude of yield loss from either biotic or abiotic stressors depends on the severity of the stress and growth stage of the plant when stress occurs. Usually, stresses at reproductive stages cause a higher yield loss than at vegetative stages. For example, soybean yield is more sensitive to moisture stress and defoliation at reproductive than vegetative plant growth stages.[55–57] Consequently, multiple stresses at crop reproductive stages can cause substantial yield losses.

In addition to their direct role on yield, abiotic stressors also can interact with biotic stressors, altering plant response. The impact of insect injury on yield loss can be reduced by provision of adequate growth requisites at reproductive stages. If nutritional and water supply are optimal, defoliated plants can produce new leaves and also delay senescence to compensate for defoliation, thus avoiding significant yield loss.[10, 44] The compensatory ability of defoliated plants, therefore, depends on optimal resource availability and growth rate for canopy recovery. Because yield responses of crop plants depend on the amount of light intercepted at reproductive stages,[37] rapid canopy recovery is crucial to minimize yield loss from insect injury.

Abiotic factors can significantly modify the compensatory response of plants to insect injury. In ecological studies, the impact of herbivory on plant fitness has been controversial.[58] Some have argued that herbivory can be beneficial to plants by stimulating growth and leading to overcompensation[59, 60] while others have maintained that herbivory can reduce plant fitness significantly.[61] One reason for such diverse opinions is the interaction of herbivory with abiotic factors. Plant compensation and overcompensation are greatly mediated by resource availability for plant growth. Therefore, the impact of herbivory on plant fitness should be viewed in conjunction with the impact of abiotic factors. The current consensus is that herbivory generally reduces plant fitness or crop yield. However, plants can compensate for herbivory depending on the intensity of injury, plant growth stage when injury occurs, and optimal resource availability for rapid plant growth.

In insect pest management programs, the impact of insect injury on yield loss can be magnified on previously stressed plants. Therefore, plant tolerance to insect injury can be increased by improving plant vigor through provision of optimum resources required for plant growth.

7.7 RESPONSES OF CULTIVARS TO INSECT INJURY

7.7.1 LIGHT INTERCEPTION

Canopy architecture determines the amount of light intercepted, plant competitive ability for light, and consequently plant fitness or yield. Because light interception, which significantly contributes to seed yield of cultivated crops, depends on plant architecture, we can infer that cultivated plants have been selected for canopy architectures with improved light interception capacities over time. Modified canopy architecture is still an important selection criterion in plant breeding programs.[16, 62–64] Similarly, in natural systems, plants with better canopy structure and hence better light interception are better adapted than plants with canopy structures that result in lower light interception.

Light interception is modified by canopy volume and canopy structure. In grain crops, yield is a function of light interception, and these two parameters usually have significant linear relationships. However, canopy volume or LAI is not linearly related to light interception or yield.[65] Consequently, whereas light interception may show the yield potential of crop plants, LAI alone does not accurately indicate canopy light interception or seed yield.[66] An exception to this rule may be in forage crops, where total biomass is the harvestable yield.[67]

Light interception, which intrinsically accounts for changes in canopy architecture and light penetration into plant canopies, can indicate canopy photosynthesis and seed yield better than LAI. The major difference between light interception and LAI may be that light interception accounts for changes in the light extinction coefficient but not LAI. Therefore, differences in the light interception of cultivars primarily can be due to differences in light extinction coefficients, if LAIs are the same. Light extinction coefficient differences primarily are due to canopy architecture differences among cultivars.

Some factors that may be responsible for differences in light extinction coefficients are differences in the angle, shape, and size of leaves. Variability in these leaf and leaflet features and their contribution to canopy light interception and yield has been observed in soybean. Cultivars with different degrees of leaflet orientation toward the sun were identified, including differences in the orientation of central and lateral leaflets.[68] Variation in leaf angle and leaf distribution in soybean canopies can be driven by differences in light intensities.[69]

Comparison of near-isogenic soybean lines showed that isolines with narrow leaflets had lower LAI but permitted better light penetration into soybean canopies compared to wide or ovate leaflets which had higher LAI.[70] There were no significant yield differences between these isolines because low LAI in narrow isolines is compensated by light penetration into the canopy.[71, 72] Examination of light extinction coefficients between these lines showed that isolines with narrow leaflets had a higher light extinction coefficient compared to isolines with ovate leaflets.[10]

In addition to LAI, light interception and light extinction coefficient may be useful when selecting cultivars tolerant to insect injury. In soybean, we found that isolines with narrow leaflets could tolerate defoliation as much as or better than isolines with wide leaflets because of improved light interception after defoliation injury was imposed.[10]

7.7.2 PHOTOSYNTHESIS

Photosynthesis directly contributes to plant biomass. Therefore, it is reasonable to expect that cultivars with a high rate of leaf photosynthesis may yield greater than cultivars with a low rate of leaf photosynthesis. However, although several cultivars with a high rate of leaf photosynthesis have been identified, there was no significant correlation between leaf photosynthesis and seed yield.[73–76] High yield usually has been achieved by improving leaf longevity, harvest index, and season length,[4, 77] not by improving leaf photosynthetic rates. This does not imply that the possibility for developing cultivars by improving their photosynthetic rates is unattainable. It has just not been a practical approach to increase seed yield.

Although the possibility for improving yield by increasing leaf photosynthesis rates seems rather remote, empirical evidence supports a significant correlation between canopy-apparent photosynthesis and seed yield.[78–82] In these studies, canopy-apparent photosynthesis partly accounted for improved seed yield. Although leaf photosynthesis directly contributes to canopy photosynthesis, it does not precisely indicate canopy photosynthesis or seed yield because of variations in leaf longevity, LAI, light interception, and canopy architecture. The contribution of canopy architecture to canopy-apparent photosynthesis and yield has been studied in cotton.[79, 83] In these studies, cotton cultivars with sub-okra leaf morphologies modified canopy architecture and increased canopy-apparent photosynthesis and cotton lint yield.

To understand yield loss from insect injury, canopy-apparent photosynthesis may provide more useful information than leaf photosynthesis. In addition, insect injury can cause changes in canopy architecture; understanding plant physiological responses at a canopy level would be desirable. However, there is limited information regarding canopy photosynthetic responses to insect injury.

Leaf photosynthetic responses to insect injury help us understand immediate plant physiological responses and may indicate plant fitness or yield response to insect injury. Relationships between cultivar photosynthetic rates and plant tolerance to insect injury have not been studied. Do cultivars with a high rate of leaf photosynthesis have a better tolerance to insect injury than cultivars with a low rate of photosynthesis? If so, can leaf photosynthesis be used as an indicator of a cultivar's tolerance level? Answers to these questions may help us understand the relationships between photosynthesis and plant tolerance.

To understand the contribution of photosynthetic rates to plant tolerance, photosynthetic responses of cultivars after insect injury may provide more useful information than rates in the absence of insect injury. Cultivars may have comparable rates of leaf photosynthesis in the absence of injury. Indeed, in a study that examined alfalfa genotypes differing in resistance to insect pests and diseases, Chatterton[84] did not find significant photosynthetic rate differences when these genotypes were healthy. However, genotypes with different resistance mechanisms can alter their photosynthetic rates after insect injury is imposed.

In a greenhouse study, we observed the photosynthetic response of wheat lines differing in resistance to Russian wheat aphid, *Diuraphis noxia.*[85] Two lines, one with antibiotic and the other with tolerance resistance mechanisms to Russain wheat aphid, were evaluated. Photosynthetic rates of these lines after aphid injury revealed that the tolerant cultivar, PI 262660, compensated for aphid injury as early as three days after aphids were removed. However, this was not the case with the antibiotic line, PI 137739. The antibiotic line showed a significant photosynthetic rate reduction by aphid injury compared to the tolerant line, suggesting a trade-off between an induced defense and photosynthesis. This study demonstrated that photosynthetic rate differences after insect injury may significantly contribute to plant tolerance in some cultivars. In addition, leaf photosynthesis after insect injury may be useful to understanding physiological and growth response of cultivars to insect injury.

7.8   CULTIVARS AND PEST MANAGEMENT DECISIONS

Differences in physiological, growth, and yield responses of cultivars to insect injury have important implications for pest management strategies. An important component of modern pest management strategy is the economic injury level (EIL). Usually, EILs are developed for a given crop, not cultivars. EILs developed based on a susceptible cultivar tend to be lower than EILs developed based on a tolerant cultivar.

Consequently, use of a single EIL for all cultivars may present a problem in making pest management decisions.

In addition to intrinsic cultivar differences, environmental factors can also affect plant tolerance to insect injury. Moisture stress can reduce plant tolerance to insect injury by reducing tillering capacity[86] or canopy recovery[10, 44] Furthermore, temperature can also affect the expression of tolerance genes to insect injury.[87] These environmental factors, affecting plant compensation and yield loss to insect injury, can also affect EILs.[88, 89] Such EIL variations based on cultivars and environmental variables suggest that the EIL is dynamic and should be treated as such.

Variation of cultivars in tolerance to insect injury and its implications in developing EILs have been examined in soybean.[44] In this study, "Dunbar," an early-maturing cultivar and the least tolerant to defoliation injury, had the lowest EIL value, compared to "Clark" and "Corsica," which are slightly late maturing and defoliation tolerant. Dunbar's EIL was half of the EIL for the other two cultivars. Therefore, the tolerance level of cultivars must be considered before developing and implementing EILs. In soybean, some of the factors that can affect yield loss and EILs of cultivars include maturity group, canopy architecture, and moisture and nutrient availability. One solution to the problem of EIL variability involves developing EILs based on maturity groups, canopy architecture, and resource availability.

7.9 CONCLUSIONS

Plant architecture is the spatial arrangement of leaves and other photosynthetic organs on stems and branches. Because leaves collect solar energy and provide surfaces for gas exchange, their arrangement in plant canopies is crucial for light interception and photosynthesis. Hence, plant architecture and canopy structures determine photosynthetic efficiencies and significantly contribute to plant fitness. In addition, plant canopies modify canopy microenvironments and can significantly alter the distribution, diversity, and fitness of arthropods, plant pathogens, and other neighboring plants. In agroecosystem, plant architecture determines plant productivity, the abundance and success of arthropod pests, diseases, and weeds, and their natural enemies, thus impacting pest management strategies.

Cultivars with modified plant architectures capable of directly suppressing pest populations can be selected. Further, cultivars capable of tolerating pest injury because of a modified canopy architecture and improved light interception following injury may be useful to manage pests. This suggests that cultivars tolerant to pest injury can be selected based on canopy architectures. Such cultivars can become ideal components of integrated pest management programs because yield loss from such cultivars is minimal and the need for pesticides can be reduced or eliminated. Consequently, evaluation of cultivars for their architecture that enables them to tolerate pest injury may be a promising approach for pest management.

Cultivars with modified architectures for better light interception and canopy photosynthesis and consequently higher yields have been selected and bred. In addition, cultivars with modified architectures to suppress weeds have been identified and used in agricultural production. However, cultivars with modified architectures have not been widely developed and used to manage insect pests. In insect pest management programs, tolerant cultivars have been least implemented compared to antibiotic and antixenotic cultivars, primarily because mechanisms that contribute to tolerance are not well known. One feasible approach, therefore, is to study plant tolerance based on canopy architectures.

REFERENCES

1. Norman, J. M., and Campbell, G. S., Canopy structure, in *Plant Physiological Ecology,* Pearcy, R. W., Ehleringer, J., Mooney, H. A., and Rundel, P. W., Eds., Chapman and Hall, London, 1989, chap.14.
2. Welles, J. M., and Norman, J. M., Instrument for indirect measurement of canopy architecture, *Agron. J.,* 83, 818, 1991.
3. Hunt, T. E., Haile, F. J., Hoback, W. W., and Higley. L. G., Indirect measurement of insect defoliation, *Environ. Entomol.,* 28, 1136, 1999.
4. Thomas, H., Canopy survival, in *Crop Photosynthesis: Spatial and Temporal Determinants,* Baker, N. R., and Thomas, H., Eds., Elsevier, Amsterdam, 1992, chap. 2.
5. Mooney, H. A., and Gulmon, S. L., Constraints on leaf structure and function in reference to herbivory, *BioScience,* 32, 198, 1982.
6. Thornley, J. H. M., *Mathematical Models in Plant Physiology,* Academic Press, London, 1976.
7. Pepper, G. E., Pearce, R. E., and Mock, J. J., Leaf orientation and yield of maize, *Crop Sci.,* 17, 883, 1977.
8. Ong, C. K., and Monteith, J. L., Canopy establishment: light capture and loss by crop canopies, in *Crop Photosynthesis: Spatial and Temporal Determinants,* Baker, N. R., and Thomas, H., Eds., Elsevier, Amsterdam, 1992, chap. 1.
9. Flent, F., Kiniry, J. R., Board, J. E., Westgate, M. E., and Reicosky, D. C., Row spacing effects on light extinction coefficients of corn, sorghum, soybean, and sunflower, *Agron. J.,* 88, 185, 1996.
10. Haile, F. J., Higley, L. G., Specht, J. E., and Spomer, S. M., The role of leaf morphology in soybean tolerance to defoliation, *Agron. J.,* 90, 353, 1998.
11. Sultan, S. E. and Bazzaz, F. A., Phenotypic plasticity in *Polygonum persicaria.* I. Diversity and uniformity in genotypic norms of reaction to light, *Evolution,* 47, 1009, 1993.
12. Rich, P. M., Holbrook, N. M., and Luttinger, N., Leaf development and crown geometry of two iriarteoid palms, *Am. J. Bot.,* 82, 328, 1995.
13. King, D. A., The adaptive significance of tree height, *Am. Nat.,* 135, 809, 1990.
14. Niklas, K. J., The role of phyllotactic pattern as a "development constraint" on the interception of light by leaf surfaces, *Evolution,* 42, 1, 1988.
15. Schimel, D. S., Kittel, T. G. F., Knapp, A. K., Seastedt, T. R., Parton, W. J., and Brown, V. B., Physiological interactions along source gradients in tallgrass prairie, *Ecology,* 72, 672, 1991.
16. Coyne, D. P., Modification of plant architecture and crop yield by breeding, *HortScience,* 15, 244, 1980.
17. Lawton, J. H., Plant architecture and the diversity of phytophagous insects, *Annu. Rev. Entomol.,* 28, 23, 1983.
18. Isichaikul, S., Fujimura, K., and Ichitkawa, T., Humid microenvironment prerequisite for the survial and growth of nymphs of the rice brown planthopper, *Nilaparvata lugens* (Stal) (Homoptera: Delphacidae), *Res. Population Ecol.,* 36, 23, 1994.

19. Jarosik, V., *Phytoseiulus persimilis* and its prey *Tetranychus utricae* on glasshouse cucumbers and peppers: key factors related to biological control efficiency, *Acta Entomologica Bohemoslovaca,* 87, 414, 1990.

20. Marcandier S., and Khachatourians, G. G., Susceptibility of the migratory grasshopper, *Melanoplus sanguinipes* (Fab.) (Orthoptera: Acrididae), to *Beauveria bassiana* (Bals.) Vuillemin (Hyphomycete): influence of relative humidity, *Can. Entomol.,* 119, 901, 1987.

21. Boyer, J. S., Plant productivity and environments, *Science,* 218, 443, 1982.

22. Bloom, A. J., Chapin, F. S, III, and Mooney, H. A., Resource limitation in plants — an economic analogy, *Annu. Rev. Ecol. Syst.,* 16, 363, 1985.

23. Meyer W. S., and Walker, S., Leaflet orientation in water-stressed soybeans, *Agron. J.,* 73, 1071, 1981.

24. Oosterhuis, D. M., Walker, S., and Eastham, D. J., Soybean leaflet movement as an indicator of crop water stress, *Crop Sci.,* 25, 1101, 1985.

25. Shackel, K. A., and Hall, A. E., Reversible leaflet movements in relation to drought adaptation of cowpeas, *Vigna unguiculata* (L.) Walp, *Aust. J. Plant Physiol.,* 6, 265, 1979.

26. Rowson, H. M., Vertical wilting and photosynthesis, transpiration, and water use efficiency of sunflower leaves, *Aust. J. Plant Physiol.,* 6, 109, 1979.

27. Ehleringer, J. R., and Forseth, I. N., Diurnal leaf movements and productivity in canopies, in *Plant Canopies, Their Growth, Form, and Function,* Russell, G., Marshall, B., and Jarvis, P. G., Eds., Cambridge University Press, Cambridge, 1989, chap. 7.

28. Fukai, S., and Loomis, R. S., Leaf display and light environments in row-planted cotton communities, *Agric. Meteorol.,* 17, 353, 1976.

29. Travis, R. L., and Reed, R., The solar tracking pattern in a closed alfalfa canopy, *Crop Sci.,* 23, 664, 1983.

30. Ventura, Y., and Mendlinger, S., Effects of suboptimal low temperatures on plant architecture and flowering in muskmelons *Cucumis melo* L., *J. Hortic. Sci. Biotechnol.,* 73, 640, 1998.

31. Shibles, R. M., and Weber, C. R., Interception of solar radiation and dry matter production by various soybean planting patterns, *Crop Sci.,* 6, 55, 1966.

32. Heitholt, J. J., Meredeth, M. R., Jr., and Willford, J. R., Comparison of cotton genotypes varying in canopy architecture in 76-cm vs. 102-cm rows, *Crop Sci.,* 36, 955, 1996.

33. Legere, A., and Schreiber, M. M., Competition and canopy architecture as affected by soybean (*Glycine max*) row width and density of redroot pigweed (*Amaranthus retroflexus*), *Weed Sci.,* 37, 84, 1989.

34. Tremmel, D. C., and Bazzaz, F. A., Plant architecture and allocation in different neighborhoods: implications for competitive success, *Ecology,* 76, 262, 1995.

35. Willcott, J., Herbert, S. J., and Liu, Z. Y., Leaf area display and light interception in short-season soybeans, *Field Crops Res.,* 9, 173, 1984.

36. Westgate, M. E., Forcella, F., Reicosky, D. C., and Somsen, J., Rapid canopy closure for maize production in the northern US corn belt: radiation-use-efficiency and grain yield, *Field Crops Res.,* 49, 249, 1997.

37. Higley, L. G., New understanding of soybean defoliation and their implication for pest management, in *Pest Management in Soybean,* Copping, L. G., Green, M. G., and Rees, R. T., Eds., Elsevier Science Publishers, London. 1992, 56.

38. Board, J. E., Wier, A. T., and Boethel, D. J., Critical light interception during seed filling for insecticide application and optimum soybean grain yield, *Agron. J.,* 89, 369, 1997.

39. Ingram, K. T., Herzog, C., Boote, K. C., Jones, J. W., and Barfield, C. S., Effects of defoliating pests on soybean canopy CO_2 exchange and reproductive growth, *Crop Sci.,* 21, 961, 1981.

40. Boote, K. J., *Concepts for Modeling Crop Response to Pest Damage,* American Society of Agricultural Engineers, St. Joseph, MI, 1981.

41. Peterson, R. K. D., Danielson, S. D., and Higley, L. G., Photosynthetic response of alfalfa to actual and simulated alfalfa weevil (Coleoptera: Curculionidae) injury, *Environ. Entomol.,* 21, 501, 1992.

42. Peterson, R. K. D., and Higley, L. G., Temporal changes in soybean gas exchange following simulated insect defoliation, *Agron. J.,* 88, 550, 1996.

43. Rubia, E. G., Hong, K. L., Zalucki, M., Gonzales, B., and Norton, G. A., Mechanisms of compensation of rice plants to yellow stem borer *Scirpophaga incertulas* (Walker) injury, *Crop Prot.,* 15, 335, 1996.

44. Haile, F. J., Higley, L. G., and Specht, J. E., Soybean cultivars and insect defoliation: yield loss and economic injury levels, *Agron. J.,* 90, 344, 1998.

45. Sadras, V. O., and Wilson, L. J., Growth analysis of cotton crops infested with spider mites. I. Light interception and radiation-use-efficiency, *Crop Sci.,* 37, 481, 1997.

46. Peterson, R. K. D., Higley, L. G., Haile, F. J., and Barrigossi, J. A. F., Mexican bean beetle (Coleoptera: Coccinellidae) injury affects photosynthesis of *Glycine max* and *Phaseolus vulgaris, Environ. Entomol.,* 27, 373, 1998.

47. Godfrey, L. D., Holtzer, T. O., Spomer, S. M., and Norman, J. M., European corn borer (Lepidoptera: Pyralidae) tunneling and drought stress: effects on corn yield, *J. Econ. Entomol.,* 84, 1850, 1991.

48. Spike, B. P., and Tollefson, J. J., Yield response of corn subjected to Western corn rootworm (Coleoptera: Chrysomelidae) infestation and lodging, *J. Econ. Entomol.,* 84, 1585, 1991.

49. Urias-Lopez, M. A., Physiological and Yield Responses of Different Types of Maize to Western Corn Rootworm Larval Injury, Dissertation, University of Nebraska – Lincoln, 1998.

50. Welter, S. C., Arthropod impact on plant gas exchange, in *Insect-Plant Interactions,* Bernays, E. A., Ed., CRC Press, Boca Raton, 1989, 135.

51. Waggoner, P. F., and Berger, R. D., Defoliation, disease, and growth, *Phytopathololgy,* 77, 393, 1987.

52. Johnson, K. B., Defoliation, disease, and growth: a reply, *Phytopathology,* 77, 1495, 1987.

53. de Koeijer, K. J., and van der Werf, W., Effect of beet yellowing viruses on light interception and light use efficiency of the sugarbeet crop, *Crop Prot.,* 14, 291, 1995.

54. Lindquist, J. H., Mortensen, D. A., and Johnson, B. E., Mechanisms of corn tolerance and velvetleaf suppressive ability, *Agron. J.,* 90, 787, 1998.

55. Ashley, D. A., and Ethridge, W. J., Irrigation effects on vegetative and reproductive development of three soybean cultivars, *Agron. J.,* 70, 467, 1978.

56. Fehr, W. R., Caviness, C. E., and Vorst, J. J., Response of indeterminate and determinate soybean cultivars to defoliation and half-plant cut-off, *Crop Sci.,* 17, 913, 1977.

57. Hunt, T. E., Higley, L. G., and Witkowski, J. F., Soybean growth and yield after simulated bean leaf beetle injury to seedlings, *Agron. J.,* 86, 140, 1994.

58. Belsky, A. J., Does herbivory benefit plants? A review of the evidence, *Am. Nat.,* 127, 870, 1986.

59. McNaughton, S. J., Compensatory plant growth as a response to herbivory, *Oikos,* 40, 329, 1983.

60. Paige, K. N., and Whitham, T. J., Overcompensation in response to mammalian herbivory: the advantage of being eaten, *Am. Nat.,* 129, 407, 1987.

61. Crawley, M. J., Insect herbivores and plant population dynamics, *Annu. Rev. Entomol.,* 34, 531, 1989.

62. Caviness, C. E., Registration of soybean germplasm lines R85–395 and R88–1259, with quantafoliolate leaves, *Crop Sci.,* 31, 495, 1991.
63. Brothers, M. E., and Kelly, J. D., Interrelationship of plant architecture and yield components in the pinto bean ideotype, *Crop Sci.,* 33, 1234, 1993.
64. Nelson, R., The inheritance of branching type of soybean, *Crop Sci.,* 36, 1150, 1996.
65. Wells, R., Soybean growth response to plant density: relationship among canopy photosynthesis, leaf area index, and light interception, *Crop Sci.,* 31, 755, 1991.
66. Robinson, T. L., and Lakso, A. N., Bases of yield and production efficiency in apple orchard systems, *J. Am. Soc. Hortic. Sci.,* 116, 188, 1991.
67. Redfearn, D. D., Moore, K. J., Vogel, K. P., Waller, S. S., and Mitchell, R. B., Canopy architecture and morphology of switchgrass populations differing in forage yield, *Agron. J.,* 89, 262, 1997.
68. Wofford, T. J., and Allen, F. L., Variation in leaflet orientation among soybean cultivars, *Crop Sci.,* 22, 999, 1982.
69. Blaine, B. L., and Baker, D. G., Orientation and distribution of leaves within soybean canopies, *Agron. J.,* 64, 26, 1972.
70. Hicks, D. R., Pendleton, J. W., Bernard, R. L., and Johnson, T. J., Response of soybean plant types to planting patterns, *Agron. J.,* 61, 290, 1969.
71. Hartung, R. C., Specht, J. E., and Williams, J. H., Agronomic performance of selected soybean morphological variants in irrigation culture with two row spacings, *Crop Sci.,* 20, 604, 1980.
72. Mandl, F. A., and Buss, G. R., Comparison of narrow and broad leaflet isolines of soybean, *Crop Sci.,* 21, 25, 1981.
73. Dornhoff, G. M., and Shibles, R. M., Varietal differences in net photosynthesis of soybean leaves, *Crop Sci.,* 10, 42, 1970.
74. Bhagsari, A. S., Ashley, D. A., Brown, R. H., and Boerma, H. R., Leaf photosynthetic characteristics of determinate soybean cultivars, *Crop Sci.,* 17, 929, 1977.
75. Palit, P., Kundu, A., Mandal, R. K., Sircar, S. M., Photosynthetic efficiency and productivity in tropical rice, *Plant Biochem. J.,* 3, 54, 1976.
76. Secor, J., McCarty, D. R., Shibles, R., and Green, D. E., Variability and selection for leaf photoysnthesis in advanced generations of soybeans, *Crop Sci.,* 22, 255, 1982.
77. Hay, R. K. M., and Walker, A. J., *An Introduction to the Physiology of Crop Yield,* Longman and John Wiley, New York, 1989.
78. Wells, R., Schulze, L., Ashley, D. A., Boerma, H. R., and Brown, R. H., Cultivar differences in canopy apparent photosynthesis and their relationship to seed yield in soybeans, *Crop Sci.,* 22, 886, 1982.
79. Wells, R., Meredith, W. R., Jr., and Willford, J. R., Canopy photosynthesis and its relationship to plant productivity in near isogenic cotton lines differing in leaf morphology, *Plant Physiol.,* 82, 635, 1986.
80. Boerma, H. R., and Ashley, D. A., Canopy photosynthesis and seed-fill duration in recently developed soybean cultivars and selected plant introductions, *Crop Sci.,* 28, 137, 1988.
81. Ashley, D. A., and Boerma, H. R., Canopy photosynthesis and its association with seed yield in advanced generations of a soybean cross, *Crop Sci.,* 29, 1042, 1989.
82. Dong, S. T., and Hu, C. H., Effect of plant population density on canopy net photosynthesis and their relation to grain yield in maize cultivars, *Photosynthetica,* 29, 25, 1993.
83. Meredeth, W. R., Jr., Registration of eight sub-okra cotton germplasm lines, *Crop Sci.,* 28, 1035, 1988.

84. Chatterton, N. J., Photosynthesis of 22 alfalfa populations differing in resistance to diseases, insect pests, and nematodes, *Crop Sci.,* 16, 833, 1976.
85. Haile, F. J., Higley, L. G., Ni, X., and Quisenberry, S. S., Physiological and growth tolerance in wheat to Russian wheat aphid (Homoptera: Aphididae) injury, *Environ. Entomol.,* 28, 787, 1999.
86. Sauphanor, B., Some factors of upland rice tolerance to stem-borers in West Africa, *Insect Sci. Appl.,* 6, 429, 1985.
87. Sosa Jr., O., and Foster, J. E., Temperature and the expression of resistance in wheat to the Hessian fly, *Environ. Entomol.,* 5, 333, 1976.
88. Ostlie, K. R., and Pedigo, L. P., Soybean response to simulated green cloverworm (Lepidoptera: Noctuidae) defoliation: progress toward determining comprehensive economic injury levels, *J. Econ. Entomol.,* 78, 437, 1985.
89. Hammond, R. B., and Pedigo, L. P., Determination of yield-loss relationships for two soybean defoliators by using simulated insect-defoliation techniques, *J. Econ. Entomol.,* 75, 102, 1982.

8 Drought Stress, Insects, and Yield Loss

Fikru J. Haile

CONTENTS

8.1 INTRODUCTION

Environmental factors, such as extremes in soil moisture and temperature, impact the fitness of plants and arthropods, and greatly mediate plant–arthropod interactions. In natural and agricultural systems, environmental resources are rarely optimal for plant growth and development and can reduce plant fitness or yield. According to Boyer,[1] overall U.S. agricultural productivity is limited by environmental stresses to 25% of its potential. Moisture stress is most likely the major abiotic stressor to which plants are subjected. Most frequently, plants encounter multiple stressors, both biotic and abiotic, at different stages in their growth and development. Consequently, studies that attempt to characterize plant response to stress conditions should account for all variables that potentially affect plant fitness.

Often, agronomic factors, such as the water status of plants, are not taken into consideration when devising a pest management program. However, such agronomic factors may play an important role directly on pest population development and indirectly by altering the suitability of host plants to insect injury or by modifying the level of plant resistance to insects. Attempts to develop comprehensive pest management plans should consider variables that may confound the response of plants to pest injury.

Several models have been developed to address the impact of temperature on the physiological and biochemical processes of plants and animals. In contrast, water-relations, including moisture stress, have received less emphasis, despite the fact that temperature range can be modified by water-relations. Water constitutes a large proportion of plants and animals and plays vital physiological and biochemical roles. Therefore, more emphasis should be given to understanding how plant–insect interactions are affected by moisture stress and other abiotic factors. In this chapter, I discuss the significance of moisture stress on plant and insect fitness, the mediation of plant–insect interaction by moisture stress, and the impact of these interactions on yield loss caused by insect pests.

8.2   IMPACT OF MOISTURE STRESS ON INSECT POPULATIONS

Arthropods are vulnerable to changing microenvironments around plants or plant parts that they inhabit. Because of their small size and poikilothermic existence, arthropods are at a disadvantage when faced with rapidly changing microenvironments. Abiotic factors, such as moisture stress and temperature, modify plant canopy size, leaf temperature, rate of plant transpiration, and rate of soil–water evaporation, and have a direct impact on the survival and fitness of herbivorous arthropods. Indirectly, abiotic factors also can impact arthropod fitness, primarily by influencing plants that shelter the arthropods, including modification of plant microenvironments and alteration of the nutritional quality of host plants.[2]

8.2.1   Direct Impacts

Arthropods may be subjected to extreme water and temperature stress in their surroundings, which may substantially reduce their fitness. By modifying the relative humidity and temperature in their nests, some social insects are capable of maintaining homeostasis to overcome the high mortality rate from unfavorable environmental conditions. Many other insects also have developed unique adaptations that enable them to live in otherwise unfavorable conditions.

Moisture stress has a direct impact on insect fitness, primarily by increasing vapor pressure deficits in the insect's immediate microenvironment, leading to desiccation. Microenvironments in moisture-stressed plants are usually low in relative humidity. Because the relative humidity is variable at different canopy heights, insect distribution can also vary in these canopies. For instance, Isichaikul et al.[3] demonstrated that rice canopies near ground level were more favorable for the distribution and survival of brown planthopper nymphs because of a relatively more humid microenvironment than upper canopies.[3]

Most soil-inhabiting insects are adapted to moist and high relative humidity environments. These insects are more affected by changes in moisture stress than foliar insects. Insects, such as Collembolans, have less control over evaporative water loss compared to foliar insects that strictly regulate water loss from their bodies. If the duration of moisture stress is extended, soil-inhabiting insects may suffer from desiccation.

The survival of corn rootworm, *Diabrotica* spp., and other soil inhabiting insects depends on soil moisture and soil texture.[4, 5] Soil texture determines the rate at which the soil moisture changes. In rapidly drying soils, larval mortality is higher because of faster desiccation than in less rapidly drying soils.[5] Consideration of soil moisture for some insect pests, such as corn rootworms, may be an important criterion before pest management action is taken.

8.2.2 INDIRECT IMPACTS

8.2.2.1 Changes in Microenvironment

An insect's microenvironment primarily is influenced by precipitation, temperature, and the type and density of vegetation.[6, 7] Changes in the level of soil moisture stress that plants experience can greatly modify insect microenvironments. Additionally, moisture stress impacts the microenvironment and insect populations by altering the growth and suitability of plants.

Moisture stress results in elevated leaf temperatures as plants fail to cool via transpiration. Closure of the stomates in moisture-stressed plants reduces transpirational water loss, limiting CO_2 uptake for photosynthesis. Because of reduced photosynthetic rates, moisture-stressed plants maintain smaller canopy size. If canopies fail to close, direct radiation can enter through plant canopies. Consequently, arthropods that live on moisture-stressed plants may experience warmer microclimate than well watered plants because of direct radiation and high leaf temperatures.

High temperatures in moisture-stressed canopies may increase the metabolic and reproductive rates of some insects, leading to population build-up.[8–11] However, elevated plant canopy temperatures resulting from moisture stress may not be conducive for insects adapted to canopies with high relative humidity. Moisture-stressed plants create increased vapor pressure deficits, causing rapid water loss from insects.

Changes in microenvironment may impact arthropods and their natural enemies in different ways. Sometimes the microenvironment is more favorable to the natural enemies than the herbivores, or vice versa. It is also possible that changes in microenvironment act on insect population indirectly by altering the fitness of natural enemies. Indeed, in most cases the direct impact on insect fitness from changes in microenvironment is less than the impact from natural enemies. Therefore, the generalization that moisture stress leads to build-up of some arthropod pests should be considered with caution, because of the indirect impact of microenvironment on natural enemies.

In agricultural ecosystems, understanding the insect microenvironment is essential to design effective pest management strategies. Microenvironments with low relative humidity generally may not be suitable to deploy fungal pathogens as biological control agents. Comparisons of canopy microenvironment and the immediate cuticular microenvironment of grasshoppers suggest that the latter may be more important in using *Beuvaria bassiana* for biological control. The role of ambient relative humidity was less important than the relative humidity at the cuticular levels of the migratory grasshopper for the development of this fungus, demonstrating that it can potentially be used even in areas with low relative humidity.[12] However, the cuticular relative humidity depends on ambient relative humidity and practical use of this pathogen for pest management has not been confirmed.

Other cultural practices that can impact insect microenvironment and also determine changes in insect pest populations are irrigation management[13] and cropping systems (i.e., monoculture vs. polyculture).[14]

8.2.2.2   Changes in Plant Nutritional Quality

As insects approach host plants, determining the suitability of a host plant and further acceptance to resume feeding are essential processes that take place at the insect–plant interface. Insect herbivores determine the nutritional quality of their host plant, and sustained feeding will only ensue if the host plant can provide essential nutrients needed for the growth and development of insects.

The nutritional suitability of host plants can be influenced by environmental factors, such as soil moisture and nutrients available for plant growth and development. Changes in the levels of nitrogen alter the suitability of plants to herbivores. Also, the level of moisture stress plants experience can alter the composition of essential elements and can affect plant nutritional quality.[15]

In moisture-stressed plants, the concentration of solutes is assumed to increase, improving their nutritional quality. Mattson and Haack[11] suggest that improved nutritional quality of plants in response to moisture stress occurs because of increases in the concentration of carbohydrates, proteins, and minerals compared to unstressed plants, which are not saturated with these compounds that are essential for arthropod growth. However, improved plant nutritional quality may not occur following moisture stress[16, 17] and may not affect insect fitness.

Although changes in the nutritional quality of host plants occur after moisture stress, high canopy temperature and low relative humidity also may occur after moisture stress. Changes in these latter parameters may have a greater impact on insect fitness and their natural enemies than changes in the nutritional quality of host plants.

8.2.3   Moisture Stress and Insect Outbreaks

Temporal and spatial changes in insect populations, including cyclic outbreaks of pest-species, have been attributed to environmental factors, primarily extremes in temperature and precipitation. Proponents of the density-independent theory[18] have held the view that environment is a major factor responsible for population regulation. Conversely, biotic factors such as competition, predation, parasitism, and disease also have been proposed to regulate population by proponents of the density-dependent theory.[19] Although there have been strong arguments between the two schools of thought, both biotic and abiotic factors most likely determine changes in herbivore populations.

Insect outbreaks have commonly occurred after drought. A "climatic release hypothesis" that attributes insect outbreaks directly to weather variables has been proposed[20, 21] to explain this phenomenon. Proponents of this hypothesis regard moisture stress as the major factor for insect outbreaks. Although the relationships are not clear, the evidence seems compelling. For instance, the largest outbreak of the saddled prominent, *Heterocampa guttivitta,* took place following droughts in the 1960s in the northeastern U.S.[21]

The climatic release hypothesis argues that the primary role of weather is to directly affect insect fitness. This hypothesis favors the density-independent theory for population regulation. However, increasing evidence suggests that weather may not have a direct impact on insect populations. Many ecologists agree that weather variables only set the maximum and minimum boundaries within which the herbivore population fluctuates. The actual change in population is primarily regulated by density-dependent factors. The direct impact of weather alone may not provide a satisfactory explanation for insect outbreak episodes following moisture stress conditions. The occurrence of drought cannot directly cause increases in insect populations. The impact of drought is indirect by altering plant solute concentrations or the amount of secondary plant metabolites that modify the suitability of host plants to herbivores.[3, 11] In addition, drought stress may not favor the activity of natural enemies that would normally keep the pest population in check.

Insect outbreaks following drought conditions may not be entirely attributed to improved plant nutritional quality. Another possible explanation is the action of natural enemies. When host plants are not limited by water, relative humidity is usually high, favoring fungal pathogens that may keep the insect populations in check. These fungal pathogens work in a density-dependent fashion; when the insect population increases, the pathogens cause significant mortality, maintaining the population at low levels. However, when drought occurs, relative humidity decreases, creating in unfavorable condition for the activity of fungal pathogens. In the absence of these natural enemies, the pest population may increase unchecked.

Environmental factors influence changes in herbivore population, but most likely only indirectly by acting upon natural enemies and to some extent by altering the suitability of host plants because of changes in plant chemistry.[2] For some cyclic insect outbreaks associated with moisture stress, consideration of the impact of natural enemies, particularly of fungal pathogens, may provide a plausible explanation. Consequently, although drought conditions may set ranges within which insect populations fluctuate, the major population regulation most likely is by density-dependent factors, such as fungal pathogens.

8.3   MOISTURE STRESS AND PLANT DEFENSE TO HERBIVORES

Ecological studies indicate that plants with limited resources tend to increase the production of defensive compounds (secondary plant compounds). Therefore, plants that experience moisture stress may increase the production of secondary plant compounds. Insect herbivores feeding on moisture-stressed plants may encounter higher concentrations of secondary plant compounds, which then can reduce insect fitness. However, plant fitness may always be reduced when insects feed on moisture-stressed plants. This is because increased plant temperature induced by moisture stress would increase the metabolic activity of herbivorous insects, including their ability to detoxify secondary plant compounds.[11]

Although little evidence exists, changes in the concentration of solutes in crop plants have been documented after moisture stress and/or arthropod injury.[22, 23] The exact role in plant defense for the solutes that increase during moisture stress has not been determined. Plants subjected to moisture stress increase the concentration of solutes to adjust water potential, reducing the deleterious impacts of moisture stress. One such solute that increases in plants subjected to moisture stress is proline.[23, 24] There is no evidence that proline is involved in plant defense. However, it has been established that increases in proline concentration during moisture stress are because of its role in plant osmoregulation. Increased proline content in some plants has been known to serve as a feeding stimulant for some herbivores. For example, moisture-stressed barley with high levels of proline was preferred by grasshoppers and locusts.[25]

Moisture stress may modify the concentration of soluble compounds in plants, including plant defense compounds. However, the role of moisture stress on plant resistance may be indirect by altering the concentration of some elements or compounds needed for the synthesis of defense compounds. Dale[26] has reviewed literature on the role of plant nutrients such as N, P, and K on plant resistance. Although the role of these elements on plant resistance has not been directly established, studies on wild plants suggest that changing the nutritional status of plants can influence the production of secondary plant compounds. For instance, in plants that depend on N for the production of defensive compounds, increased N supply can promote plant defense. It is therefore likely that moisture stress conditions that alter plant solute concentration also may impact the level of plant defense or resistance to herbivores.

In cultivated plants, moisture stress may affect the level of plant resistance expressed to insect pests. Some of the studies that have examined the role of moisture stress on the level of plant resistance are on soybean. These studies suggest that soybean resistance to insect pests can be altered by moisture stress. Resistance to Mexican bean beetle, *Epilachna varivestis,* was reduced when soybeans were provided with sufficient water.[27, 28] Although changes in the concentration of secondary plant metabolites involved in soybean resistance against the Mexican bean beetle were not determined in these studies, the high level of resistance in moisture-stressed soybeans could be due to high concentrations of defense compounds. In addition, moisture-stressed plants are less suited nutritionally to Mexican bean beetle compared to well watered soybeans. Therefore, a high mortality rate of the beetles on moisture-stressed soybeans could partly be due to a low nutritional quality.[27, 28]

8.4 PLANT PHYSIOLOGICAL RESPONSES TO DROUGHT AND INSECT INJURY

8.4.1 CHANGES IN PLANT–WATER POTENTIAL

Plants undergo diurnal changes in water potential depending on the atmospheric temperature, wind speed, and soil moisture. Generally, the highest leaf water potential (less negative) occurs at midnight and early morning when the atmospheric temperature is low and the wind speed is minimal.[29] Late in the morning and in the afternoon, because of a relatively higher temperature and wind, plants lose water via transpiration and can experience low water potential speed. These changes in plant–water potential can affect insect microenvironments. Consequently, insects experience a warmer microenvironment when plants are at low water potential.

Diurnal changes in water potential can influence consumption rates of insect herbivores. Warmer microenvironments in the afternoon that partly cause low water potential may increase consumption rate because of higher insect metabolic rates. In addition, insects that feed on plant sap may experience some changes in their feeding. This is because changes in plant–water potential also may impact the amount of plant sap available to sap-feeding insects. Although low water potential may increase solute concentration in plant sap, thus improving plant nutritional quality, the sap volume that would be available to insects may be reduced. This may be the case in insects such as aphids, which depend on sap pressure to pump plant sap into their alimentary canal. Low water potential may reduce the sap pressure, reducing sap flow. Conversely, these insects can take advantage of the high water potential of plants during midnight, which increases the volume of sap available for insect consumption because of increased sap pressure.

Increased sap viscosity resulting from low water potential may have reduced numbers of cereal aphids raised on water stressed cereals.[30] However, similar experiments using the Russian wheat aphid, *Diuraphis noxia,* did not cause reductions in aphid populations, suggesting that this aphid is tolerant to sap viscosity of moisture-stressed plants.[31] On the contrary, Russian wheat aphid populations increased on moderately stressed wheat most likely due to increases in soluble nitrogen and starch of moisture-stressed wheat.[31]

Plant–water potential is a more accurate measurement of plant water status and a better indicator of physiological and growth-rate processes than soil–water content. Therefore, measurement of water potential, along with other physiological processes, would indicate the growth and yield potential of plants. Although the amount of soil moisture can directly affect water potential, insect herbivores also can alter the water relations of their host plants. Studies indicate that either defoliating or sap-removing arthropods can alter water potential. By imposing similar levels of moisture stress, defoliated plants maintain higher leaf water potentials than undefoliated plants.[32] This difference occurs because, in defoliated plants, there are fewer leaves through which evaporative water loss takes place (i.e., reduced transpirational water loss). Conversely, undefoliated plants experience a higher transpirational water loss than defoliated plants. Consequently, defoliated plants experience improved water status and may have reduced leaf senescence compared to undefoliated plants. Improved water status is one of the physiological bases for delayed senescence and supports previous findings that defoliation does not cause a reduction in photosynthesis. In undefoliated plants, normal senescence causes reductions in the rate of photosynthesis.

Because piercing and sucking arthropods remove plant sap, it is possible that they also can reduce plant–water potential.[33] However, a study that examined the interaction of moisture stress and mite injury showed that although water potential was reduced by moisture stress, mite injury alone did not cause a significant reduction in water potential.[34] Arthropods such as mites cause a substantial injury to plants only when populations are high. Several studies indicate that moisture stress, which creates a warmer microclimate for mites, induces an increase in mite populations.[8, 9] Such a large number of mites potentially can remove a significant amount of plant sap and can reduce plant–water potential. Reduced photosynthetic rates by mites and other sap-removing arthropods may partly be attributed to low water potential, particularly under moisture-stress conditions. However, the major physiological mechanism for reduced photosynthesis in plants injured by piercing and sucking arthropods is mechanical injury to plant cells and in some cases the release of toxic saliva that impairs chlorophyll.

8.4.2 CHANGES IN PLANT HORMONES

Under moisture stress, plants undergo changes in hormone concentration. Abscisic acid is the major hormone that increases in plants experiencing moisture stress.[24, 35–37] Some legumes increase proline levels by 30% in response to moisture stress.[24, 38] High concentrations of abscisic acid or proline in moisture-stressed plants reduce stomatal conductance, preventing water loss via transpiration. In plants adapted to drought conditions, the concentration of abscisic acid is usually high. In plants not adapted to drought conditions, moisture stress induces increases in abscisic acid concentration transiently until there is ample moisture.

Production of abscisic acid occurs in roots. It is then transported to leaves, where it is responsible for closing the stomates. It is hypothesized that the initiation of abscisic acid production in roots only occurs following a decline in soil moisture content that results in a threshold root water potential.[39, 40] Decline in root water potential below a threshold initiates abscisic acid synthesis. Once synthesized, abscisic acid is rapidly translocated to the leaves, minimizing transpirational water loss. Although abscisic acid activity may reduce shoot production because of the closure of the stomates to CO_2 and consequent reduction in the rate of photosynthesis, studies have shown that abscisic acid promotes root growth.[39, 41] This results in a high root:shoot ratio in moisture-stressed plants. The continued root growth is an adaptation in moisture-stressed plants to access more water.

Changes in plant hormones, such as abscisic acid, represent an adaptive trait to stress conditions. Plant hormones are known to mediate the response of plants to abiotic stresses such as moisture stress, salinity, and extremes in temperature. However, the role of plant hormones in mediating plant response to injury by arthropods is not well known. Plant hormones may play an important role in arthropod–plant interactions. Higley et al.[42] suggest that different stress agents may produce similar plant physiological changes. A plant cannot differentiate between biotic and abiotic stressors. If arthropod injury produces similar physiological changes to those of moisture stress, perhaps the same hormone, such as abscisic acid, may be responsible for adaptive response of plants to arthropod injury. However, research must be conducted to understand if hormones mediate arthropod–plant interactions.

8.4.3 CHANGES IN PHOTOSYNTHESIS

Because plants reduce transpirational water loss by closing their stomates when subjected to moisture stress, the rate of photosynthesis also is reduced because of CO_2 limitation. The impact of insect feeding on photosynthetic rates has been well documented. Photosynthetic responses depend on the nature of insect feeding. For defoliating insects, the rate of photosynthesis in the remaining leaves does not commonly decline.[43–46] However, for piercing and sucking insects, because of cellular damage and/or removal of chlorophyll, the rate of photosynthesis declines.[34, 43, 47–49]

The impact of insect herbivores on photosynthesis is greatly mediated by the amount of moisture available to plants. Consequently, the photosynthetic response of plants subjected to both insect injury and moisture stress can be different from the response of plants subjected to either insect injury or moisture stress. Injury by piercing and sucking insects, such as spider mites, caused a significantly greater photosynthetic rate reduction when plants were subjected to moisture stress compared to an injury on well-watered plants.[34]

For defoliating insects, changes in photosynthetic rate relate to the proportion of injury. Generally, removal of a portion of a leaf or leaflet does not alter photosynthetic rates in the remaining leaves or portions of leaves.[32, 45, 46, 50] However, removal of a few leaves may alter plant–water relations at the plant level and may impact photosynthetic rates of the remaining leaves. Usually, delayed leaf senescence can occur without a significant photosynthetic rate reduction in the remaining leaves following defoliation. In defoliated plants, water and light conditions reduce leaf senescence so that the remaining leaves maintain a greater photosynthetic rate than undefoliated plants.

The impact of defoliation on photosynthesis can be explained primarily in relation to plant–water potential. Defoliation does not seem to significantly affect photosynthetic rates of well-watered soybeans. However, photosynthetic rates of defoliated soybeans were greater than those of the undefoliated check in moisture-stressed soybeans. Defoliation of well-watered soybeans did not significantly improve the water potential because the soybeans already exhibited greater water potential. In contrast, defoliation significantly improved the water potential of moisture-stressed soybeans and caused higher photosynthetic rates.[32]

8.5   CHANGES IN CANOPY SIZE AND LIGHT INTERCEPTION BY DROUGHT

The level of moisture stress plants experience impacts plant height, leaf size, and the angle of leaf orientation. These changes in turn alter plant canopy size and the amount of canopy light interception. Under moisture stress conditions, plant height and leaf size are reduced, resulting in a small canopy. Conversely, under optimal water supply, plants tend to be taller and produce relatively larger leaves, maintaining larger canopies compared to moisture-stressed ones.

Plants have developed different adaptive strategies to overcome the undesirable effects of moisture stress. One adaptation is the modification of canopy architecture. Plants alter leaflet orientation in response to moisture stress to reduce the incident solar radiation, minimizing canopy heat load. Shackel and Hall[51] demonstrated that when drought was imposed in dry beans, *Phaseolus vulgaris,* the leaflets tracked the sun (diaheliotropism) in the morning and avoided the sun (paraheliotropism) in the afternoon by vertical orientation, thus reducing the incident light on the leaflets. Well-watered plants were diaheliotropic, with horizontal orientation in the afternoon for maximum light interception. In soybean, moisture stress exposes the lower leaf surface (which is lighter than the upper leaf surface) to the sun to reflect some of the incident radiation away from the leaves, thereby reducing heat load.[52, 53] Plants employ these strategies when soil moisture content is too low to employ transpirational cooling.

Canopy size and canopy light interception can indicate crop yield potential. In soybean, canopy size at reproductive stage determines the amount of light interception, which directly contributes to yield. If light interception is reduced below 90%, a significant soybean yield loss can occur. Insect defoliation and moisture stress can alter canopy size of crop plants. A large canopy with greater light interception can attain a greater apparent canopy photosynthesis, and can result in increased crop yield. However, excessive increase in plant canopy size does not necessarily reflect increased crop yield because of limitation by harvest index.[54]

Crop plants that experience arthropod injury to nonreproductive plant parts may benefit from excessive canopy size. For instance, in areas where defoliating insect pests are prevalent, cultivars that maintain more leaf tissue would be preferred because of their potential to tolerate loss of leaf tissue to defoliating insects without incurring a significant yield loss. Maintenance of large canopy size can be influenced by environmental factors such as optimal supply of water. See Chapter 7 for a detailed discussion on the role of canopy size and yield loss from defoliation.

8.6   ROLE OF MOISTURE STRESS ON PLANT TOLERANCE TO INSECT INJURY

Plant tolerance or compensation to insect injury may represent an alternative defense strategy in addition to chemical (antibiotic) or structural (antixenotic) defenses to reduce damage to plants. Obviously, the evolution of compensatory plant response plays a key role in promoting plant fitness. Plants that have compensatory response mechanisms have a fitness advantage in the presence of severe herbivore injury. Compensatory response may have evolved in response to the selection pressure by herbivores, although some suggest that it could have also evolved in response to plant competition.[55] Regardless, compensatory response is a desirable plant trait both in natural and agricultural ecosystems.

Plant tolerance to insect injury is an inherited trait. Generally, tolerant cultivars have a rapid growth rate and recover from insect injury[56–58] without significant yield loss. Although tolerance is genetically controlled, its expression may be confounded by environmental factors. If appropriate growing conditions are not met, plant tolerance may not be expressed.

Availability of moisture and nutrients in the soil affects the source-sink relationships in plants, directly impacting plant compensation to arthropod injury. Bloom et al.[59] suggested that plants allocate resources to new biomass to acquire resources that are in short supply and potentially limit plant growth. For instance, plants growing in soils deficient with moisture would have to allocate more resources to roots to absorb more moisture. Because most of the photosynthates are directed to roots, such plants would be susceptible to injury by defoliating herbivores because there are limited photosynthates to replace defoliated leaves. Therefore, plants with limited resources are more susceptible to arthropod injury compared to plants with an optimum supply of resources. Likewise, plants adapted to growing in suboptimal environments have a slow growth rate and are less able to replace tissues lost to herbivory.[60, 61] Consequently, such plants tend to deploy active chemical defenses to deter generalist herbivores.

Plant compensation to herbivore injury depends on the level of injury, plant phenology when injury occurs, plant parts injured, and also the resources available for growth and development of plants. Because plant compensation to insect injury depends on the plant growth rate, optimal resource supply can directly influence the level of compensation plants exhibit. Moisture stress modifies plant tolerance to insect injury. Assuming that all other resources for plant growth and development are met, provision of optimum soil moisture can increase plant tolerance to insect injury. For example, although significant defoliation was imposed, soybeans grown in a year with ample rainfall replaced leaf tissue lost by defoliation in 11 days to attain a critical leaf area index of 3.5. Ample moisture supply initiated rapid soybean growth, delayed maturity, and delayed senescence with sustained photosynthesis. However, during a year with suboptimal moisture, there was no canopy recovery and defoliation caused significant yield reduction.[62, 63]

8.7   METABOLIC COSTS TO MAINTAIN PLANT COMPENSATION AND MEDIATION BY MOISTURE

Compensatory growth of plants to arthropod injury usually occurs in environments rich with resources required for plant growth and development.[64] Plant compensation to herbivory depends on the plant's primary metabolism, and it seems that there are limited or no metabolic costs associated with compensation if resources are available to plants in sufficient amount. In contrast, chemical defense against herbivores is more strongly developed in plant species that are adapted to low resource environments, including drought conditions.[60, 65] Chemical plant defense usually requires additional metabolic pathways to produce a compound used for defense. Therefore, chemical defense is considered more expensive than compensatory growth.

Regarding metabolic costs involved in tolerance, ecological studies suggest that compensatory growth is not costly if plants are grown in resource-rich environments, such as optimal water supply. Some ecologists argue that herbivory is advantageous and leads to overcompensation.[66, 67] However, in agroecosystems, available evidence suggests that compensatory growth is costly. For example, the high level of tolerance in wild tomatoes was a trade-off for less yield compared to cultivated tomatoes that are less defoliation tolerant but high yielding.[68] A tolerant barley cultivar supported equal greenbug population density as did the susceptible cultivar, but gave higher grain yield. But in the absence of greenbug infestation, the tolerant barley cultivar yielded lower than the susceptible cultivar, thus demonstrating a yield-drag associated with plant tolerance.[69]

Plants deploying compensatory growth as a defense strategy generally have a rapid growth rate and allocate resources to leaves and roots for more resource acquisition.[61] In suboptimal environments, plants would not exhibit a compensatory response. It may be inaccurate to generalize that plants under natural conditions do compensate for herbivory but not plants in agricultural systems. If resources are sufficiently provided, plants in an agricultural system also could exhibit some level of compensatory growth to herbivory. This has been demonstrated with compensatory regrowth of soybean to defoliation when moisture and nutrient were in optimum supply.[62, 63] Consequently, the answer to the question of whether compensatory growth is costly to the plant, like chemical or structural defenses, should be considered with caution. It seems that all defenses are costly, including compensatory growth. In compensatory growth, it is costly because plants need an adequate supply of resources to exhibit compensatory responses. In an agricultural ecosystem, provision of adequate resources to increase the level of plant compensation to arthropod injury may entail costs associated with farming inputs, for instance, costs for irrigation water where precipitation is marginal.

8.8   INTERACTION OF DROUGHT STRESS, INSECT INJURY, AND PLANT YIELD LOSS

Moisture stress reduces photosynthetic rates, plant biomass, and yield.[70–72] Moisture-stressed plants have smaller canopies, are more susceptible to insect injury, and suffer heavier yield loss compared to unstressed plants. Consequently, the economic injury levels of drought-stressed soybeans were different from those that were not drought-stressed.[73, 74] These studies showed that drought-stressed soybeans had lower economic injury levels compared to drought-unstressed soybeans.

In soybeans, a leaf area index (LAI) of 3.5 at reproductive stage is considered critical for maximum yield. This critical LAI corresponds to about 90% canopy light interception. Reduction of the LAI below 3.5, or light interception below 90%, results in significant yield loss.[44, 75] When growing conditions are optimal for soybeans, including sufficient precipitation or irrigation water, soybeans can produce a leaf area index as high as 7. Consequently, these soybeans with ample water supply can tolerate a significant defoliation injury without a significant yield loss. For instance, if the canopy has attained an LAI of 7, removal of half of the leaf area may not result in a significant yield reduction.

Although optimal moisture supply may enhance the level of tolerance to insect injury, excess moisture can be harmful for plant growth.[76, 77] In some areas where there is excessive precipitation, drainage becomes a challenge to overcome water logging. Therefore, optimal water supply that favors rapid plant growth is desirable. The more vigorous the plants are, the more tolerant they can be to injuries by insect pests. Therefore, optimizing the supply of resources required for plant growth is one alternative to increasing the level of plant tolerance to insect injury and consequently reducing yield loss.

In agricultural ecosystems, moisture stress can interact with arthropod injury to affect plant gas exchange, dry-matter, and yield.[62, 63, 77, 78] Agronomic factors, such as soil moisture available for plant growth, affect plant vigor altering the level of tolerance to arthropod injury. Therefore, consideration of agronomic inputs in a pest management plan may be essential to increase plant tolerance to arthropod injury.

8.9 CONCLUSIONS

Seldom are environmental requisites in optimal supply for plants to achieve their genetic yield potential. In natural and agricultural systems, plants experience stress from both biotic and abiotic factors that can disrupt normal plant physiological processes and growth. As a result, stressed plants experience reduced fitness or yield.

Moisture stress is one of the major abiotic factors that limit agricultural productivity. Plants experiencing moisture stress have reduced physiological processes such as photosynthesis, transpiration, and reduced biomass and reduced yield. If there is extreme moisture stress, particularly at early growth stages, plants may not recover from moisture stress and can die. However, if the level of moisture stress is not severe, plants deploy different strategies to maintain fitness. Some of the adaptations to moisture stress include increases in some hormone concentration to regulate stomatal conductance limiting transpirational water loss. In addition, moisture-stressed plants tend to increase root growth to absorb more water from the soil.

Plants face multiple stressors that can limit their growth, development, and fitness. Because little research has focused on the interactions of abiotic stress and insect injury, the broad importance of the interaction is not clear. Few studies indicate that abiotic factors can alter plant response to insect injury. In agricultural ecosystems, yield loss occurs from both moisture stress and insect injury. However, the magnitude of yield loss from the interaction of the two stressors could be different from the magnitude of yield loss if these stressors were acting independently. This implies that, at least in some systems, the impact of insect injury depends on moisture stress and other abiotic factors. Yield loss from insect injury is most likely greater in plants subjected to moisture stress than in unstressed plants. This is because plants provided with optimum resources may tolerate insect injury more than plants without optimum resources. Consequently, there may be opportunities to increase plant tolerance to insect injury and reduce yield losses by cultural practices that increase plant vigor.

REFERENCES

1. Boyer, J. S., Plant productivity and environments, *Science,* 218, 443, 1982.
2. Roades, D. F., Herbivore population dynamics and plant chemistry, in *Variable Plants and Herbivores in Natural and Managed Systems,* Denno, R. F., and McClure, M. S., Eds., Academic Press, New York, 1983, chap. 6.
3. Isichaikul, S., Fujimura, K., and Ichikawa, T., Humid microenvironment prerequisite for the survival and growth of nymphs of the rice brown planthopper, *Nilaparvata lugens* (Stal) (Homoptera: Delphacidae), *Res. Popul. Ecol.,* 36, 23, 1994.
4. Lummus, P. F., Smith, J. C., and Powell, N. L., Soil moisture and texture effects on survival of immature Southern corn rootworm, *Diabrotica undecimpunctata howardi* Barber (Coleoptera: Chrysomelidae), *Environ. Entomol.,* 12, 1529, 1983.
5. Burst, G. E., and House, G. J., Effects of soil moisture, texture, and rate of soil drying on egg and larval survival of the Southern corn rootworm (Coleoptera: Chrysomelidae), *Environ. Entomol.,* 19, 697, 1990.
6. Rosenberg, N. J., Blad, B. L., and Verma. S. B., *Microclimate, The Biological Environment,* 2nd Ed., Wiley-Interscience, New York, 1983.
7. Stoutjesdijk, P. H., and Barkman, J. J., *Microclimate Vegetation and Fauna,* Opulus Press, Uppsala, 1992.
8. Perring, T. M., Holtzer, T. O., Toole, J. L., and Norman, J. M., Temperature and humidity effects on ovipositional rates, fecundity, and longevity of adult female Banks grass mites (Acari: Tetranychidae), *Ann. Entomol. Soc. Am.,* 77, 581, 1984.
9. Perring, T. M., Holtzer, T. O., Toole, J. L., and Norman, J. M., Relationships between corn-canopy microenvironments and Banks grass mite (Acari: Tetranychidae) abundance, *Environ. Entomol.,* 15, 79, 1986.
10. Toole, J. L., Norman, J. M., Holtzer, T. O., and Perring, T. M., Simulating Banks grass mite population dynamics as a subsystem of a crop canopy-microenvironment model, *Environ. Entomol.,* 13, 329, 1984.
11. Mattson, W. J., and Haack, R. A., The role of drought in outbreaks of plant-eating insects, *Bioscience,* 37, 110, 1987.
12. Marcandier S., and Khachatourians, G. G., Susceptibility of the migratory grasshopper, *Melanoplus sanguinipes* (Fab.) (Orthoptera: Acrididae), to *Beauveria bassiana* (Bals.) Vuillemin (Hyphomycete): influence of relative humidity, *Can. Entomol.,* 119, 901, 1987.
13. Trichilo, P. J., Wilson, L. T., and Grimes, D. W., Influence of irrigation management on the abundance of leafhoppers (Homoptera: Cicadellidae) on grapes, *Environ. Entomol.,* 19, 1803, 1990.
14. Biladdawa, C. W., The effect of crop microenvironment on the pea leaf weevil's behavior: an explanation for weevil departure from diverse cropping systems, *Insect Sci. Appl.,* 9, 509, 1988.

15. Schulze, E. D., Water and nutrient interactions with plant water stress, in *Response of Plants to Multiple Stresses,* Mooney, H. A., Winner, W. E., and Pell, E. J., Eds., Academic Press, London, 1991, chap. 4.

16. Holtzer, T. D., Archer, T. L., and Norman, J. M., Host plant suitability in relation to water stress, in *Plant Stress-Insect Interactions,* Wiley, New York 1988, chap. 3.

17. Stone, C., and Bacon, P. E., Relationships among moisture stress, insect herbivory, foliar cineole content, and the growth of river red gum *Eucalyptus camaldulensis, J. Appl. Ecol.,* 31, 604, 1994.

18. Andrawartha, H. G., and Birch, L. C., *The Distribution and Abundance of Animals,* University of Chicago Press, Chicago, 1954.

19. Nicholson, A. J., An outline of the dynamics of animal populations, *Aust. J. Zool.,* 2, 9, 1954.

20. Wellington, W. G., Atmospheric circulation processes and insect ecology, *Can. Entomol.,* 86, 312, 1954.

21. Martinat, P. J., The role of climatic variation and weather in forest insect outbreaks, in *Insect Outbreaks,* Barbosa, P., and Schultz, J., Eds., Academic Press, New York, 1987, chap. 10.

22. Singh, B. B., and Gupta, D. P., Proline accumulation and relative water content in soybean (*Glycine max*) varieties under water stress, *Ann. Bot.,* 52, 109, 1983.

23. Schroeder, P. C., Brandenburg, R. L., and Nelson, C. J., Interaction between moisture stress and potato leafhopper (Homoptera: Cicadellidae) damage in alfalfa, *J. Econ. Entomol.,* 81, 927, 1988.

24. Harborne, J. B., *Ecological Biochemistry,* Academic Press, London, 1993.

25. Bright, S. W. J., Lea, P. L., Kueh, J. S. H., Woodcock, C., Hollomon, C. W., and Scott, G. C., Proline content does not influence pest and disease susceptibility of barley, *Nature,* 295, 592, 1982.

26. Dale, D., Plant-mediated effects of soil mineral stresses on insects. in *Plant Stress-Insect Interactions,* Heinrichs, E. A., Ed., Wiley-Interscience, New York, 1988, chap. 2.

27. Hammond, R. B., Bledsoe, L. W., and Nazim Anwar, M., Maturity and environmental effects on soybeans resistant to Mexican bean beetle (Coleoptera: Coccinellidae), *J. Econ. Entomol.,* 88, 175, 1995.

28. Jenkins, E. B., Hammond, R. B., St. Martin, S. K., and Cooper, R. L., Effect of soil moisture and soybean growth stage on resistance to Mexican bean beetle (Coleoptera: Coccinellidae), *J. Econ. Entomol.,* 90, 697, 1997.

29. Kramer, P. J., *Water Relations of Plants,* Academic Press, San Diego, CA, 1983.

30. Pons, X., and Tatchell, G. M., Drought stress and cereal aphid performance, *Ann. Appl. Biol.,* 126, 19, 1995.

31. Johnson, G. D., Ni, X., Mclendon, M. E., Jacobson, J. S., and Wraith, J. M., Impact of Russian wheat aphid (Homoptera: Aphididae) on drought stressed spring wheat, in *Response Model for an Introduced Pest — The Russian Wheat Aphid,* Quisenberry, S. S., and Peairs, R. B., Eds., Entomological Society of America, Lanham, MD, 1998, chap. 6.

32. Haile, F. J., Physiology of Plant Tolerance to Arthropod Injury, Ph.D. dissertation, University of Nebraska, Lincoln, 1999.

33. Dorschner, K. W., Johnson, R. C., Eikenbary, R. D., and Ryan, J. D., Insect–plant interactions: greenbugs (Homoptera: Aphididae) disrupt acclimation of winter wheat to drought stress, *Environ. Entomol.,* 15, 118, 1986.

34. Youngman, R. R., and Barnes, M. M., Interaction of spider mites (Acari: Tetranychidae) and water stress on gas-exchange rates and water potential of almond leaves, *Environ. Entomol.,* 15, 594, 1986.

35. Aspinall, D., Role of abscisic acid and other hormones in adaptation to water stress, in *Adaptation of Plants to Water and High Temperature Stress,* Turner, N. C., and Kramer, P. J., Eds., Wiley-Interscience, New York, 1980, chap. 11.

36. Farquhar, G. D., Wong, S. C., Evans, J. R., and Hubick, K. T., Photosynthesis and gas exchange, in *Plants Under Stress: Biochemistry, Physiology, and Ecology and their Applications to Plant Improvement,* Jones, H. G., Flowers, T. J., and Jones, M. B., Eds., Cambridge University Press, Cambridge, 1989, chap. 4.

37. Dubey, R. S., Photosynthesis in plants under stressful conditions, in *Handbook of Photosynthesis,* Pessarakli, M., Ed., Marcel Dekker, Inc., Tucson, 1997.

38. Stewart, C. R., and Hanson, A. D., Proline accumulation as a metabolic response to water stress, in *Adaptation of Plants to Water and High Temperature Stress,* Turner, N. C., and Kramer, P. J., Eds., Wiley-Interscience, New York, 1980, chap. 12.

39. Shamshad, S., and Naqvi, M., Plant/crop hormones under stressful conditions, in *Handbook of Photosynthesis,* Pessarakli, M., Ed., Marcel Dekker, Inc., Tucson, 1997, chap. 30.

40. Sharp, R. E., and Davies, W. J., Regulation of growth and development of plants growing with a restricted supply of water, in *Plants Under Stress: Biochemistry, Physiology, and Ecology and their Applications to Plant Improvement,* Jones, H. D., Flowers, T. J., and Jones, M. B., Eds., Cambridge University Press, Cambridge, 1989, chap. 5.

41. Saab, I. N., Sharp, R. E., Pitchard, J., and Voetberg, G. S., Increased endogenous abscisic acid maintains primary root growth and inhibits shoot growth of maize seedlings at low water potentials, *Plant Physiol.,* 93, 1329, 1990.

42. Higley, L. G., Browde, J. A., and Higley, P. M., Moving towards new understanding of biotic stress and stress interactions, in *International Crop Science I,* Buxton, D. R., Shibles, R., Forseberg, R. A., Blad, B. L., and Asay, K. H., Eds., Crop Science Society of America, Madison, WI, 1993, 749.

43. Welter, S.C, Arthropod impact on plant gas exchange, in *Insect-Plant Interactions,* Bernays, E. A., Ed., CRC Press, Boca Raton, 1989, 135.

44. Higley, L. G., New understanding of soybean defoliation and their implications for pest management, in *Pest Management in Soybean,* Copping, L. G., Green, M. B., and Rees, R. T., Eds., Elsevier Science Publishers, London, 1992, 56.

45. Peterson, R. K. D., Danielson, S. D., and Higley, L. G., Photosynthetic response of alfalfa to actual and simulated alfalfa weevil (Coleoptera: Curculionidae) injury, *Environ. Entomol.,* 21, 501, 1992.

46. Peterson R. K. D., and Higley, L. G., Temporal changes in soybean gas exchange following simulated insect defoliation, *Agron. J.,* 88, 550, 1996.

47. Miller, H., Porter, D. R., Burd, J. D., Mornhinweg, D. W., and Burton, R. L., Physiological effects of Russian wheat aphid (Homoptera: Aphididae) on resistant and susceptible barley, *J. Econ. Entomol.,* 87, 493, 1994.

48. Burd, J. D., and Elliott, N. C., Changes in chlorophyll a fluorescence induction kinetics in cereals infested with Russian wheat aphid (Homoptera: Aphididae), *J. Econ. Entomol.,* 89, 1332, 1996.

49. Haile, F. J., Higley, L. G., Ni, X., and Quisenberry, S. S., Physiological and growth tolerance in wheat to Russian wheat aphid (Homoptera: Aphididae) injury, *Environ. Entomol.,* 28, 787, 1999.

50. Klubertanz, T., Pedigo, L. P., and Carlson, R. E., Soybean physiology, regrowth, and senescence in response to defoliation, *Agron. J.,* 88, 577, 1996.

51. Shackel, K. A., and Hall, A. E., Reversible leaflet movements in relation to drought adaptation of cowpeas, *Vigna unguiculata* (L.) Walp, *Aust. J. Plant Physiol.,* 6, 265, 1979.

52. Meyer W. S., and Walker, S., Leaflet orientation in water-stressed soybeans, *Agron. J.,* 73, 1071, 1981.

53. Oosterhuis, D. M., Walker, S., and Eastham, J., Soybean leaflet movement as an indicator of crop water stress, *Crop Sci.,* 25, 1101, 1985.

54. Hay, R. K. M., and Walker, A. J., *An Introduction to the Physiology of Crop Yield,* Longman and John Wiley, New York, 1989.

55. Trumble, J. T., Kolodny-Hirsch, D. M., and Ting, I. P., Plant compensation to arthropod herbivory, *Annu. Rev. Entomol.,* 38, 93, 1993.

56. Zuber, M. S., Musick, G. J., and Fairchild, M. L., A method of evaluating corn strains for tolerance to the Western corn rootworm, *J. Econ. Entomol.,* 64, 514, 1971.

57. Owens, J. C., Peters, D. C., and Hallauer, A. R., Corn rootworm tolerance in maize, *Environ. Entomol.,* 3, 767, 1974.

58. Riedell, W. E., and Evenson, P. D., Rootworm feeding tolerance in single-cross maize hybrids from different eras, *Crop Sci.,* 33, 951, 1993.

59. Bloom, A. J., Chapin III, F. S., and Mooney, H. A., Resource limitation in plants — an economic analogy, *Annu. Rev. Ecol. Syst.,* 16, 363, 1985.

60. Bryant, J. P., Chapin III, F. S., and Klien, D. R., Carbon/nutrient balance of boreal plants in relation to herbivory, *Oikos,* 40, 357, 1983.

61. Chapin III, F. S., Integrated responses of plants to stress: a centralized system of physiological response, *BioScience,* 41, 29, 1991.

62. Haile, F. J, Higley, L. G., and Specht, J. E., Soybean cultivars and insect defoliation: yield loss and economic injury levels, *Agron. J.,* 90, 344, 1998a.

63. Haile, F. J, Higley, L. G., Specht, J. E., and Spomer, S. M., The role of leaf morphology in soybean tolerance to defoliation, *Agron. J.,* 90, 353, 1998b.

64. Herms, D. A., and Mattson, W. J., The dilemma of plants: to grow or defend, *Q. Rev. Biol.,* 67, 283, 1992.

65. Coley, P. D., Herbivory and defensive characteristics of tree species in lowland tropical forest, *Ecol. Monogr.,* 53, 209, 1983.

66. McNaughton, S. J., Compensatory plant growth as a response to herbivory, *Oikos,* 40, 329, 1983.

67. Paige, K. N., and Whitham, T. G., Overcompensation in response to mammalian herbivory: the advantage of being eaten, *Am. Nat.,* 129, 407, 1987.

68. Welter, S. C., and Steggall, J. W., Responses of tomato to simulated and real herbivory by tobacco hornworm (Lepidoptera: Sphingidae), *Environ. Entomol.,* 20, 1537, 1993.

69. Castro, A. M., Rumi, C. P., and Arriaga, H. O., Influence of greenbug root growth of resistant and susceptible barley genotypes, *Environ. Exp. Bot.,* 28, 61, 1988.

70. McPherson, H. G., and Boyer, J. S., Regulation of grain yield by photosynthesis in maize subjected to water deficiency, *Agron. J.,* 69, 714, 1977.

71. Ashley, D. A., and Ethridge, W. J., Irrigation effects of vegetative and reproductive development of three soybean cultivars, *Agron. J.,* 70, 467, 1978.

72. Souza P. I. De., Egli, D. B., and Bruening, W. P., Water stress during filling and leaf senescence in soybean, *Agron. J.,* 89, 807, 1997.

73. Hammond, R. B., and Pedigo, L. P., Determination of yield-loss relationships for two soybean defoliators by using simulated insect-defoliation techniques, *J. Econ. Entomol.,* 75, 102, 1982.

74. Ostlie, K. R., and Pedigo, L. P., Soybean response to simulated green cloverworm (Lepidoptera: Noctuidae) defoliation: progress toward determining comprehensive economic injury levels, *J. Econ. Entomol.,* 78, 437, 1985.

75. Shibles, R. M., and Weber, C. R., Interception of solar radiation and dry matter production by various soybean planting patterns, *Crop Sci.,* 6, 55, 1966.
76. Guidi, L., and Soldatinit, G. F., Chlorophyll fluorescence and gas exchange in flooded soybean and sunflower plants, *Plant Physiol. Biochem.,* 35, 713, 1997.
77. Linkemer, G., Board, J. E., and Musgrave, M. E., Waterlogging effects on growth and yield components in late-planted soybean, *Crop Sci.,* 38, 1576, 1998.
78. Godfrey, L. D., Holtzer, T. O., Spomer, S. M., and Norman, J. M., European corn borer (Lepidoptera: Pyralidae) tunneling and drought stress: effects on corn yield, *J. Econ. Entomol.,* 84, 1850, 1991.

9 The Impact of Herbivory on Plants: Yield, Fitness, and Population Dynamics

Kevin J. Delaney and Tulio B. Macedo

CONTENTS

9.1 INTRODUCTION: THE IMPORTANCE OF HERBIVORY TO PLANTS

Terrestrial herbivorous insects alone represent about 25% of all animal and plant species,[1] and many other organisms (invertebrate and vertebrate groups in marine, aquatic, and terrestrial habitats) have herbivorous lifestyles as well. The large impact

135

of plants on herbivores seems clear, but the impact of herbivores on plants (especially at plant distribution and community levels) has not been studied as extensively.[2] The impact of herbivory on plants has been a controversial topic,[3] proponents of the main view suggesting that herbivory is detrimental to plants while others maintain that herbivory generally has minor effects compared to abiotic factors, no impact, or even benefits plants.[4-6]

It is clear that herbivory has exerted a strong evolutionary impact affecting plant evolution over time, so that how we see plants today is at least partly due to herbivory. For example, plant chemicals have been suggested to drive the coevolution of plant–insect interactions,[7] either as pairwise[8] or diffuse[9, 10] coevolutionary plant–herbivore interactions. Several lines of evidence suggest that herbivory has not generally had a favorable impact on plant fitness. This evidence includes the widespread presence of secondary plant defensive compounds, plant morphological defensive structures, and induced plant defenses, all potentially defensive mechanisms against herbivores.[11-16]

In addition, considering plants and herbivores in the context of tritrophic interactions with primary predators, plant responses can interact with the effects of predators (and parasitoids) in several ways to affect herbivores.[17-18] Plants may have lower nutritional quality to delay herbivore development so that predators have more time to consume herbivores.[17] It also seems that plants often provide predators more effective physical access to herbivores,[18] and plants will sometimes release compounds following herbivory that attract predators of the herbivores.[17-20]

Many of the authors who have written about the impact of herbivory on plants work with plants in temperate, seasonal environments. Coley and Barone[16] offer a different perspective, suggesting that herbivory pressure is most intense when herbivores and plants have a continuous, year-round association. In support of this idea, tropical plants tend to be more chemically, mechanically, and/or spatially defended than temperate plants.[16] Also, the pattern of resource investment to plant defenses tends to differ between seasonal and aseasonal plant species. Because leaves often remain on aseasonal plants longer than on seasonal plants, leaves appear to be more valuable to aseasonal plants. These data suggest that the degree of plant defenses tends to increase as the severity of herbivory pressure increases. This in turn suggests that many plants "resist" herbivory injury not only because of a general negative evolutionary impact of herbivory on plant fitness, but also because of the current negative effects of herbivory on many plants.[2, 16]

The impact of herbivory on plants has importance for both applied and theoretical reasons. Farmers, plant breeders, and scientists need to understand how herbivores affect plants if they are to maximize crop yields, while selecting plants maximally resistant to herbivores and thus suffering reduced yield losses. Useful insights can be obtained by reviewing how plants respond to herbivory injury by drawing on studies from agricultural systems as well as basic ecology studies. In natural systems, i.e., environments without human interference, plants are subjected to a broad array of biotic and abiotic stresses. Compared to standard agricultural settings, natural systems are diverse, with high genetic variation. Also, natural systems may experience greater degrees of abiotic stresses than agricultural systems. These conditions of stress are themselves unfavorable to plants and may compound with insect injury,[21] in space and time.[22]

Unfortunately, how the impact of herbivory on plants is measured varies by discipline or focus. With managed systems, interest usually concerns plant yields, while studies in natural systems usually attempt to estimate the impact of herbivory on plant fitness by measuring survival and reproduction parameters. Some ecologists take the issue a step further, and consider how herbivores affect plant population dynamics and plant distributions. Thus, how researchers measure herbivory impact is not a trivial issue and influences the level at which we can understand plant–herbivore interactions.

9.2 PLANT EVOLUTIONARY RESPONSES TO HERBIVORY

Plant responses to herbivore attacks are dependent on interactions among plant characteristics, herbivores, and environmental factors.[23–28] Plants have several strategies that appear to be evolutionary responses, at least in part, to the impact of herbivory. Plants can avoid herbivory by living near different species more preferred by herbivores, by hiding (texture theory), or by avoiding herbivores spatially or temporally.[22] Alternatively, plants can address herbivory more directly via resistance. In host plant resistance, these resistance mechanisms (antixenosis, antibiosis, and tolerance) can occur at different times relative to the occurrence of herbivory.[22, 29] Antixenosis serves as a strategy whereby plants avoid herbivores by changing in some manner, which causes herbivores to not prefer or even avoid consuming that plant relative to other plant species. Thus, antixenosis generally serves as a deterrent strategy to prevent herbivory prior to consumption. Antibiosis, where the plant affects (usually negatively) the biology of herbivores that attempt to consume it, often occurs through secondary chemicals,[12] low plant tissue nutrient value,[30] and mechanical defenses.[31] Tolerance is a strategy where the plant endures or compensates for injury without appreciable reduction to individual fitness.

Herbivores exhibit a range of feeding strategies from monophagy (specialized consumption on one plant host species) to polyphagy (general consumption on many different plant host species). Diet breadth is a key trait most important in determining an herbivore's reaction to vegetation texture. Because species-specific feeding habits determine an herbivore's perception of vegetation texture (food vs. nonfood), it follows that diet breadth shapes each herbivore's relationship with vegetation texture.

Kareiva states that herbivore number can be influenced by three primary routes: (1) vegetation texture may limit the number of parasites and predators in a habitat and, consequently, determine the degree of herbivore regulation by their natural enemies; (2) changes in vegetation texture might alter suitability of individual food plants for herbivore growth and reproduction; and (3) vegetation texture may shape the movement and searching behavior of herbivores and affect their host-finding success.[32] For example, Karban[33] reported patterns of tobacco hornworm, *Manduca quinquemaculata,* abundance on wild tobacco plants, *Nicotiana attenuata. Manduca quinquemaculata* was more likely to be found on large tobacco plants, and on tobacco plants with flowering neighbors like *Eriastrum densifolium.* Thus, larger patches can support greater numbers of a herbivorous species because of increased habitat heterogeneity. In many cases, therefore, plants will not be able to avoid herbivores and will instead resist herbivory.

Constitutive defenses exist in a plant before herbivory, and generally seem to serve to prevent herbivory injury. Early studies on plants have shown elevated concentrations of antifeedants and tannins when the infestations of herbivores were either increased or their distribution on parts of a plant was greater.[34] These defenses consist of toxic chemicals that repel herbivore attack. Berenbaum[35] further discusses the importance of secondary compounds in determining patterns of host-plant utilization. She found that primary compounds produced by plants are especially important in defense against oligophagous herbivores of limited mobility.

Induced resistance in plants is characterized by low specificity, which is different from the mode of action of antigens in vertebrates.[36, 37] Plant fitness can be higher in early-induced plants because plant damage from herbivory injury by a focal herbivorous species is reduced at an earlier point in the plant's life,[38] or subsequent herbivory by other species can be reduced.[39, 40] Chemical production can take place at or near the wound site, with the apparent objective to arrest the herbivore's initial attack. In other instances, responses can take place centimeters away from the wound site, which seems to work directly against a persistent attack or against future attacks.[41]

Finally, plants have varying degrees of tolerance to injury; some plants can compensate for certain levels of herbivory injury by reducing net damage to plant yields or fitness.[42, 43] Tolerance serves as a mechanism in response to herbivory, and hence also occurs following herbivory injury. Unlike induced defenses, tolerance as a strategy does not attempt to prevent further injury, but instead plants are able to recover from injury received. The major difference between plant defensive strategies is that antixenosis and antibiosis exert selective pressures on plants to defend themselves and their herbivores to overcome plant defenses, whereas tolerance only exerts selective pressure on the plants' ability to recover from injury.[44]

Plants do not equally express all defenses because it appears that some forms of defense are expensive in terms of resource allocation, so a trade-off exists between the capacity for defense and the ability to tolerate injury (a cost of defense) for many species.[45–53] Indeed, a corollary to this cost-of-defense argument is the idea that induced defenses exist because it is too expensive to maintain a constitutive form of defense, and thus induced defenses are activated only as a result of sufficiently powerful negative stimuli like herbivory injury.[14, 47, 54] However, trade-offs have not been detected between plant defensive capability and plant ability to tolerate herbivory injury in other situations.[47, 49, 54–57]

The cost-of-defense argument has led to the suggestion that tolerance via compensation to herbivory injury serves as a viable evolutionary strategy when sufficient nutrient resources are available, and that under conditions where limited resources are common, defenses to avoid herbivory injury better serve plants.[58] Again, this supports the argument that many plants over evolutionary time have evolved defenses because herbivory has negative impact on individual plant fitness. Yet, the importance of herbivory on plant evolution has been suggested to be weak relative to other biotic selective agents like pathogens and allelochemicals from competing plant species.[49] A difficult problem with assessing the impact of herbivory on plants and the relative importance of herbivory impact vs. the impact of other biotic factors on plants is that researchers can only directly study the impact of present herbivores on present plants. We can only infer the past impact of herbivory on plant evolution based on evidence from the fossil record or with comparative method approaches, and attempt to estimate the present importance of herbivory to plants relative to other biotic factors.

9.3   PLANT TOLERANCE TO HERBIVORY

9.3.1   DIFFERENT COMPENSATION RESPONSES

A continuum of tolerance responses theoretically is possible that indicates how herbivory injury can translate to plant damage in terms of reductions in units of yield, fitness, or distribution area.[42] It is possible that small amounts of injury result in large amounts of plant damage, a nonlinear relationship that would reflect a plant's hypersensitive response to herbivory. It is, however, more common for a linear relationship to exist between herbivory injury and plant damage where each unit of injury results in a constant amount of plant damage. This linear injury to damage relationship tends to exist when plants have no compensatory mechanisms in response to injury.

When plants have compensatory abilities, several forms of tolerance to herbivory injury can exist. With partial compensation, a nonlinear relationship exists between herbivory and plant damage such that each unit of injury initially results in very low damage and only with increasing levels of injury do damage levels increase. With full compensation, which usually only occurs at low injury levels, no plant damage is received for each unit of injury. (Many often consider herbivores to not affect plants if this response is observed.) Overcompensatory plant responses are expected to occur only for low levels of herbivory,[43, 58] (but see Paige and Whitham[5]) where for each unit of herbivory injury there is a negative amount of plant damage. Therefore, with overcompensation the plant is better off after the injury and actually "benefits" by regrowth following herbivory injury. At the other extreme, after very high levels of herbivory injury, plant damage will either increase very slowly (nonlinear-desensitization) or not at all (inherent impunity) with each unit of injury. The key point about plant tolerance is that the consideration of plant compensatory abilities is critical when studying the impact of herbivory on plants.[43]

9.3.2   PLANT OVERCOMPENSATION: BENEFICIAL HERBIVORY?

Some authors have advanced the view that herbivory can benefit some plants because of plant overcompensatory responses in growth at low[58] or even high herbivory injury levels.[5, 60, 61] McNaughton[58] pointed out that herbivory injury often does not translate into linear forms of damage to plants because of plant compensation, in which case herbivory generally does cause some plant damage but not as much as expected based on the injury received. However, with overcompensatory plant responses, plant vegetative growth and/or reproductive production increases following herbivory injury, in which plant fitness would be expected to be higher in plants subject to herbivory relative to plants not subject to herbivory. However, overcompensation to herbivory may not be the result of herbivory selection pressure on plants, but rather plant regrowth responses to other factors that happen to also be stimulated by herbivory.[61] Hence, Belsky et al.[61] suggest that overcompensation may not indicate a plant–herbivore mutualistic coevolutionary relationship,[62] and several assumptions behind the idea that herbivory benefits plants may not be true.

Yet, several examples of overcompensation have been suggested. Some examples include a history of management pressures by humans on a biennial, *Gentianella campestris*,[63] possibly between herbivores and a cyanogenic grass, *Cynodon plectostachyus*,[59] between grazing herbivores and some grasses,[3, 5, 59] and with the biennial scarlet gilia, *Ipomopsis aggregata*.[5, 61, 64–66] Additionally, it has been suggested that overcompensation may not be quite so rare a form of plant response to herbivory.[66] Yet, despite finding overcompensatory responses in *I. aggregata* following shoot consumption by deer, other researchers with the same study organism have not been able to detect overcompensation[67, 68] to the same types of injury. Bergelson et al.[69] reported partial plant compensation (not overcompensation) for herbivory injury in *I. aggregata*, and injury from each of several different herbivores could have negative fitness consequences.[69] With this example, *I. aggregata* may have geographical variation in both compensatory responses to herbivory injury and pressure from different numbers of herbivores,[68, 69] and interactions between different types of injury that can result in either plant damage or plant overcompensation.[66]

The debate about plant overcompensation remains unclear because variation in environmental conditions and genetic variation in plant ability to compensate for injury seem to exist even within a single plant species. Thus, making general statements about plant responses to herbivory may be difficult even at the level of a species, let alone for general plant responses to herbivory. It seems that herbivores usually have negative effects on plant growth and fitness,[2, 3, 16, 62] although in some limited number of situations herbivory seems to benefit plant fitness.[66] At a broader level, some herbivores may serve as controlling factors of plant population dynamics and distributions.[2, 46, 70–73]

9.4 HERBIVORY: DIFFERENT PLANT TISSUES

Several different plant tissues can be consumed directly by herbivores. These tissues include roots, stems, leaves, flowers, fruits, and seeds. In addition, multiple herbivores may consume tissue or fluids from all of these plant tissues from one plant species.[70, 71] Some herbivores consume plant tissues, other herbivores suck plant fluids (phloem or xylem sap) from roots, leaves, and stems, and some insects develop in plant tissue, resulting in gall formation. Thus, many of the effects of herbivory exert indirect consequences on plant fitness, while other forms of herbivory damage plant reproductive tissues and/or reduce plant survivorship. However, whether these negative herbivory influences on individual plant fitness also result in effects on population dynamics (e.g., herbivores regulate population sizes or distributions of plant species) is a separate issue. Just because herbivores harm individual plants does not mean that they necessarily affect plant population dynamics.[2]

At a mechanistic level, herbivory can impair photosynthetic processes by decreasing photosynthetic rates of leaves following injury,[74–80] though the effect of herbivory on photosynthesis can depend on other factors like nitrogen level,[80] and CO_2 levels can affect subsequent leaf area.[81] Also, some herbivores may cause leaf photosynthesis rate decreases in a plant whereas injury from other herbivores does not.[79] In addition, compensation can occur by increasing photosynthetic rates following injury,[75] such as following stem-mining injury to rice by the yellow stem borer, *Scirpophaga incertulas* (Walker),[82] and simulated herbivory injury to *Nicotiana sylvestris.*[83]

9.5   MANAGED SYSTEMS: HERBIVORY AND CROP YIELDS

In managed systems, herbivory is important because it affects the yields of crop plants or growth of animals fed on rangelands. Yields can come from several different plant tissues including seeds, fiber content of tissue, fruits, roots, sugar levels in tissue, leaves, or stems.[84] Thus, many examples of yield do not measure plant reproductive effort (and hence an estimate of fitness), but measure highly exaggerated allocations to reproductive tissues atypical of plants from natural systems. Indirect plant injury does not necessarily result in yield reductions. For example, tomato plants with jasmonic acid induction had 60% less leaf injury and fewer flowers than control plants with no herbivory, but plants with induced jasmonic acid as a result of herbivory injury did not suffer yield reductions compared to control plants.[85] However, a negative relationship was found between leaf-feeding lepidopterans and tomato yield, where indirect feeding resulted in yield loss.[86] Comparing the impact of each of several different herbivores on a plant can point out key time periods or particular herbivores that cause yield losses to one plant species. Also, the same herbivore can have different impacts on yield across different plant hosts.[84]

Different forms of herbivory also may differentially influence yield quality. For example, glycoalkaloid content in potato roots (yield), *Solanum tuberosum*, increased following leaf herbivory (gross tissue removal) by Colorado potato beetles, *Leptinotarsa decemlineata*, but did not change following leaf herbivory (sucking injury) by potato leafhoppers, *Empoasca fabae.*[87] Also, root feeding on *Brassica* species by insects increased toxic glucosinolate levels in foliage,[88] and concentrations of foliar alkaloids increased in a Solaneaceous plant, *Atropa acuminata,* following insect feeding and simulated injury.[89] Thus, herbivory may have indirect effects that do not necessarily reduce yield, but reduce yield quality and hence usable yield.

Yield losses also can vary from injury caused by different herbivores, in degree of yield loss, duration of plant damage following herbivory, and timing of yield loss. For example, on alfalfa, *Medicago sativa*, different trends in yield losses were found from injury by several different herbivores including: a weevil, *Hypera brunneipennis,* sucking aphids, *Acyrthosiphon pisum* and *A. condoi,* and several Lepidopteran species (alfalfa caterpillar, *Colias eurytheme,* beet armyworm, *Spodoptera exigua,* and western yellowstriped armyworm, *S. praefica*).[90] The weevil caused yield losses for several cuttings following injury, the aphids caused yield losses only while at damaging densities, and yields increased following low injury levels from the lepidopteran species in this study. Also, an interaction between weevil and aphid feeding resulted in yield loss 30 to 60 days following peak numbers. Although feeding injury resulted in yield losses, there were no reductions in either alfalfa stand densities or in individual alfalfa plant mortality rates.

Control of weevils and slugs via insecticides allowed white clover, *Trifolium repens,* to have higher yields than when these two herbivores were not controlled.[91] Simulated defoliation on sugarbeets resulted in a linear decrease in 0.5% sucrose yield for each 1% defoliation applied.[92] Thus, it is important to observe the intensity of the attack by herbivores, each plant part attacked, and timing of the attack.[21, 93] With soybean, *Glycine max,* plants can tolerate 40% pre-bloom but only 25% post-bloom leaf defoliation by herbivores without suffering yield loss. Thus, age of plant can affect its susceptibility to tolerating herbivory injury. In one study, deciduous forest trees tolerated a rate of 30% annual leaf defoliation without having any effects detected on survival or growth rates during the three years of observations.[70]

Herbivory can affect crop plant architecture, such as leaf canopy structure. For example, with soybean, herbivory appears to cause yield losses only if canopy leaf area falls below a leaf area index (LAI) level of 3.5.[94] This is because photosynthetic efficiencies are about 95% at LAIs above 3.5, but efficiencies drop below LAIs of 3.5.[44] Although soybeans tend to have a high tolerance for herbivory injury, several factors like soil moisture, nutrient levels, and soybean genotypes affect actual yield losses observed. Even 50 to 70% leaf defoliation may not result in yield losses under optimal water and soil nutrient conditions, while under suboptimal conditions soybean tends to experience yield losses following defoliation.[44, 95] Also, canopy architecture changes when cotton experiences simulated reproductive organ injury, resulting in reduced cotton photosynthetic capacity.[96]

The attack of herbivores can be avoided, in some instances, by an interesting strategy where a plant lives in association with other plant species. For example, Rish[97] analyzed the densities of five herbivorous beetle species on monocultural squash and a polyculture of squash, maize, and beans. He found that beetles were less abundant in maize–bean–squash polycultures, partially because the beetles avoid host plants shaded by corn. Bach[98] focused on the response of one specialist herbivore, the striped cucumber beetle, *Acalymma vittata,* to cucumber monocultures vs. cucumber–broccoli–maize polycultures. By controlling total plant density, host-plant density, and plant diversity, Bach was able to distinguish the effect of these three confounding variables. He reported a significant effect of both plant density and diversity on *A. vittata* abundance, where fewer cucumber beetles per cucumber plant were found on cucumbers in polycultures compared to cucumber monocultures.

A similar result was observed with cassava, where yields were affected by injury from whiteflies, but yields were also reduced by interspecific competition with other plants, like cowpea.[99] Thus, monocultural cassava did not experience yield losses from interspecific competition, but suffered yield losses from whitefly injury. When cassava was grown with cowpea and whiteflies were eliminated, cassava again suffered yield losses from interspecific competition. However, when cassava was grown with cowpea and whiteflies were not eliminated, cassava had the best yields of the experiment. This interaction was not observed when cassava was intercropped with maize, as cassava yields were similar with and without whiteflies. Thus, several comparisons may be needed to determine when herbivory with other biotic factors will cause yield losses. On cabbage grown with and without "living mulches" (creeping bentgrass, red fescue, Kentucky bluegrass, white clover), populations of multiple herbivores were lower when cabbage was with a living mulch.[100] However, the benefits of reduced herbivory may trade off with cabbage yield reductions due to competition with living mulches. The impact of herbivory on plant yield depends on the presence of other plant species, whether the herbivores prefer the other plants, and whether yield losses are greater due to interspecific plant competition (with or without the presence of herbivores) or herbivory (with or without plant competition).

The interesting aspect of the interaction between crops and insects is that phenotypic variations in plant adaptations to herbivorous insects may be used as an approach to decrease insect herbivory injury on crop plants, permitting yield increments and subsequent stabilization of plant crop yields. Plant genetic engineering has been used broadly to develop plant varieties with elevated resistance to insect herbivore injury. This technique utilizes genetic material found in wild or crop species, conferring the desirable characteristic to yield increment, and will be an important area of research in the future for agricultural plants.[84]

9.6 NATURAL SYSTEMS: HERBIVORY AND PLANT FITNESS

In natural systems, how herbivory affects a plant is usually measured in terms of plant growth, survival, and/or reproduction. These parameters, when combined, indicate individual plant fitness. Thus, ecologists and evolutionary biologists tend to be interested in how herbivory serves as a selective force driving the evolution of plant traits when considering the plant side of plant–herbivore interactions. In this section we will consider the effects of sucking herbivory and tissue consumption, and then the effects of multiple herbivores on their plant hosts.

A point worth noting here is that most of the studies cited in this chapter report the effects of herbivory on plant growth, survival, and/or reproductive (sexual or asexual) traits. With annual (and perhaps biennial) plants, how herbivory affects plant survival and reproduction directly translates to plant fitness. However, perennial plants offer a serious challenge because reductions in plant survival or reproduction in one or multiple years may or may not lower the lifetime fitness of a perennial. Energy reserves and plant compensatory mechanisms may allow a perennial to not suffer or benefit from herbivory during part of its life. Studies that follow perennials through their entire life can discuss how herbivory affects plant fitness, but it is rarely feasible for researchers to pursue such long-term studies.

Until such long-term studies are performed, only limited or partial inferences can be drawn from many studies on how herbivory affects perennial lifetime fitness. These inferences may or may not be correct, but can suggest hypotheses and predictions to be tested in long-term studies on how herbivory affects perennial lifetime fitness. As an example, consider a study with scrub oak, *Quercus ilicifolia,* and herbivory by gypsy moth larvae, *Lymantria dispar.*[126] In years with complete defoliation by *L. dispar,* scrub oak acorn production dropped by 88% and stem growth by 49%, relative to control plants sprayed with an insecticide. Despite these results, it is nearly impossible to relate the effects on such high herbivory in some years of the scrub oak life to lifetime fitness. The observed reproductive output reductions may result in lower plant reproductive lifetime fitness, but acorn production in other years (especially mast years) may be more important to plant fitness, or the oaks may tolerate a certain amount of injury (across years) before lifetime fitness is affected. Thus, leaf herbivory might depress a perennial plant's reproductive output in some years, and yet not be sufficient to depress perennial lifetime fitness.

9.6.1 ONE HERBIVORE SPECIES

9.6.1.1 Mining/Sucking/Gall Injury

The effects of a root-boring ghost moth, *Hepialus californicus,* and seed-feeding insects were examined on bush lupine, *Lupinus arboreus,* their host plant.[101] Suppression of seed-feeding insects resulted in a 78% increase in seed output over three years of *L. arboreus,* while suppression of the ghost moth increased seed production by 31% and there was 18% higher survival of such plants. The effects were different by year, because suppression of seed-feeding insects reduced seed output only during the first two years, while the suppression of root-boring ghost moth only reduced seed production in the third year. Yet, the study showed that both above- and below-ground feeding injury can have negative fitness impact on *L. arboreus.*

A shoot-galling sawfly, *Euura lasiolepis,* reduced reproductive bud formation on arroyo willow, *Salix lasiolepis,* on single shoots with galls compared to shoots without galls, and on whole plant reproductive bud formation.[102] Most willows experienced low reductions because of low sawfly densities, but a sufficiently large (27.5%) group of plants had a 10% or greater reduction in reproductive bud formation as a result of sawfly shoot galls, and sawfly galls seemed to negatively impact their willow hosts. Another study examined the impact of a stem gall-forming fly, *Resseliella clavula,* on flowering dogwood, *Cornus florida.*[6] In the first three years of their study, inflorescence and fruit numbers decreased on shoots with fly galls compared to ungalled shoots. However, in the fourth year, galled shoots were longer and almost doubled the number of inflorescences relative to ungalled shoots, an overcompensatory response. Also, more vegetative shoots were produced in the third and fourth years of the study, which led to more reproductive buds. Thus, whether positive or negative impact results from gall formation on a long-lived plant's fitness depends on the time scale considered.

9.6.1.2   Consumption Injury

Feeding on inner bark layers of ponderosa pine, *Pinus ponderosa,* by Abert's squirrels, *Sciurus aberti,* resulted in a decrease of incremental growth, male strobilus formation, female cone production, and seed quality, and hence a general decrease in several traits that should affect plant fitness.[103] Several *P. ponderosa* characteristics in trees Abert's squirrels target are under genetic control and possibly have heritable additive genetic variation; hence, these herbivores may impose selective pressure on the pines. High levels of herbivory on leaves of a fast-growing annual, *Rudbeckia hirta,* by caterpillars of a nymphalid, *Chlosyne nycteis (Charidryas nycteis),* reduced seed set in flowers.[104] Those plants with more rapid growth experienced reduced seed set reductions relative to plants with slower growth rates. Foliar herbivory also can affect both male and female reproductive traits, as with wild radish, *Raphanus raphanistrum.*[105] Simulated leaf herbivory by unidentified scarab beetles, chrysomelid beetles, and grasshoppers also resulted in fewer flowers and lower seed production with a perennial herb, *Oenothera macrocarpa.*[106] Fruit predation by a noctuid moth, *Hadena bicruris,* on two species of plants, *Silene alba* and *S. dioica,*[107] had a large negative impact on fecundity of both plants, but flower and fruit production were reduced in *S. alba* by a factor of two compared to *S. dioica.* Thus, a single herbivore species can have different degrees of impact on plant fecundity depending on the plant host the herbivore consumes.

9.6.2   MULTIPLE HERBIVORES ON PLANTS

Many plants face multiple herbivore species during their life. In such situations, the effects of each herbivore and interactions between effects of different herbivores on their plant host need to be considered in order to gain a complete picture of the selective pressure of herbivory on the plant in question.[108] Studies with two herbivores form an obvious starting point for studying the impact of multiple herbivores on one plant species. For example, the effects of herbivory injury by flea beetles, *Phyllotreta cruciferae,* and diamondback moths, *Plutella xylostella,* are not independent on *Brassica rapa [B. campestris].*[109] When diamondback moths are common, *B. rapa* performs best when there is a large amount of injury from flea beetles. When diamondback moth injury is low, plants perform best when flea beetle injury also is low. In a different study, the presence of branch-causing herbivores results in more branches on a goldenrod, *Solidago altissima,* and an aphid, *Philaenus spumarius,* and a spittlebug, *Lepyronia quadrangularis,* are found more abundantly on the goldenrod as branch number increases.[110] Hence, the presence or absence of a herbivore causing one type of injury to a plant can influence selection pressure that other herbivores impose on the same host plant. Thus, the interactions of injury by multiple herbivores may be required to assess the complete impact of herbivory on plant performance and fitness.[109, 110]

Another study on multiple herbivores involves examining the separate effects of three herbivores: a xylem sap-feeding spittlebug, *P. spumarius,* a phloem sap-feeding aphid, *Uroleucon caligatum,* and leaf-feeding beetle, *Trirhabda* sp., on a goldenrod plant host, *S. altissima*.[111] Plant growth rate was reduced most by spittlebug feeding, slightly by beetle feeding, and was unchanged as a result of aphid feeding. Seed number decreased as a result of feeding from each of the three herbivores (order of importance: spittlebug, beetle, aphid), but only under high soil fertility conditions. In addition, spittlebug and beetle feeding resulted in flowering delays, but again only under high soil fertility conditions. Oddly, *S. altissima* sexual reproduction was harmed by herbivory only under "favorable" soil conditions, whereas asexual reproductive growth was not as harmed by herbivory injury under any soil conditions and appeared to be less susceptible to injury than sexual reproductive growth. Thus, an interaction was detected between herbivory injury and an abiotic environmental condition, soil quality.[111]

In a different six-year study on *S. altissima,* herbivore levels tended to be low, causing some reductions in inflorescence numbers but no detectable effects on stems.[112] Thus, these goldenrods tended to order the impact of herbivory more on seed production than on maintenance of present ramets, and the effects of herbivory in one year never affected plant performance in a following year. Occasionally, outbreaks of a few dominant insect herbivores completely inhibited flowering and stunted stem growth of *S. altissima*.[112] Thus, herbivory on *S. altissima* had mild chronic effects on plant performance most of the time, but occasional herbivore outbreaks strongly impose selection pressure on the goldenrods. For example, catastrophic mortality resulted from herbivory injury by a specialist chrysomelid beetle, *Rhabdopterus praetexus,* on an annual plant, *Impatiens pallida*.[113] Also, *I. pallida* living in areas with *R. praetexus* had early reproduction relative to plants in beetle-free regions, suggesting the influence of the beetles on a plant life history trait.

Another study on the separate effects of three herbivores involved a thrips, *Apterothrips apteris,* a spittlebug, *P. spumarius,* and plume moth caterpillars, *Platyptilia williamsii,* on the perennial plant, *Erigeron glaucus*.[115] Thrips feeding resulted in reduced seedling root biomass, and damaged ray petals of flowers which was associated with a reduced likelihood of visitation by pollinators. Fewer flowerheads were produced by *E. glaucus* only in the third year after three years of feeding by either the plume moth caterpillar or by spittlebugs compared to control plants. In addition, once herbivory ended, only plants injured by spittlebugs produced fewer flowerheads than control plants for an additional year before returning to normal flower production levels. Also, feeding from any of these three insect herbivores did not reduce *E. glaucus* survival, but survival was reduced by the impact of gophers, *Thomomys bottae.* Thus, feeding injury by all four herbivores reduced reproductive or survival parameters that could negatively impact the fitness of *E. glaucus.* Other studies also have found that herbivores can affect plant reproductive fitness traits in subsequent years following herbivory injury.[115, 116]

Leaf herbivores also have a negative effect on fitness of tropical plants, for example *Piper arieianum*.[117] This plant faces herbivory from a suite of 95 species in Costa Rica, and there are several genotypes conferring resistance to different herbivorous species.[118] Thus, one genotype would have resistance to herbivore-A but not herbivore-B, while another genotype would have resistance to herbivore-B but not herbivore-A. Clones that experienced greater degrees of herbivory injury grew

less than clones that received less damage, which should result in fitness consequences to individual plants. The species of herbivore causing the most injury to *P.
arieianum* changed over time, so intensity and quality of selection pressure from herbivory changed over time as well.[118]

Injury that is concentrated spatially on certain tissues (such as reproductive
branches) of *P. arieianum* can result in up to an 80% reduction in seed production,
and the plant has certain times when it is most vulnerable to concentrated injury. The
same amount of injury spread throughout a plant rather than concentrated on vulnerable tissues resulted in plants that could not be distinguished from control plants.[119]
A similar result occurred with wild radish, *R. sativus,* where flower number, reproductive biomass, and total biomass were higher in no-injury control plants and plants
that received 25% simulated herbivory injury on four leaves, than on leaves with 50%
injury on two leaves, or 100% injury on a new or mature single leaf.[120] Both pattern
(concentrated or diffuse) and timing of herbivory injury can be critical to the degree
(if any) of resulting plant damage. On a side note, it has been suggested that induced
plant defenses do not always reduce herbivore densities and hence herbivory pressure. Instead, induced defenses may sometimes help spread out future injury on plant
tissues, where diffuse herbivory should result in less harm on a plant compared to
concentrated herbivory.[121, 122]

Removal of herbivores was performed with annual morning glory, *Ipomoea purpurea,* to assess the selection pressure imposed by four different insect herbivores.[55]
Seed number increased by 20% when insect herbivores were removed, and genetic
variation for seed number was eliminated. Thus, these herbivores seem to impose
selection pressure on some traits of *I. purpurea* by having a negative impact on individual plant fitness based on seed number. Herbivory on flowers and seeds can also
impose selection pressure on flower timing in wild sunflower, *Helianthus annuus.*[123]
Five insects were studied that feed on developing sunflower seeds. Primary injury
from some herbivores (the head-clipping weevil, *Haplorhynchites aeneus,* the sunflower moth, *Homoeosoma electellum,* and the sunflower bud moth, *Suleima
helianthana*) occurred early, injury was low early and increased over a season from a
seed fly, *Gymnocarena diffusa,* and injury was constant from two seed weevils,
Smivronyx fulvus and *S. sordidus.* The impact of seed herbivory generally had negative effects on plant reproductive fitness, but selection for late flowering by *H. annuus*
seemed to be driven from two of the herbivores, *S. helianthana* and *H. electellum.*
This finding is based on two phenotypic selection analyses, one analysis examining
effects from all herbivores, and a second analysis examining effects of all herbivores
but the two sunflower moths, *S. helianthana* and *H. electellum.*[123]

Although herbivory injury can impose selection pressure on plant size and resistance to particular herbivores, without genetic variation no evolutionary plant
responses can occur. This was observed in a study examining the selective pressures
of the tobacco flea beetle, *Epitrix parvula,* and a grasshopper, *Sphenarium pur-
purascens,* on an annual weed, *Datura stramonium.*[124] Both herbivores imposed
selection pressures on plant size and/or plant resistance to herbivores. Yet, there was
no detectable genetic variation on any reproductive traits that might affect plant fitness as assessed with paternal half-sibling family analysis. Thus, information on the

heritability of the plant traits is necessary to determine where herbivores can have present evolutionary selection pressure on a plant host.

9.7 HERBIVORE INJURY: INDIRECT EFFECTS ON PLANTS

9.7.1 HERBIVORY AND PLANT COMPETITIVE ABILITY

Much of the debate around the evolutionary importance of herbivory to plants has focused on evidence for direct impact of herbivores on individual plant performance or on plant population dynamics. However, herbivory can influence plant fitness through various indirect mechanisms, including altering plant competitive ability relative to conspecific or heterospecific plants, and how plants interact with other biotic stresses.

Louda et al.[125] reviewed data on the influence of herbivory on plant performance and competitive relationships. From their review, they concluded that herbivory can influence competition in two ways. First, competition was altered by indirectly changing plant access to resources through changes in plant growth and morphology as a result of herbivory. Second, competition was altered because herbivory can change the distribution and abundance of some plant species, which again can change resources available to competing plants. The review and theoretical work by Louda et al.[125] led to a concluding observation that "herbivory was particularly important where constraints in resources, growing season, or growth strategies limited plant compensation for losses, and diminished the species capacity to maintain itself against competitors."

9.7.2 LEAF QUALITY

Defoliation by herbivores can alter the quality of replacement leaves. On a scrub oak, *Quercus ilicifolia,* that was completely defoliated by gypsy moth caterpillars, *Lymantria dispar,* plants produced a second set of leaves each summer after defoliation occurred.[126] These secondary leaves had lower nitrogen, copper, and zinc levels than primary leaves of control plants, so it is possible that leaf quality changed. Yet, how the change in leaf nutrient content might affect leaf functioning or attractiveness to herbivores was not studied, so a connection to plant performance or fitness could not be estimated.

9.7.3 FLORAL ATTRACTIVENESS TO POLLINATORS

Leaf herbivory also can result in reduced flower attractiveness to pollinators, and differentially affect male and female fitness parameters.[127, 128] For example, leaf herbivory from white cabbage butterfly caterpillars, *Pieris rapae,* on an obligate outcrossing wild radish species, *R. raphanistrum,* resulted in reduced floral attraction and reward characters, and thus reduced pollinator visitation to wild radish.[127] Herbivory by *P. rapae* larvae resulted in fewer and smaller flowers on *R. raphanistrum,* and injured radishes had fewer bee visits as a result.[129] In this study, when the number of flowers was equalized between injured and uninjured plants, bee visits did not differ and hence bees seemed to use flower number as a cue. However, syrphid flies visited uninjured flowers more often even when controlling for flower numbers, and seemed to not use flower number as a cue.[129]

In a different study involving *B. rapa,* high and low resistant lines (to flea beetle attack) were studied in response to herbivory by larval *P. rapae.*[130] The leaf herbivory injury resulted in later injury to flower petals, and high resistance flowers had reduced petal size even in the absence of injury (a cost of resistance), so leaf herbivory and resistance to herbivory both affected plant floral features. Although there were no differences in number of open flowers of damaged low and high resistance plants, pollinators spent more time at flowers of injured low resistance than high resistance plants, perhaps because palatability or reward levels of high resistance flowers may have decreased following injury. Longer pollinator visits result in higher fitness for *B. rapa,* so it seems that a trade-off exists between chemical defense and floral attractiveness, yet injury also reduces floral attractiveness.[130] Herbivory impact on reduced flower attractiveness to pollinators has also been demonstrated for *O. macrocarpa,* as leaf herbivory results in fewer and smaller flowers.[106] Also, two hawk moth species (*Dolba hyloeus* and *Paratraea plebeja*) preferred flowers with larger corollas, and flowers with larger corollas were more likely to produce fruits. Thus, indirect effects of herbivory injury on flower number and/or quality can reduce a plant's floral attractiveness to pollinators and lower plant fitness.[106]

9.7.4 PLANT SIZE AND ARCHITECTURE

On a larger scale, a plant's size can affect whether it will be heavily attacked by herbivores. For example, 64% of individuals of a perennial herbaceous plant, *Vicia cracca,* had inflorescence injury from herbivores.[131] The injury levels on plants were classified as none, partial, and totally browsed. A large proportion of the inflorescence injury occurred on small plants, while larger plants suffered partial or no injury. Partially browsed plants compensated for numbers of inflorescences, fruits, seeds, and seed weight, but not for flower number. Totally browsed plants experienced almost complete reproductive failure, and hence only heavy herbivory injury had large negative effects on plant fecundity. Hence, smaller plants were at much greater risk for negative impact of herbivory than larger plants.[131]

In contrast, herbivory also can affect plant architecture and sexual expression. The effects of herbivory by a cone boring moth, *Dioryctria albovitella,* over a three-year period on pinyon pine, *Pinus edulis,* affected plant architecture, reduced shoot production and growth rate, and resulted in functionally male plants as female cone-bearing ability was lost.[132] However, when the moths were removed, normal growth and reproduction patterns resumed.

9.7.5 OTHER FACTORS

Simulated injury (cutting shoots) on *I. aggregata* had different consequences depending on whether or not plants received fertilizer (nitrogen limited) and hand pollination (pollen limited), as plants had (partial) compensation abilities only when not under nitrogen and pollen limiting conditions.[68] Those plants receiving simulated injury were under strong selection pressure for early flowering and increased plant height, so injury affected plant architecture, phenology, and reduced plant fitness. Injury may reduce flower pollination of *I. aggregata*, and, depending on available resources, may increase or reduce fitness of scarlet gilia plants.[65] Clipping injury reduced plant reproductive effort (flowers, fruits, seeds) while emasculation increased production of these same traits. When both injuries were considered together, plant fitness increased with emasculation and no clipping injury, while plant fitness decreased with emasculation and clipping injury.[65] The two forms of injury interacted to influence plant reproductive effort and fitness. Other factors of importance include good nutritional conditions and water availability so that unstressed plants can tolerate herbivory injury better than stressed plants.[21]

On a different note, variation in a biotic factor—size of different ant species—was related to the impact of a beetle's, *Stethobaris* sp., herbivory on an orchid, *Schomburgkia tibicinis*.[133] As the body size of the ant species increased, herbivory by the beetle decreased, and plant fruit production increased, so plant reproductive efforts increased with size of associated ant species. Plant reproductive output for *Opuntia stricta* increased by 50% for plants when a suite of nine ant species was allowed to remain compared to a treatment where ants were removed despite continuous nectar collection by these ants.[134] The ants helped reduce bud damage from a pyralid moth, while damage from sucking bugs and mining dipterans were not different between ant-excluded and antincluded treatments. Thus, biotic factors may have specific interactions with other biotic and abiotic factors in altering how herbivory will impact plants.

Both internal and external conditions can influence plant responses to herbivory. For example, two perennial grass species replace each other along a topographic/resource gradient. Partial defoliation had negative, neutral, or positive effects on plant mass of each species, depending on where a species fell along a geographic gradient.[135] The pattern of response was different for each plant species, yet each species had overcompensatory responses to defoliation in edaphic environments where each species was most abundant. Thus, the combination of several factors may need to be considered in some cases before the general impact of herbivory on a plant species can be assessed.

9.8   HERBIVORY AND PLANT POPULATION DYNAMICS

The influence of herbivory on plant population dynamics is often assessed with the use of pesticides for small herbivores, where cleared vs. control comparisons can be made, while larger vertebrate herbivores tend to be physically excluded from herbivore-free plots. Several effects of insect herbivory were inferred from this approach with early succession communities containing the plant *Trifolium pratense*.[136] Plants grew taller and had more leaves (controlling for leaf area of individual leaves) when insect herbivores were cleared with an insecticide, while herbivory reduced total leaf area and increased variation in plant size across individual plants. Individual fitness of plants was reduced by herbivory, as seed number/plant and individual seed weights decreased, and significant mortality was imposed on seedlings in two of three study sites. Thus, herbivory affects individual plant fitness and in interaction with plant competition also determines changes in numbers of *T. pratense* populations. Herbivory also affected size variability in eight species of annual and perennial grasses and forbs, again inferred by comparing plots with insects cleared by insecticides having large size variation with control plots that have low size variation.[137]

A guild of inflorescence insect herbivores had several effects on native Platte thistle, *Circium canescens,* both on plant fitness and population dynamic parameters.[138] Using an insect exclusion design by insecticide, thistles in exclusion plots had higher seed output, seedling density was higher, later blooming flowers contributed more to the seed pool, and more seedlings led to higher numbers of flowering adults than thistles in plots with inflorescence insect herbivores. This study shows that an inflorescence guild of herbivores reduces Platte thistle seed production and maternal fitness, but population dynamic parameters were also affected, including the reduction of seedling recruitment and thistle density.[138] An additional insect exclusion study with a tall thistle, *C. altissium,* determined that insect herbivores reduce thistle growth and survival.[139] It appears that the thistle is limited by the impact of insect herbivory injury, and thistle can also be limited via seed herbivory by an introduced weevil.[140] A different study on a cruciferous plant, *Moricandia moricandiodes,* indicated that seed herbivory by sheep reduced seed dispersal and swamped out the effects of other seed herbivores, while pollinators did not have any detectable effects on seed dispersal.[141] Thus, herbivory has the potential to affect plant populations and limit plant distributions and/or densities in natural settings.

The consequences of herbivory can be such that plant fitness and abundance are affected, but these effects do not necessarily lead to changes in species diversity. Bach[142] observed the impact of a flea beetle, *Altica subplicata,* on the sand-dune willow, *Salix cordata,* in terms of growth and mortality on the willow and any changes in dune successional plant species. Using mesh cages to exclude flea beetles in some plots and allow flea beetles in other plots with no cage, she found that flea beetles reduced plant height and diameter by a factor of two, while plant mortality was three times higher on one duneside and six times higher on another dune side when flea beetles had access to the sand-dune willow. The abundance of other plant species differed between exclusion and inclusion plots, with herbaceous monocots becoming more prevalent and woody plants becoming less abundant, even though the flea beetle only fed on the sand-dune willow. Thus, the indirect effects of flea beetle herbivory injury on sand-dune willow altered sand-dune plant successional patterns, but did not alter overall species diversity (richness or evenness) patterns in a three-year study.[142]

Different herbivores will have different levels of importance in affecting plant population dynamics. Following mass mortality of a sea urchin, *Stronglyocentrotus droebachiensis,* once again *L. longicruris* became a dominant canopy plant in its rocky subtidal habitat, suggesting that the sea urchin had played a powerful role in suppressing kelp population levels. A mesogastropod, *Lacuna vincta,* became the only major herbivore on the kelp, while other abundant herbivorous grazers like a chiton, *Tonicella rubra,* and a limpet, *Notoacmaea testudinalis,* could not consume the chemically defended kelp tissues. The impact of *L. vincta* reduced kelp canopy area, but this did not provide understory plant species sufficient time to take advantage of temporary light availability. Unlike *S. droebachiensis, L. vincta* could only consume kelp tissues that did not have high chemical defense concentrations, and hence did not consume tissue that would cause kelp mortality. Thus, some potential herbivores were entirely deterred by kelp defenses and had no present impact on the plants. The mesogastropod reduced plant canopy area, but did not cause kelp mortality or affect community structure. Only the presence of the sea urchin caused major kelp mortality and therefore altered the plant community structure of the subtidal zone.[45]

9.9 CONCLUSIONS

Over evolutionary time, it seems clear that herbivory has exerted a negative, perhaps large, impact on plant performance. An array of plant defenses against herbivory suggests this point, while mutualistic plant–herbivore interactions seem unlikely even for plants that presently overcompensate from herbivory. Herbivory tends to have negative effects on plant performance traits (growth, survival, sexual, and vegetative reproduction) that generally combine to determine plant fitness. However, in some cases plant compensation can occur such that plants suffer no net damage or even benefit following injury in natural settings. Several factors make it difficult to clearly state how herbivory will lower plant yields in managed systems or reduce traits related to plant fitness (fecundity and survival) in natural settings. Plant compensation to injury can vary across gradients of abiotic conditions (water availability, nutrient availability) and interact with biotic factors as well (such as plant competition levels). In addition, plants often do not experience injury from a single herbivore species only, and interactions between injuries from multiple herbivores can make the issue of estimating herbivory impact on plant performance quite complex. Also, the timing and concentration of herbivory on plant tissues can influence impact of herbivory on the plant. A caution to consider is that estimating the impact of herbivory on lifetime fitness of perennial plant species is quite difficult, especially when studies do not follow a perennial throughout its entire life. Since most studies do not examine herbivory over a perennial's entire life, such studies can make few statements about how herbivory affects perennial plant fitness. Instead, statements are limited to how herbivory affects plant performance in perennial species.

Herbivory often will affect plant reproductive effort and/or survivorship, but can also have indirect effects on plant fitness, such as with reduced floral attractiveness to pollinators. Subtle effects of herbivory may not be obvious but still have important consequences to plant fitness, while chronic effects of herbivory may take years to detect. Finally, herbivores that feed on annuals, biennials, and long-lived plants, along a gradient may have temporary to continuous pressure on their plant host(s). Thus, the present selection pressure that herbivores impose on plants varies greatly,[61, 66, 135] and plants will evolve only when they possess additive genetic variation in traits that affect plant fitness.

Just because herbivory injury affects individual plant performance does not necessarily mean that injury affects plant population dynamics like recruitment, density, distribution, and community structure of plant species. High seed predation can limit plant populations,[71, 143] but effects on seedling or adult survivorship strongly affect plant recruitment in a habitat. Although multiple herbivores may affect individuals of a plant species, not all of the herbivores will necessarily affect population dynamics of the plant species. Thus, certain keystone herbivores[45, 143] can have large effects even on plant community structure while other herbivores may have minor or negligible effects on present-day plant performance parameters.

The effect of herbivory on plants is not a simple issue. A number of factors often require study before a complete picture can be made for the effects of herbivory on a single plant species. However, researchers have examined a number of levels of plant performance parameters. These levels include plant physiological responses, growth changes, yields, effects on survivorship and fecundity traits that affect plant fitness, the selection pressure of multiple herbivores on a plant species, the responses of a plant species across a variety of environmental conditions, and several aspects of plant population and community level parameters. Studies at all of these levels will continue to help in how we understand plant–herbivore interactions from the perspective of how herbivores affect plants.

REFERENCES

1. Strong, D. R., Lawton, J. H., and Southwood, T. R. E., *Insects on Plants,* Blackwell Scientific, Oxford, U.K., 1984.
2. Crawley, M. J., Insect herbivores and plant population dynamics, *Annu. Rev. Entomol.,* 34, 531, 1989.
3. Belsky, A. J., Does herbivory benefit plants?, *Am. Nat.,* 127, 870, 1986.
4. Boyer, J. S., Plant productivity and environment, *Science,* 218, 443, 1982.
5. Paige, K. N., and Whitham, T. G., Overcompensation in response to mammalian herbivory: the advantage of being eaten, *Am. Nat.,* 129, 407, 1987.
6. Sacchi, C. F., and Conner, E. F., Changes in reproduction and architecture in flowering dogwood, *Cornus florida,* after attack by the dogwood club gall, *Resseliella clavula, Oikos,* 86, 138, 1999.
7. Ehrlich, P. R., and Raven, P. H., Butterflies and plants: a study in co-evolution, *Evolution,* 18, 586, 1964.
8. Hougen-Eitzman, D., and Rausher, M. D., Interactions between herbivorous insects and plant-insect coevolution, *Am. Nat.,* 143, 677, 1994.
9. Rausher, M. D., Genetic analysis of coevolution between plants and their natural enemies, *Trends Ecol. Evol.,* 12, 212, 1996.
10. Iwao, K., and Rausher, M. D., Evolution of plant resistance to multiple herbivores: quantifying diffuse coevolution, *Am. Nat.,* 149, 316, 1997.
11. Whittaker, R. H., and Feeny, P. P., Allelochemics: chemical interactions between species, *Science,* 171, 757, 1971.
12. Levin, D. A., The chemical defenses of plants to pathogens and herbivores, *Annu. Rev. Ecol. Syst.,* 7, 121, 1976.
13. Rosenthal, G. A., and Janzen, D. H., Editors, *Herbivores: Their Interaction with Secondary Plant Metabolites,* Academic Press, New York, 1979.

14. Karban, R., and Myers, J. H., Induced plant responses to herbivory, *Annu. Rev. Ecol. Syst.*, 20, 331, 1989.
15. Sagers, C. L., and Coley, P. D., Benefits and costs of defense in a shrub, *Ecology*, 76, 1835, 1995.
16. Coley, P. D., and Barone, J. A., Herbivory and plant defenses in tropical forests, *Annu. Rev. Ecol. Syst.*, 27, 305, 1996.
17. Price, P. W., Bouton, C. E., Gross, P., McPheron, B. A., Thompson, J. N., and Weis, A. E., Interactions among three trophic levels: influence of plants on interactions between insect herbivores and natural enemies, *Annu. Rev. Ecol. Syst.*, 11, 41, 1980.
18. Schmitz, O. J., Resource edibility and trophic exploitation in an old-field food web, *Proc. Natl. Acad. Sci., USA*, 91, 5364, 1994.
19. Marquis, R. J., and Whelan, C. Plant morphology and recruitment of the third trophic level: subtle and little-recognized defenses?, *Oikos*, 75, 330, 1996.
20. Paré, P. W., Alborn, H. T., and Tumlinson, J. H., Concerted biosynthesis of an insect elicitor of plant volatiles, *Proc. Natl. Acad. Sci. USA*, 95, 13971, 1998.
21. Denno, R. F., and McClure M. S., Introduction variability: a key to understanding plant-herbivore interactions, in *Variable Plants and Herbivores in Natural and Managed Systems*, Denno, R. F., and McClure, M. S., Eds., Academic Press, New York, 1983, 1.
22. Kennedy, G. G., and Barbour, J. D., Resistance variation in natural and managed systems, in *Plant Resistance to Herbivores and Pathogens: Ecology, Evolution, and Genetics*, Fritz, R. S., and Simms, E. L., Eds., University of Chicago Press, Chicago, 1992, 13.
23. Gallun, R. L., and Khush, G. S., Genetic factors affecting expression and stability of resistance, in *Breeding Plants Resistant to Insects*, Maxwell, F. G., and Jennings, P. R., Eds., Wiley, New York, 1980, 63.
24. Tingey, W. M., and Singh, S. R., Environmental factors influencing the magnitude and expression of resistance, in *Breeding Plants Resistant to Insects*, Maxwell, F. G., and Jennings, P. R., Eds., Wiley, New York, 1980, 87.
25. Gould, F., Genetics of plant-herbivore systems: interactions between applied and basic studies, in *Variable Plants and Herbivores in Natural and Managed Systems*, Denno, R. F., and McClure, M. S., Eds., Academic Press, New York, 1983, 599.
26. Diehl, S. R., and Bush G. L., An evolutionary and applied perspective of insect biotypes, *Annu. Rev. Entomol.*, 29, 471, 1984.
27. Futuyama, D. J., and Peterson S. C., Genetic variation in the use of resources by insects, *Annu. Rev. Entomol.*, 30, 217, 1985.
28. Smith, C. M., *Plant Resistance to Insects: A Fundamental Approach*, Wiley, New York, 1989.
29. Horber, E., Types and classification of resistance, in *Breeding Plants Resistant to Insects*, Maxwell, F. G., and Jennings, P. R., Eds., John Wiley, New York, 1980, 15.
30. Moran, N., and Hamilton, W. D., Low nutritive quality as defense against herbivores, *J. Theor. Biol.*, 86, 247, 1980.
31. Fernandes, G. W., Plant mechanical defenses against insect herbivory, *Revista Brasileira de Entomol.*, 38, 421, 1994.
32. Kareiva, P., Influence of vegetation texture on herbivore populations: resource concentration and herbivore movement, in *Variable Plants and Herbivores in Natural and Managed Systems*, Denno, R. F., and McClure, M. S., Eds., Academic Press, New York, 1983, 259.
33. Karban, R., Neighborhood affects a plant's risk of herbivory and subsequent success, *Ecol. Entomol.*, 22, 433, 1997.

34. Feeny, P., Plant apparency and chemical defense, *Recent Adv. Phytochem.*, 10, 1, 1976.
35. Berenbaum, M. R., Turnabout is fair play: secondary roles for primary compounds, *J. Chem. Ecol.*, 21, 925, 1995.
36. Kuc, J., Multiple mechanisms, reaction rates, and induced resistance in plants, in *Plant Disease Control*, Staples, R. C., Ed., Wiley, New York, 1981, 259.
37. Kuc, J., Plant immunization and its applicability for disease control, in *Innovative Approaches to Plant Disease Control*, I. Chet, Editor, Wiley, New York, 1987, 255.
38. Agrawal, A. A., Induced responses to herbivory and increased plant performance, *Science*, 279, 1201,1998.
39. Wold, E. N., and Marquis, R. J., Induced defenses in white oak: effects on herbivores and consequences for the plant, *Ecology*, 78, 1356, 1997.
40. Agrawal, A. A., Induced responses to herbivory in wild radish: effects of several herbivores and plant fitness, *Ecology*, 80, 1713, 1999.
41. Ryan, C. A., Insect-induced chemical signals regulating natural plant protection responses, in *Variable Plants and Herbivores in Natural and Managed Systems*, Denno, R. F., and McClure, M. S., Eds., Academic Press, New York, 1983, 243.
42. Higley, L. G., Browde, J. A., and Higley, P. M., Moving towards new understandings of biotic stress and stress interactions, in *International Crop Science I*, Buxton, D. R., Shibles, R., Forseberg, R. A., Blad, B. L., and Asay, K. H., Eds., Crop Science Society of America, Madison, WI, 1993, 749.
43. Trumble, J. T., Kolodny-Hirsch, D. M., and Ting, I. P., Plant compensation for arthropod herbivory, *Annu. Rev. Entomol.*, 38, 93, 1993.
44. Haile, F. J., Physiology of Plant Tolerance to Arthropod Injury, Ph.D. Dissertation, University of Nebraska – Lincoln, 1999.
45. Johnson, C. R., and Mann, K. H., The importance of plant defense abilities to the structure of subtidal seaweed communities: the kelp, *Laminaria longicruris*, survives grazing by the snail, *Lacuna vincta*, at high population densities, *J. Exp. Mar. Biol. Ecol.*, 97, 231, 1986.
46. Simms, E. L., and Fritz, R. S., The ecology and evolution of host-plant resistance to insects, *Trends Ecol. Evol.*, 5, 356, 1990.
47. Simms, E. L., Costs of plant resistance to herbivory, in *Plant Resistance to Herbivores and Pathogens: Ecology, Evolution, and Genetics*, Fritz, R. S., and Simms, E. L., Eds., The University of Chicago Press, Chicago, 1992, 392.
48. Fineblum, W. L., and Rausher, M. D., Trade-off between resistance and tolerance to herbivore damage in a morning glory, *Nature*, 377, 517, 1995.
49. Bergelson, J., and Purrington, C. B., Surveying patterns in the cost of resistance in plants, *Am. Nat.*, 148, 536, 1996.
50. Baldwin, I. T., Jasmonate-induced responses are costly but benefit plants under attack in native populations, *Proc. Natl. Acad. Sci. USA*, 95, 8113, 1998.
51. Mauricio, R., Costs of resistance to natural enemies in field populations of the annual plant *Arabidopsis thaliana, Am. Nat.*, 151, 20, 1998.
52. Stowe, K. A., Experimental evolution of resistance in *Brassica rapa:* correlated response of tolerance in lines selected for glucosinolate content, *Evolution*, 52, 703, 1998.
53. Elle, E., van Dam, N. M., and Hare, J. D., Cost of glandular trichomes, a "resistance" character in *Datura wrightii* Regel (Solanaceae), *Evolution*, 53, 22, 1999.
54. Agrawal, A. A., Strauss, S. Y., and Stout, M. J., Cost of induced responses and tolerance to herbivory in male and female fitness components of wild radish, *Evolution*, 53, 1093, 1999.

55. Simms, E. L., and Rausher, M. D., The evolution of resistance to herbivory in *Ipomoea purpurea:* II. Natural selection by insects and costs of resistance, *J. Ecol.,* 77, 537, 1989.
56. Shen, C. S., and Bach, C. E., Genetic variation in resistance and tolerance to insect herbivory in *Salix cordata, Ecol. Entomol.,* 22, 335,1997.
57. Chapin, F. S., and McNaughton, S. J., Lack of compensatory growth under phosphorus deficiency in grazing-adapted grasses from the Serengeti Plains (Tanzania), *Oecologia,* 79, 551, 1989.
58. McNaughton, S. J., Compensatory plant growth as a response to herbivory, *Oikos,* 40, 329, 1983.
59. Georgiadis, N. J., and McNaughton, S. J., Interactions between grazers and a cyanogenic grass, *Cynodon plectostachyus, Oikos,* 51, 343, 1988.
60. Paige, K. N., Overcompensation in response to mammalian herbivory: from mutualistic to antagonistic interactions, *Ecology,* 73, 2076, 1992.
61. Belsky, A. J., Carson, W. P., Jensen, C. L., and Fox, G., Overcompensation in plants: herbivore optimization or red herring?, *Evol. Ecol.,* 7, 109, 1993.
62. Jaremo, J., Tuomi, J., Nilsson, P., and Lennartsson, T., Plant adaptations to herbivory: mutualistic versus antagonistic coevolution, *Oikos,* 84, 313, 1999.
63. Lennartsson, T., Tuomi, J., and Nilsson, P., Evidence for an evolutionary history of over-compensation in the grassland biennial *Gentiantella campestris* (Gentianaceae), *Am. Nat.,* 149, 1147, 1997.
64. Gronemeyer, P. A., Dilger, B. J., Bouzat, J. L., and Paige, K. N., The effects of herbivory on paternal fitness in scarlet gilia: better moms also make better pops, *Am. Nat.,* 150, 592, 1997.
65. Juenger, T., and Bergelson, J., Does early season browsing influence the effect of self-pollination in scarlet gilia?, *Ecology,* 81, 41, 2000.
66. Paige, K. H., Regrowth following ungulate herbivory in *Ipomopsis aggregata:* geographic evidence for overcompensation, *Oecologia,* 118, 316, 1999.
67. Juenger, T., and Bergelson, J., Pollen and resource limitation of compensation to herbivory in scarlet gilia, *Ipomopsis aggregata, Ecology,* 78, 1684, 1997.
68. Juenger, T., and Bergelson, J., Pairwise versus diffuse natural selection and the multiple herbivores of scarlet gilia, *Ipomopsis aggregata, Evolution,* 52, 1583, 1998.
69. Bergelson, J., Juenger, T., and Crawley, M. J., Regrowth following herbivory in *Ipomopsis aggregata:* compensation but not overcompensation, *Am. Nat.,* 148, 744, 1996.
70. Kulman, H. M., Effects of insect defoliation on growth and mortality of trees, *Annu. Rev. Entomol.,* 16, 289, 1971.
71. Harper, J. L., *Population Biology of Plants,* Academic Press, London, 1977.
72. Crawley, M. J., *Herbivory: The Dynamics of Animal-Plant Interactions,* University of California Press, Berkeley, 1983.
73. Waloff, N., and Richards, O. W., The effect of insect fauna on growth, mortality, and natality of broom, *Sarathamnus spoparius, J. Appl. Ecol.,* 14, 787, 1977.
74. DeAngelis, J., Berry, R. E., and Krantz, G. W., Photosynthesis, leaf conductance, and leaf chlorophyll content in spider mite (Acari: Tetranchyidae) injured peppermint leaves, *Environ. Entomol.,* 12, 345, 1983.
75. Welter, S. C., Arthropod impact on plant gas exchange, in *Insect–Plant Interactions,* Bernays, E. A., Ed., CRC Press, Boca Raton, 1989, 135.
76. Welter, S. C., Farnham. D. S., McNally, P. S., and Freeman, R., Effect of Willamette mite and Pacific spider mite (Acari: Tetranychidae) on grape photosynthesis and stomatal conductance, *Environ. Entomol.,* 18, 953, 1989.

77. Buntin, G. D., Braman, S. K., Gilbertz, D. A., and Phillips, D. V., Chlorosis, photosynthesis, and transpiration of azalea leaves after azalea lace bug (Heteroptera: Tingidae) feeding injury, *J. Econ. Entomol.,* 89, 990, 1996.

78. Schaffer, B., Peña, J. E., Colls, A. M., and Hunsberger, A., Citrus leafminer (Lepidoptera: Gracillariidae) in lime: assessment of leaf damage and effects on photosynthesis, *Crop Prot.,* 16, 337, 1997.

79. Peterson, R. K. D., Higley, L. G., Haile, F. J., and Barrigossi, J. A. F., Mexican bean beetle (Coleoptera: Chrysomelidae) injury affects photosynthesis of *Glycine max* and *Phaseolus vulgaris, Environ. Entomol.,* 27, 373, 1998.

80. Jackson, P. R., and Hunter, P. E., Effects of nitrogen fertilizer level on development and populations of the pecan leaf scorch mite (Acari: Tetranychidae), *J. Econ. Entomol.,* 76, 432, 1983.

81. Hughes, L., and Bazzaz, F. A., Effect of elevated CO_2 on interactions between the western flower thrips, *Frankliniella occidentalis* (Thysanoptera: Thripidae) and the common milkweed, *Asclepias syriaca, Oecologia,* 109, 286, 1997.

82. Rubia, E. G., Heong, K. L., Zalucki, M., Gonzales, B., and Norton, G.A., Mechanisms of compensation of rice plants to yellow stem borer *Scirpophaga incertulas* (Walker) injury, *Crop Prot.,* 15, 335, 1996.

83. Baldwin, I. T., and Ohnmeiss, T. E., Coordination of photosynthetic and alkaloidal responses to damage in uninducible and inducible *Nicotiana sylvestris, Ecology,* 75, 1003, 1994.

84. Duck, N., and Evola, S., Use of Transgenes to increase host plant resistance to insects: opportunities and challenges, in *Advances in Insect Control: The Role of Transgenic Plants,* Carozzi, N., and Koziel, M., Eds., Taylor and Francis, NY, 1997, 1.

85. Thaler, J. S., Induced resistance in agricultural crops: effects of jasmonic acid on herbivory and yield in tomato plants, *Environ. Entomol.,* 28, 30, 1999.

86. Welter, S. C., Johnson, M. W., Toscano, N. C., Perring, T. M., and Varela, L., Herbivore effects on fresh and processing tomato productivity before harvest, *J. Econ. Entomol.,* 82, 935, 1989b.

87. Hlywka, J. J., Stephenson, G. R., Sears, M. K., and Yada, R.Y., Effects of insect damage on glycoalkaloid content in potatoes (*Solanum tuberosum*), *J. Agric. Food Chem.,* 42, 2545, 1994.

88. Birch, A. N. E., Griffiths, D. W., Hopkins, R. J., Smith, W. H. M., and McKinlay, R. G., Glucosinolate responses of swede, kale, forage and oilseed rape to root damage by turnip root fly (*Delia floralis*) larvae, *J. Sci. Food Agric.,* 60, 1, 1992.

89. Khan, M. B., and Harborne, J. B., A comparison of the effect of mechanical and insect damage on alkaloid levels in *Atropa acuminata, Biochem. Syst. Ecol.,* 19, 529, 1991.

90. Summers, C. G., Effects of selected pests and multiple pest complexes on alfalfa productivity and stand persistence, *J. Econ. Entomol.,* 82, 1782, 1989.

91. Mowat, D. J., and Shakeel, M. A., The effect of some invertebrate species on persistence of white clover in ryegrass swards, *Grass and Forage Sci.,* 44, 117, 1989.

92. Capinera, J. L., Effect of simulated insect herbivory on sugarbeet yield in Colorado, USA, *J. Kansas Entomol. Soc.,* 52, 712, 1979.

93. Pedigo, L.P., *Entomology and Pest Management,* 2nd ed., Prentice-Hall, Upper Saddle River, New Jersey, 1996.

94. Higley, L. G., New understandings of soybean defoliation and their implications for pest management, in *Pest Management of Soybean,* Copping, L. G., Green, M. B., and Rees, R. T., Eds., Elsevier, Amsterdam, Netherlands, 1992, 56.

95. Hammond, R. B., and Pedigo, L. P., Determination of yield-loss relationships for two soybean defoliators by using simulated insect defoliation techniques, *J. Econ. Entomol.,* 75, 102, 1982.

96. Sadras, V. O., Cotton responses to simulated insect damage: radiation-use efficiency, canopy architecture and leaf nitrogen content as affected by loss of reproductive organs, *Field Crops Res.,* 48, 199, 1996.

97. Risch, S., Insect herbivore abundance in tropical monocultures and polycutures: an experimental test of two hypotheses, *Ecology,* 62, 1325, 1981.

98. Bach, C. E., Effects of plant diversity and time of colonization on an herbivore–plant interaction, *Oecologia,* 44, 319, 1980.

99. Gold, C. S., Altieri, M. A., and Bellotti, A. C., Effects of intercrop competition and differential herbivore numbers on cassava growth and yields, *Agric. Ecosystems Environ.,* 26, 131, 1989.

100. Andow, D. A., Nicholson, A. G., Wien, H. C., and Wilson, H. R., Insect populations grown with living mulches, *Environ. Entomol.,* 15, 293, 1986.

101. Maron, J. L., Insect herbivory above- and belowground: individual and joint effects on plant fitness, *Ecology,* 79, 1281, 1998.

102. Sacchi, C. F., Price, P. W., Craig, T. P., and Itami, J. K., Impact of shoot galler attack on sexual reproduction in the arroyo willow, *Ecology,* 69, 2021, 1988.

103. Snyder, M. A., Interactions between Abert's squirrel and ponderosa pine: the relationship between selective herbivory and host plant fitness, *Am. Nat.,* 141, 866, 1993.

104. Paulissen, M. A., Exploitation by, and effects of, caterpillar grazers on the annual, *Rudbeckia hirta, Am. Midland Nat.,* 117, 439, 1987.

105. Lehtila, K., and Strauss, S.Y., Effects of foliar herbivory on male and female reproductive traits of wild radish, *Raphanus raphanistrum, Ecology,* 80, 116, 1999.

106. Mothershead, K., and Marquis, R. J., Fitness impacts of herbivory through indirect effects on plant-pollinator interactions in *Oenothera macrocarpa, Ecology,* 81, 30, 2000.

107. Biere, A., and Honders, S. J., Impact of flowering phenology of *Silene alba* and *S. dioica* on susceptibility to fungal infection and seed predation, *Oikos,* 77, 467, 1996.

108. Simms, E. L., Examining selection on the multivariate phenotype: plant resistance to herbivores, *Evolution,* 44, 1177, 1990.

109. Pilson, D., Two herbivores and constraints on selection for resistance in *Brassica rapa, Evolution,* 50, 1492, 1996.

110. Pilson, D., Aphid distribution and the evolution of goldenrod resistance, *Evolution,* 46, 1358, 1992.

111. Meyer, G. A., and Root, R. B., Effects of herbivorous insects and soil fertility on reproduction of goldenrod, *Ecology,* 74, 1117, 1993.

112. Root, R. B., Herbivore pressure on goldenrods (*Solidago altissima*): its variation and cumulative effects, *Ecology,* 77, 1074, 1996.

113. Schemske, D. W., Population structure and local selection in *Impatiens pallida* (Balsaminaceae), a selfing annual, *Evolution,* 38, 817,1984.

114. Karban, R., and Strauss, S.Y., Effects of herbivores on growth and reproduction of their perennial host, *Erigeron glaucus, Ecology,* 74, 39, 1993.

115. Klopfenstein, W. G., Effect of European Red Mite Feeding on Growth and Yield of "Red Delicious" Apple, Ph.D. Dissertation, Ohio State University, Columbus, 1977.

116. McClure, M. S., Competition between herbivores and increased resource heterogeneity, in *Variable Plants and Herbivores in Natural and Managed Systems,* Denno, R. F., and McClure, M. S., Eds., Academic Press, New York, 1983, 125.

117. Marquis, R. J., Leaf herbivores decrease fitness of a tropical plant (*Piper arieianum*), *Science*, 226, 537, 1984.
118. Marquis, R. J., Genotypic variation in leaf damage in *Piper arietinum* [sic] (Piperaceae) by a multispecies assemblage of herbivores, *Evolution*, 44, 104, 1990.
119. Marquis, R. J., A bite is a bite is a bite? Constraints on response to folivory in *Piper arietinum* [sic] (Piperaceae), *Ecology*, 73, 143, 1992.
120. Mauricio, R., Bowers, M. D., and Bazzaz, F. A., Pattern of leaf damage affects fitness of the annual plant *Raphanus sativus*, *Ecology*, 74, 2066, 1993.
121. Lehtila, K., Optimal distribution of herbivory and localized compensatory responses within a plant, *Vegetatio*, 127, 99, 1996.
122. Marquis, R. J., Plant architecture, sectorality and plant tolerance to herbivores, *Vegetatio*, 127, 85, 1996.
123. Pilson, D., Herbivory and natural selection on flowering phenology in wild sunflower, *Helianthus annuus*, *Oecologia*, 122, 72, 2000.
124. Nunez, F. J., and Dirzo, R., Evolutionary ecology of *Datura stramonium* L. in central Mexico: natural selection for resistance to herbivorous insects, *Evolution*, 48, 423, 1994.
125. Louda, S. M., Keeler, K. H., and Holt, R. D., Herbivore influences on plant performance and competitive interactions, in *Perspectives on Plant Competition*, Grace, J. B., and Tilman, D., Eds., Academic Press, New York, 1990, 413.
126. May, J. D., and Killingbeck, K. T., Effects of herbivore-induced nutrient stress on correlates of fitness and on nutrient resorption in scrub oak (*Quercus ilicifolia*), *Can. J. Forest Res.*, 25, 1858, 1995.
127. Strauss, S. Y., Conner, J. K., and Rush, S. L., Foliar herbivory affects floral characters and plant attractiveness to pollinators: implications for male and female plant fitness, *Am. Nat.*, 147, 1098, 1996.
128. Strauss, S. Y., Floral characters link herbivores, pollinators, and plant fitness, *Ecology*, 78, 1640, 1997.
129. Lehtila, K., and Strauss, S. Y., Leaf damage by herbivores affects attractiveness to pollinators in wild radish, *Raphanus raphanistrum*, *Oecologia*, 111, 396, 1997.
130. Strauss, S. Y., Siemens, D. H., Decher, M. B., and Mitchell, O. T., Ecological costs of plant resistance to herbivores in the currency of pollination, *Evolution*, 53, 1105, 1999.
131. Kang, H., Consequence of floral herbivory in *Vicia cracca* (Leguminosae), *Korean J. Biol. Sci.*, 2, 55, 1998.
132. Whitham, T. G., and Mopper, S., Chronic herbivory: impacts on architecture and sex expression of pinyon pine, *Science*, 228, 1089, 1985.
133. Rico, G. V., and Thien, L. B., Effect of different ant species on reproductive fitness of *Schomburgkia tibicinis* (Orchidaceae), *Oecologia*, 81, 487,1989.
134. Oliveira, P. S., Rico, G. V., Diaz, C. C., and Castillo, G. C., Interaction between ants, extrafloral nectaries and insect herbivores in neotropical coastal sand dunes: herbivore deterrence by visiting ants increases fruit set in *Opuntia stricta* (Cactaceae), *Functional Ecol.*, 13, 623, 1999.
135. Alward, R. D., and Joern, A., Plasticity and overcompensation in grass responses to herbivory, *Oecologia*, 95, 358, 1993.
136. Gange, A. C., Brown, V. K., Evans, I. M., and Storr, A. L., Variation in impact of insect herbivory on *Trifolium pratense* through early plant succession, *J. Ecol.*, 77, 537, 1989.
137. Gange, A. C., and Brown, V. K., Insect herbivory affects size variability in plant populations, *Oikos*, 56, 351, 1989.

138. Louda, S. M., and Potvin, M. A., Effect of inflorescence-feeding insects on the demography and lifetime fitness of a native plant, *Ecology,* 76, 229, 1995.

139. Guretsky, J. A., and Louda, S. M., Evidence for biological control: insects decrease survival and growth of native thistle, *Ecol. Appl.,* 7, 1330,1997.

140. Louda, S. M., Kendall, D., Connor, J., and Simberloff, D., Ecological effects of an insect introduced for the biological control of weeds, *Science,* 277, 1088, 1997.

141. Gomez, J. M., Predispersal reproductive ecology of an arid land crucifer, *Moricandia moricandioides*: effect of mammal herbivory on seed production, *J. Arid Environ.,* 33, 425, 1996.

142. Bach, C. E., Effects of a specialist herbivore (*Altica subplicata*) on *Salix cordata* and sand dune succession, *Ecol. Monogr.,* 64, 423, 1994.

143. Brown, J. H., and Heske, E. J., Control of a desert-grassland transition by a keystone rodent guild, *Science,* 250, 1705, 1990.

10 Contrasting Plant Responses to Herbivory in Wild and Domesticated Habitats

Stephen C. Welter

CONTENTS

10.1 INTRODUCTION

The study of plant–insect interactions has long intrigued evolutionary biologists, ecologists, and agriculturists alike. Many of the questions asked are the same among

0-8493-1145-4/01/$0.00+$.50
© 2001 by CRC Press LLC

161

the research groups despite apparent differences in the native and agricultural settings. Kogan[1] suggested that the area of host-plant resistance in cropping systems developed by "pragmatically oriented" entomologists evolved independently in a parallel fashion to the area of insect–plant interactions developed by ecologists working in native systems. However, the recognition of shared interests has become progressively more widespread as evidenced in literature reviews[2–10] and the development of journals encompassing wild and agricultural settings (e.g., *Ecological Applications*).

Each area clearly has made contributions and has special advantages for particular questions. However, the acceptance or rejection of data from either wild or domesticated settings without question may prove premature. Plant life history theory for natural settings predicts optimization of various resource allocation strategies such that plant fitness is maximized for a particular set of environmental conditions.[11, 12] In contrast, Kogan[1] points out that humans have played a very important determinant role in the evolution of plant–insect interactions in agricultural settings by serving as a conscious or unconscious evolutionary sieve for crop plants as their primary consumer, moderator of multi-trophic interactions, and ultimate judge of success. Selection is often for a few key criteria, e.g., maximal yield, resistance to a specific pest, or some agronomic trait necessary for cultivation. Therefore, the concept of optimization of resource allocation to maximize fitness is not appropriate for agricultural settings.

Each system provides opportunities for addressing questions that might prove difficult in other settings. For example, agricultural settings with a high degree of control over plant genetics, environmental uniformity, physical layouts and the subsequent ability to randomize more easily provide wonderfully robust models for statistical testing of specific questions. However, these same traits make the extrapolation of agricultural data questionable or perhaps inappropriate to native systems that often are with reduced resource bases, high genetic diversity, and multiple interacting stresses. The legitimacy of crossing the agricultural/wild border ultimately will depend on the question and how it is framed.

The overall question addressed in this review is not "Do domesticated cultivars differ qualitatively in their response to herbivory compared to their wild progenitors?" The types of responses found in wild plants are paralleled in agricultural plants, such that from a qualitative perspective, the answer to the previous question is no. Instead, the question addressed in this review is "Should the changes in the plant physiology, architecture, and system characteristics resulting from crop domestication alter plant response to herbivory?" Therefore, the review is laid out in the following framework:

a) What are the changes in crop architecture and physiology that have resulted from crop domestication? What are the potential consequences of these changes on plant responses to herbivory?

b) What are the major differences in system characteristics between wild and domesticated habitats that might influence their relative responses to herbivory and do current data and theories suggest that these differences are important for predicting plant responses?

Although literature is drawn from a series of managed ecosystems, e.g., rangelands, this review emphasizes the changes associated with plant domestication in agriculture.

Examination of the contrasts between wild and domesticated genotypes ultimately leads to a progressively entangling set of hypotheses that, like many hypotheses in ecology or agroecology, have mixed support. Difficulties quickly arise as each hypothesis builds on the next until a potential house of cards is erected which may or may not have a solid foundation. Alternatively, approaching each hypothesis independently carries the risk of failing to understand the interplay of the different factors.

10.2 CROP DOMESTICATION

Given that plant physiological and morphological traits have been shown to affect plant response to herbivory, contrasting wild and domesticated plants requires an understanding of the changes associated with domestication that might prove important for predicting responses to herbivory. The domestication of wild plants was initiated approximately 10,000 years ago and has been followed by a rapid spread from their sites of origin by humans. Changes in almost all aspects of crop phenology, tolerance to environmental stress, physiology, and form have been reported, as well as significant changes in the cropping ecosystem (Table 10.1).[13–22]

Overviews of crop domestication and the associated changes in genetic, physiological, and morphological attributes are discussed relative to their effects on yield potential, stability, and adaptiveness.[23–28] Kennedy and Barbour[10] provide a review of the genetically based changes in resistance to herbivory among wild and domesticated plants. Because the authors provide a thorough review of phytochemical and

---

TABLE 10.1
Changes of Wild and Crop Plants and Their Environments Relevant to Herbivory

Character	Crop	Wild
Life History Traits		
Reproductive allocation	High	Low
Genetic diversity	Low	High
Determinism	Increased	Variable
Population Level		
Intraspecific competition	High	Variable
Density dependent mortality	Little	Significant
Community Level		
High resource availability		
Nutrient	High	Low
Water	High	Low
Soil structure	Homogeneous	Heterogeneous
Interspecific competition	High	Variable

morphological changes that influence rates of herbivore attack, this area will not be addressed within this review. Similarly, phytochemical inductions have already been recently reviewed.[29, 30] A shortened listing of potential changes from a wild to domesticated state has been compiled from these sources:

a) Shifts in life history strategies[31–33]
b) Decreased sensitivity to photoperiod length[34]
c) Increased allocation to the harvested portion of the plant[28]
d) Changes in plant canopy including generally higher leaf area indices (LAI) and leaf area duration (LAD), alterations in leaf position and inclination and corresponding rates of exposure
e) Reductions in specific plant parts such as stem, root, or reserve allocation[35]
f) General size increase[36–38]
g) Increased rates of polyploidy[24]
h) Increased uniformity in germination, synchronization of flower, and maturation[39, 40]
i) Shifts in duration of specific stages of plant development (either lengthened or shortened)[36, 41]
j) Loss of bitter and toxic substances[10, 28, 42]
k) Adaptiveness to cultivation[43]
l) Seed retention enhanced[44]
m) Reduced seed coat thickness[45]
n) Increased sensitivity to system inputs (e.g., nutrient enhancements)

Several factors have not been changed in any consistent pattern or to any great degree:

a) Maximum carbon dioxide exchange rate (CER)[24]
b) Relative growth rates[24]
c) Sensitivity to low nutrient conditions[46]

To date, crop yield increases have not resulted from increased maximum carbon exchange rates (CER). However, more recent advances using molecular approaches may produce significant changes as illustrated by the transfer of genes from C-4 maize to C-3 rice plants that alter rates of expression of enzymes important for higher photosynthetic rates.[47, 48] Higher CERs have been reported in wild progenitors or species of barley, sorghum, millets, soybean, cotton, cassava, wheat, rice, sugar cane, *Brassicas,* and sunflower. Generally, selection for increased CER in crop plants has resulted in no increase or decreases in yield potential. Similarly, no systematic increases in relative growth rates have been associated with crop domestication if plant size is corrected, especially for initial seed size.[24]

Whereas crops have not become more sensitive to low levels of nutrient stress, they are more capable of exploiting and utilizing the higher inputs of modern, high-intensity agriculture. Nitrogen use in the U.S. has increased by 355% from 1960 to 1995, with almost all acres of corn, fall potatoes, and rice as well as 75% of cotton and wheat acres receiving some type of commercial fertilizer.[49] Whereas crop yields have surged with increased nitrogen inputs, there is also a strong genotype by fertilization interaction selected for in agricultural genotypes. Agricultural genotypes are selected to respond aggressively to higher nutrient inputs compared to their wild counterparts. However, the responses are highly variable with many "high input" cultivars proving no more sensitive to nutrients stress than their wild progenitors or older cultivars.

10.2.1 DIRECT EFFECTS OF PLANT FORM AND PHYSIOLOGY ON PLANT RESPONSES

10.2.1.1 General Responses to Herbivory

In terms of the types of responses, wild and agricultural plants have demonstrated comparable responses to herbivory including all directions of change for regrowth rates, compensatory physiological responses, reallocation of photosynthates, alterations in leaf demography, and changes in architecture (for reviews, see[3–5, 50]). Reviews of herbivore impact on agricultural productivity[51–54] generally show strong yield depressions associated with levels of herbivory found in agricultural systems. However, the selection of plant–insect interactions studied in agriculture is nonrandom and generally focused on those pairings that are perceived to be of greatest economic importance. Similarly, the levels of herbivory are often quite high due to both changes in crop settings and the lack of natural enemies due to the accidental importation of crop pests. Variability has been observed in agricultural cultivars for plant tolerance as part of breeding program efforts to development tolerant crop cultivars.[55–59] Extensive research on crop tolerance has been done in field crops such as maize, sorghum, rice, or barley, but the evaluation of tolerance and its associated mechanisms is less clear relative to other forms of resistance, antixenosis or antibiosis.[47, 49, 50] Evidence for tolerance in wild and domesticated genotypes of rice also has been reported.[60–62] Whereas Panda and Heinrichs[62] developed a means to identify tolerant cultivars, the mechanisms for achieving improved tolerance were not within the scope of the paper. While Jung-Tsung et al.[52] did find two ascensions of wild rice with moderate levels of tolerance to herbivory to the brown planthopper, difficulties with incorporating resistant wild lines into domesticated cultivars using conventional breeding techniques also were discussed.

10.2.1.2 Life History Tradeoffs

One of the most dramatic changes associated with crop domestication has been the shift in allocation patterns to the harvestable portion of the plant that often contains the reproductive structures. So, questions associated with the higher yields of domesticated crops might be framed as either "What are the interactions of reproduction and tolerance to herbivory?" or "What are the trade-offs between allocation to various plant structures (e.g., leaf, root, or reproductive) and a plant's ability to respond to herbivore injury?" Life history theory predicts that plants will allocate resources so that overall lifetime fitness is optimized, such that growth and reproduction are competing for a limited pool of resources.[11, 12, 63, 64] In some cases, these traits may be in conflict with each other, the subject of trade-offs being fairly controversial.[13–22] The notion of trade-offs has resulted in various optimality arguments using economic analogies.[65–67] However, the notion of direct trade-offs, as measured by relative dry-weight allocations, is complicated because factors other than energy may be limiting[68, 69] or reproductive structures may themselves contribute to resource acquisition.[70–72] This may be uniquely important for crops with photosynthetically active fruit structures, e.g., strawberries, given the tremendous increases in fruiting associated with domestication. Problems with developing the costs of reproduction are discussed by various authors.[73–76] Therefore, use of relative biomass distribution patterns as indices of cost needs to be interpreted with some caution.

Trade-offs have been postulated between reproduction and growth rates, growth and defense, reserves and short-term growth, and ultimately the interactions of these trade-offs with plant response to herbivory. Inverse relationships between growth and non-growth processes have been documented for many plant species.[65, 77–85] Trade-offs between growth rates and allocation to secondary plant metabolites associated with defense have been found in wild and domesticated plants,[86–93] but the existence of measurable cost appears to depend on the mechanism.[94]

Natural variation in reproductive effort has been used to test the relative cost of reproduction on plant growth by examining natural patterns of change in reproduction over time,[95] by looking at fruiting and non-fruiting individuals,[96–99] or by experimental manipulation. Hand pollination of flowers to enhance fruit set demonstrated that each fruit cost the plant approximately 2% of its future leaf area, whereas a non-pollinated inflorescence was approximately half as costly.[75] Using giberrellic acid to manipulate reproductive effort, Saulnier and Reekie[100] showed that increased reproduction resulted in decreased nitrogen allocation to roots and increased allocation to shoots, increased leaf area, and generally decreased photosynthesis, but these results were modified by nutrient availability and phenology of plant. Using photoperiod to manipulate allocation to reproduction, multiple genotypes of two plant species indicated substantial variation in the cost of reproduction and variation between species.[100]

The higher relative allocation to reproduction and decreased allocation to vegetative growth in many crop plants would be predicted to decrease tolerance to herbivory if only a single factor is considered. Contrasts of the tolerance of wild and domesticated lines using equal levels of simulated herbivory within artificial settings demonstrated decreased tolerance to herbivory in the domesticated cultivars,[101, 102] but opposite results also have been observed.[103]

Positive correlations between growth and reproduction also have been obtained in some cases[104–106] or reproduction is not always correlated with a cost to growth.[76, 103] No significant cost was associated with a four-fold increase in pistillate flower formation in the bunchgrass, *Tripsacum dactyloides,* within a naturally occurring wild mutant. Similarly, Boeken[107] demonstrated no carry-over effect of reproductive rates on growth, whereas current reproduction was more strongly correlated with current environmental conditions, e.g., rainfall.

Tolerance to grazing has been correlated with rapid growth rates in some species,[107, 108] but contrasts of other pairs of species have not always shown this pattern to be true.[109] Because relative growth rates have not been consistently altered with crop domestication other than with additional inputs, innate differences in relative growth rates do not seem to be an important differentiating character for response to herbivory.

Within a recent review, Mole[61] raised a series of concerns that appear particularly relevant to the contrasts across systems. One issue raised is the notion that trade-offs between response to herbivory may lack an underlying genetic linkage, but may reflect only phenotypic plasticity in response to environmental conditions. However, the cost of defense may be distributed differentially to "third" party traits in different species. Cost of allocation to one plant structure may be "paid" for by losses in different traits by different species or genotypes.

As will be discussed later in the chapter, issues of phenotypic plasticity, genotype by environment interactions, and differential resource bases exist between many domesticated settings and wild habitats. Various authors have looked at the possibility of examining resource allocation shifts in crop cultivars as potential models for examining life history tradeoffs with long-term objectives of understanding and improving crop tolerance to herbivory.[101–103] Assuming that the trade-offs reported between tolerance and fitness occur frequently, we would have to assume that the increases in crop yields by breeders should be correlated in some cases with decreases in tolerance to yield.

10.2.1.3 Phenotypic Plasticity, Changes in Determinism, and Apical Dominance

Plants operate as a series of interconnected, but competing modules that can alter resource allocation patterns depending on the site of limitation.[110–113] Competition within modules may vary between species, with time, or between environments. Some of the more common changes and potential tradeoffs associated with crop domestication include a shift from perennial to annual life cycles, changes in relative allocation patterns, and increased apical dominance for some crop genotypes, e.g., maize.[113] These same factors have been identified as important determinants for predicting the effects of herbivory for some wild species.[115, 116]

Changes in crop life history from indeterminate to determinate growth that is sometimes associated with shifts from perennial to annual life cycles may prove important for predicting a plant's capacity to respond to herbivory. Given that a determinate growth form has a reduced growth rate and heavy allocation to reproduction after some point in its life cycle, then you would hypothesize that determinate crop cultivars would be less tolerant than their indeterminate counterparts. Welter[5] demonstrated that an indeterminate growth form of tomato was able to compensate more readily for pre-harvest fruit loss than a determinate form.

Bilbrough and Richards[110] have suggested that the differences in tolerance to browsing between the two species may reflect the tradeoff in root:shoot allocation. The ability of *Purshia tridentata* to allocate to above-ground growth following defoliation was presumed to have occurred at the expense of below-ground growth. This presumed reduced allocation below-ground is correlated with the lower ability to withstand drought in *Purshia* compared to the less defoliation tolerant *Artemesia*. Enhanced developmental plasticity was reported as a mechanism important for grazing tolerance in graminoid species.[117] Increasing rates of defoliation were associated with increasingly reduced allocation to storage reserves in a perennial spring ephemeral while minimizing the effects of herbivory on current reproduction.[117] Alfalfa was subjected to a series of defoliation regimes and the subsequent dry matter accumulation and partitioning examined. Increased allocation to foliage as evidenced by increased leaf area ratio, leaf specific weight, and leaf weight ratio supported the hypothesis of maintained allocation to leaf growth at the expense of support structures.[118]

Tolerance to herbivory has been correlated with the ability to initiate new growth from inactive meristems.[101, 108, 119–121] Loss of actively growing meristems due to later clipping resulted in substantially increased losses due to grazing,[122] whereas removal of older leaf tissue was less damaging than removal of more active, younger leaves. The selective advantage of increased ability to produce lateral branches also is heavily dependent on the growing conditions and resource limitations.[114]

The question of apical dominance and tillering as a response to herbivory is important to understanding the effects of crop domestication on herbivore tolerance for many important crop species, such as maize, barley, wheat, and rice. Crop breeding for high input systems has often targeted the development of single-stalk cultivars with less flexibility in the production of additional tillers than their wild counterparts.[123, 124] Coincident with the loss of a propensity for branching and tillering in the agricultural genotypes has been a decreased competitiveness which in turn has been compensated for in agronomic growing conditions and increased inputs.[24] Therefore, changes in crop architecture such as reduced tillering are likely to impact herbivore tolerance to herbivory, but also are likely to influence indirectly through alterations of the interplant competitive interactions discussed below.

Many plants respond to herbivore injury with increased tiller production,[125] but not all species possess this flexibility.[126, 127] Because some tillers are capable of reallocating resources between tillers following defoliation, tillering has the additional advantage of distributing the cost of defoliation to more than one tiller. Welker et al.[128] demonstrated increasing transport rates of N-15 to defoliated tillers from adjacent tillers with increasing frequency of defoliation, while Gifford and Marshall[129] also demonstrated increased transport of photosynthates. In addition, reductions in tillering have been observed in response to defoliation.[126, 130]

There is considerable difficulty in interpreting the relative costs of herbivory in wild and domesticated counterparts, in that in agriculture, single-year yields are often the key criteria; thus short-term gains that carry long-term costs may never be realized. Increased rates of tillering were correlated with increased ability to withstand infestations of a stem-boring lepidopteran in a species of perennial maize,[101] but the study was restricted to a single year. Given that increased tillering rates do not necessarily produce more tillers over longer periods and may be associated with higher tiller mortality rates during the winter,[126] the interpretation of benefit must be tied to the system and the life history of the plant species. Research with other species has demonstrated already that both the temporal scale and level of plant architecture used in the experiment influenced the perceived importance of herbivory.[131, 132] Whereas single-shoot compensation by increased branch tissue production ranged from 89 to 583%, overall reductions greater than 75% were observed for clumps of interconnected shoots of dwarf fireweed.

10.2.2 ENVIRONMENTAL EFFECTS ON PLANT RESPONSES

Differences in the environment or setting are clear between wild and agricultural systems. Humans have altered the resource base available to the plant to achieve as optimum a growing condition as biologically and economically possible in many cases. While Boyer[133] estimates that environmental stress limits 25% of all U.S. agricultural productivity, Chapin[79] suggests that "most natural environments are continuously suboptimal with respect to one or more environmental parameters, such as water or nutrient availability." Factors such as water or nutrients that are limiting in many native habitats are provided sometimes to excess in modern, high input systems. Native plants adapted to low resource environments (e.g., infertile soils, deserts, or tundra) often share common life history traits that appear adaptive to that site (e.g., low photosynthetic rates, low capacity for nutrient acquisition, and slow growth rates). Plants from low fertility sites often are associated with fixed shoot:root ratios and prove less flexible in their ability to reallocate to growth.[134] Therefore, many of the traits associated with plants adapted to low fertility sites are negatively correlated with tolerance to herbivory. In contrast, agricultural plants have been selected to respond to increased nutrient availability.[24]

However, just as gradients exist in native habitats, so do gradients exist between agricultural settings. Discussion of agriculture as a single set of conditions fails to recognize the strong differences in resources available in low vs. high input systems. Therefore, this discussion is focused on the resources as independent variables along a gradient of availability that in general runs from native habitats to traditional farming to high-input farming operations. The degree to which the gradient follows this direction is dependent on both the systems and variables under discussion (e.g., some native habitats exhibit high levels of specific nutrients).[135]

The following discussion is divided into the direct effects of environmental factors on plant response to herbivory, then followed by a brief discussion of environmental factors that influence the indirect effects of intra- and interspecific competition among plants. The objective of this brief discussion is not to provide an extensive review of environmental effects, but to cast the contrast of wild and managed ecosystems as two environments with dramatically different resource bases.

10.2.2.1 Direct Effects

10.2.2.1.1 Temperature and light

Temperature effects on plant–insect interactions can occur via three basic pathways: (1) direct effects on plant processes affecting plant susceptibility to herbivores, (2) direct effects on plant growth or physiological processes, and (3) direct effects on herbivore behavior and developmental biology.[136] The effects of temperature on plant resistance can be either positive, negative, or neutral. In general, temperature effects are more important at the extremes, but often the effects appear reversible as conditions change. Sousa and Foster[137] showed with the Hessian fly, *Mayetiola destructor,* that increasing temperatures resulted in reduced resistance in three of four tested cultivars, whereas the fourth remained relatively unchanged with temperature. For the three temperature sensitive cultivars, increased temperatures were correlated with increased rates of tillering, a trait often associated with increased tolerance in grass species.[124] The direction of change associated with crop domestication and temperature will depend on the sensitivity of the trait and differences between sites.

Ambient light levels have been correlated with shifts in resistance expression in terms of growth or phytochemistry. Stem solidity, an important component of resistance in wheat to the wheat stem sawfly, *Cephus cinctus,* was positively correlated with light levels. However, there were significant genotype-by-light interactions for various ascensions. Oka[16] reports that wild plants and their domesticated counterparts vary in their responses to environmental conditions with wild rice races having greater plasticity and more sensitivity to environmental daylength and temperature, but less sensitivity to increased fertilization regimes. Although temperature and daylength may prove important for determining plant responses to herbivory, no consistent difference between agricultural and wild systems would be predicted.

10.2.2.1.2 *Nutrient availability*

Nitrogen has proven one of the stronger variables that modify plant response to herbivory.[138] Nitrogen can be a strong determinant of a plant's ability to respond to defoliation with increased growth.[121] The authors also demonstrated interactions of nitrogen levels and delays in leaf senescence rates.[121] At high nitrogen levels, only the clipped plants exhibited delays in leaf senescence. Differential responses in allocation patterns have been modified by nutrient status. Stafford[139] demonstrated that under conditions of single-factor stress, plants allocated to the parts necessary for acquiring the limiting resource. (If carbon-limited by defoliation, then plants allocated to foliage, and if nutrient-limited, then allocations were to root development.) The consequence of concurrent stresses was to balance allocation to each structure type.

Nutrient assimilation by roots was increased by clipping of a sedge, *Kyllinga nervosa,* such that green leaf concentrations were increased for potassium, phosphorus, copper, manganese, sodium, zinc, iron, and/or calcium depending on nitrogen source.[140] However, other studies have shown decreased nitrate absorption in response to defoliation.[141] Given that herbivory has been shown to alter rates of nutrient acquisition following defoliation events[141] as well as changes in root respiration rate, the ability of a plant to compensate for herbivory is in part a function of nutrient availability such that compensatory responses are possible.

Similarly, plant mortality in response to herbivory has been positively correlated with the proportion of plant nutrients removed through herbivore feeding[119, 141] and the resulting inability to respond with increased foliar production. Although more recent literature has suggested that carbohydrate reserves are less accessible for rapid growth responses to herbivory,[120, 142–144] nutrient availability does seem important as a moderator of plant response. Plants grown under agricultural settings with artificially elevated nitrogen levels would not be predicted to exhibit similar levels of mortality. In addition, the fact that plant allocation to nitrogen reserves diminishes under conditions of high nutrient availability[24] further confounds the contrast of wild and domesticated plants.

Decreased root growth and increased root mortality can be associated with foliar injury.[145] The effects of simulated root herbivory were not ameliorated by the addition of fertilization, but shifts in allocation priorities to rapidly reestablish root:shoot ratios were observed at the expense of above-ground growth.[146] If below-ground resources are readily available, then little energy and biomass are allocated to roots and allocation to shoots can be high.[76]

Given that (1) nitrogen may ameliorate the effects of defoliation, (2) agricultural systems have elevated levels of nitrogen due to artificial inputs, (3) genotypes that are responsive to increased nutrient inputs have been systematically selected in crop breeding programs, and (4) the establishment of significant genotype by nutrient stress level by defoliation interactions has been demonstrated, then the effects of herbivory should be reduced in an agricultural setting compared to native habitats assuming all other factors are equal.

10.2.2.1.3 Water stress

Interactions between water stress and herbivore effects have been documented.[147] However, significant interactions were observed between water stress and plant allocation to vegetative biomass after stem borer injury.[148] Similarly, the effects of stem boring on corn yield were more severe under conditions of water stress in dryland farming conditions.[149] Similar results have been reported for three grass species found in Australian savannas, but the three species did not respond consistently.[150] Water stress induced by a root-feeding herbivore resulted in reduced vegetative biomass, but this condition was reduced under high water availability. In general, agricultural systems with consistently less stressful conditions would be expected to be more likely to compensate for herbivore injury than their wild counterparts.

10.2.2.1.4 Multiple stress interactions

Interspecific variation in photosynthetic compensatory responses to herbivory has been demonstrated both to exist and to vary by intensity and frequency of clipping for three species of African grasses[151] as well as with agricultural cultivars.[152] However, what is more important for this review is that significant interactions were also detected for maximum photosynthesis between frequency of defoliation, water stress, and nitrogen fertilization and that the interactions were often species-specific. Interactions between herbivore losses, nutrients, and water stress were sufficiently significant that some authors have suggested experiments focusing on herbivore effects should be conducted in the projected environment.[152] Therefore, use of single-factor experiments would not only have presented a limited perspective, but would have potentially generated erroneous patterns for conclusions on the degree of compensation observed within and among the three species. Again, the strong, systematic differences in wild and agricultural habitats would be expected to generate strong differences in plant responses.

10.2.2.2 Competition Mediated Interactions

Herbivory has long been recognized as a potential moderator of intra- and interspecific competition between plants (for review of plant competition, see Maschinski and Whitham[153] and Reader et al.[154]). Many authors have cautioned that responses to herbivory often may reflect a release from competition rather than direct effects of herbivory.[155, 156] In general, it has been suggested that indirect interactions, defined as the "pre-emptive exploitation of limited resources," are more directly affected by herbivory. As such, competition for resources such as light, nutrients, or water can all be adversely affected by either the direct effects of herbivory to resource-acquiring modules or through the indirect effects of herbivory (e.g., changes in root:shoot ratio via alterations in resource allocation patterns).

Louda et al.[157] suggest that the impact of herbivores on competitive interactions can be influenced by: (1) differential rates of herbivory on different plant species or due to environmental gradients, (2) the innate ability of the consumed plant to tolerate or respond to herbivory (see above discussion), (3) the availability of resources to the plant that determine or limit compensatory responses, and (4) the strength of the competition. All four traits have been influenced by crop domestication as discussed below. While the four traits are listed independently, often the interactions of the traits are significant and/or asymmetric between plant species (see Weiner[158] for a discussion of competitive symmetry). Weiner[158] argues that the level of asymmetry is a product of both plant characteristics and the limiting resource type (e.g., carbon vs. nitrogen).

10.2.2.2.1 Density effects

At low densities with low levels of competition, early defoliation treatments delayed tiller replacement more than later treatments, whereas at higher densities, all clipping regimes delayed tiller replacement. Effects on reproductive output by the clipping regimes were more severe at higher densities.[159] Intra-population differences in response to herbivory were reported in response to mammalian herbivory in native grasses, and these responses were modified by grazing history which in turn interacted with the consequences of defoliation and plant competitive ability.[160] Competition, resource limitation (nutrient and water), and herbivory all strongly interact for their impact on herb establishment and vegetative cover development.[161] The effects of defoliation on *Abutilon theophrasti* were insignificant on seed production at low planting densities, but reductions of approximately 50% were observed for equal levels of defoliation when the plants were at 6.25 greater densities and mutual shading proved significant.[162] Simulated injury of the seedcorn maggot adversely affected the competitive abilities of soybean, the consequences of which were exacerbated by increasing plant density.[163] In contrast, the effects of herbivory and competition were additive for three plant species.[164]

These data raise the potential risk of deriving conclusions about plant responses to herbivory under noncompetitive conditions. For example, Rosenthal and Welter[102] demonstrated that under noncompetitive conditions of a lath house, the wild perennial maize, *Zea diploperrenis,* was more tolerant to stem boring damage than the wild annual, landrace cultivars, or modern cultivars of maize due to higher rates of tiller in the wild perennial taxa and risk spreading across modules such that cost per stem boring event was compartmentalized and minimized. However, maize tillering in some phenotypes also decreases under high-density conditions typical of agricultural fields or due to herbivory.[165, 166] Therefore, these conclusions also will need to be repeated under higher density conditions typical of agricultural systems if extrapolations to agricultural settings are desired.

10.2.2.2.2 Indirect effects of resource availability

In agriculture, elimination of suboptimal conditions within a field is one of the key objectives. As such, gradients in water levels, soil condition, light levels, and nutrients are kept to a minimum, whereas such variability is the norm rather than exception in native habitats. Variability in rates of herbivory due to environmental variability is kept to a minimum, thus making the potential impact of selective herbivory on intraspecific competition more uniform across a site.

Below-ground competition for nutrients is also modified by above-ground defoliation of competitors. Dramatically increased capture of phosphorus by *Artemesia tridentata* was detected when adjacent competing plants were clipped.[167] Factors influencing the potential regrowth rate also influence their competitive ability. Root herbivory on *Centaurea maculosa* affected plant height and reproductive biomass at intermediate levels, but this effect was minimized if plant density was low and the plants were grown in a nutrient-rich environment. The effects of the herbivory were greatest in the presence of competition from a grass species and high plant density, whereas low levels of herbivory were reported to increase shoot number and biomass per area.

10.3 DIFFERENTIAL RATES OF HERBIVORY

Rates of herbivory in agriculture have been altered positively through the unintentional selection for more palatable cultivars or possibly by the selection for higher yields,[1] but see Simms[95] for an alternative perspective on costs of certain types of resistance. Many wild progenitors possess resistance mechanisms not found in cultivated genotypes.[10] Rates of herbivory in native habitats are typically listed between 5 to 15%,[168] whereas outbreaks of insect herbivores in agriculture are common due to the varieties of factors, including the absence of regulating natural enemies in the system and alteration of system characteristics (e.g., mean leaf nitrogen levels or agronomic practices).

As both ecologists and agriculturists have pointed out recently,[17, 169] understanding the effects of herbivory requires understanding its effects across a range of injury. Given that plant responses to herbivory are nonlinear in many cases,[53] comparisons that either do not control for levels of herbivore injury or do not encompass similar ranges of injury are difficult at best. Thus, differences in levels of susceptibility to damage between wild and domesticated genotypes potentially confound our ability to compare the responses to herbivory unless appropriate experimental techniques are used. Nault and Kennedy[169] also discuss how the relationships between herbivory and damage are often not described by the same regressional function even within multiple cultivars of the same species, thus making the contrasts even more difficult. Agricultural studies that do not actively ensure that the same ranges of injury are covered may not provide useful comparisons across taxa. Similarly, studies attempting to contrast plant genotypes in different habitats rarely are able to ensure equal levels of herbivory.

10.4 DATA ASYMMETRY

The effects of herbivory have been well documented in wild systems,[50, 168, 170] agricultural systems,[52, 169] or in both.[3, 4] Inclusion of wild progenitors into agricultural studies has long been a staple of crop-breeding programs looking for sources of resistance,[10, 55–57, 171] but the converse is not true for the inclusion of agricultural plants into native habitats (but see Clement and Quisenberry[172]). Because of the types of questions commonly asked by ecologists looking at the ecological consequences and evolution of plant–insect interactions, the inclusion of agricultural genotypes would in general only serve to confound their studies. As such, direct contrasts of wild and agricultural plants typically are only found in agricultural systems and hence are incapable of partitioning the genotype by environmental conditions if stress levels typical of native habitats are not included in the range of stress.

Given the wealth of data demonstrating the interactions of genotype by environmental effects and the outcome on plant phenotype, the observed effects of herbivory on wild genotypes placed into agricultural systems are suspect as predictors for the response of the wild genotype in its native habitat. Until there are common garden plots with exchange of genotypes across all habitats, no general conclusion about true differences in plant response can be determined. This is not to suggest that contrasts of specific elements cannot be made (e.g., secondary plant distribution, herbivore loads in specific habitats, contrasts within a more narrow range of conditions), but that our ability to generalize is hampered. Conclusions generated by wild genotypes placed in agricultural settings may in fact be making predictions for a phenotype never observed in native habitats as allocation patterns to both above-ground and below-ground biomass may be dramatically altered, thus skewing the results for interpretation.

Finally, perhaps the greatest difficulty for presenting definitive conclusions is the limitations of our knowledge for predicting and understanding plant response to an herbivore in even one setting. As Smith[47] concludes for agricultural plants that have received extensive continued studies on far fewer species, "very little is known about the mechanisms of plant tolerance to insect feeding." This perception is echoed by Velausamy and Heinrichs[171] for crop breeding: "In most cases, the scientists have not understood the phenomenon of tolerance and there has been no attempt to conduct experiments for separating tolerance from non-preference and antibiosis." These conclusions were repeated as recently as 1993.[173] By the time that differences among plant genotypes, types of herbivory, ranges of environmental stress, differential levels of competition, and the scarcity of data on agricultural phenotypes placed in wild settings are coupled with genotype by environmental interactions, the contrast becomes quite hazardous.

10.5   CONCLUSION

In the end, the question, "Do wild and agricultural plants vary in their response to herbivore damage?" is not appropriate. This question implies that (1) the differences observed between settings are due to plant characteristics, (2) a "typical" set of conditions exist for either agricultural or native habitats, and (3) it ignores the fact that testing conditions for many contrasts are different between wild progenitors and their domesticated counterparts. In fact, almost every variable that researchers have identified as an important modifier for the effects of herbivory has been altered with crop domestication. Therefore, the answer to the first question is "Yes, but it depends," which is a relatively unsatisfying answer.

If we are interested in answering the question, "Are the observed patterns in wild and agricultural systems different?" then we will need to consider the following factors. The datasets for understanding the responses of agricultural plants in agriculture settings and native plants in native habitats are well represented in the literature. Some information on native plants under agricultural settings is available from plant breeding literature, but usually only under optimal agricultural conditions until more recent efforts in international breeding programs. Virtually no information exists for the fourth square in the 2 × 2 matrix: agricultural plants in wild habitats. Although studies manipulating single variables such as water stress are extremely useful, they are not able to simulate the conditions of a wild habitat.

An alternative way to approach this contrast is "Does the contrast of wild and agricultural plants' response to herbivory provide insights into previously unexplored areas or independent variables?" and "Does either setting provide unique opportunities to look at specific variable(s) and their interactions?"

Understanding these differences is becoming more important than ever as we see the pooling of literature across traditional boundaries between native and agricultural habitats increase over time.[3, 4, 174] The initial resistance to pooling of data sets[1, 4] seems to have been replaced by entomologists from traditional agricultural roots sharing common interests in ecology and by ecologists looking for larger datasets and for application of their research to applied issues.

If researchers are interested in the quantitative response of plants to herbivory to determine the relative importance to plant fitness, then an unconstrained pooling of literature has the potential to introduce data with a systematic bias because of system differences and history of selection. For example, since the negative impact of herbivory often is reduced when resources are not limiting (e.g., nitrogen, water, light), then agriculture with its higher inputs of nutrients and water reduced interspecific competition due to cultivation and herbicide use would be predicted to have less proportional changes. Conversely, although the data are variable, the increase in allocation reproductive structures seems to be correlated with decreased tolerance to herbivory. Thus, there are many changes associated with agricultural plants and the associated habitats that are predicted to alter the impact of herbivory in both negative and positive directions. In addition, given the strong changes in genotype by environment interactions selected for both within and between species comparisons, then the potential for changes in plant phenotype due to domestication also alters the predicted responses.

Appropriate uses of pooled data might include looking for types of responses to herbivory in a qualitative sense or looking for potentially novel changes. Finally, the variation in phenotypes generated by artificial selection also allows for examination of specific plant traits that may have minimal variation for other traits (e.g., contrasts of cultivars with different allocation patterns or contrasts of determinate and indeterminate phenotypes). Similarly, environmental conditions can be modified with precision while maintaining uniformity of other factors such as soil type, water stress levels, and topography. Therefore, the pooling of literature remains a rich resource for developing and addressing particular questions while also potentially generating erroneous patterns by inclusion of the systematic biases across literature bases.

Within agricultural studies, correlates among general vigor, inter- and intraspecific compensatory growth, compensatory gas exchange responses, mechanical strength of various plant tissues, differential allocation patterns, and plant tolerance have been observed.[175] Whereas many crop breeding programs have successfully enhanced or incorporated resistant traits into modern cultivars,[55–57] it has been suggested that incorporation of many traits is hindered by difficulties of separating positive resistant traits from associated negative agronomic traits.[176]

REFERENCES

1. Kogan, M., Plant defense strategies and host-plant resistance, in *Ecological Theory and Integrated Pest Management,* Kogan, M., Ed., John Wiley & Sons, New York, 1986, 83.
2. Fritz, R. S., and Simms, E. L., Ecological genetics of plant-phytophage interactions, in *Plant Resistance to Herbivores and Pathogens: Ecology, Evolution and Genetics,* Fritz, R. S., and Simms, E. L., Eds., University of Chicago Press, Chicago, 1992, 1.
3. Herms, D. A., and Mattson, W. J., The dilemma of plants: to grow or defend, *Q. Rev. Biol.,* 67, 283, 1992.
4. Trumble, J. T., Kolodny-Hirsch, D. M., and Ting, I. P., Plant compensation for arthropod herbivory, in *Annual Review of Entomology,* Mittler, T. E., Radovsky, F. J., and Resh, V. H., Eds., Annual Reviews, Palo Alto, CA, 1993, 93.
5. Welter, S. C., Arthropod impact on plant gas exchange, in *Plant-insect Interactions,* Bernays, E. A., Ed., CRC Press, Boca Raton, 1989, 135.
6. Baldwin, I. T., and Preston, C. A., The eco-physiological complexity of plant responses to insect herbivores *Planta,* 208, 137, 1999.
7. Stowe, K. A., Marquis, R. J., Hochwender, C. G., and Simms, E. L., The evolutionary ecology of tolerance to consumer damage *Annu. Rev. Ecol. Syst.,* 75, in press.
8. Strauss, S. Y., Levels of herbivory and parasitism in host hybrid zones *Trends Ecol. Evol.,* 9, 209, 1994.
9. Strauss, S. Y., Siemes, D. H., Decher, M. B., and Mitchell-Olds, T., Ecological costs of plant resistance to herbivores in the currency of pollination *Evolution,* 53, 1105, 1999.

10. Kennedy, G. G., and Barbour, J. D., Resistance variation in natural and managed systems, in *Plant Resistance to Herbivores and Pathogens: Ecology, Evolution and Genetics,* Fritz, R. S., and Simms, E. L., Eds., University of Chicago Press, Chicago, 1992, 13.
11. Stearns, S. C., Trade-offs in life-history evolution *Funct. Ecol.,* 3, 259, 1989.
12. Stearns, S. C., *The Evolution of Life Histories,* Oxford University Press, Oxford, U.K., 1992, 249.
13. Baker, H. G., Human influences on plant evolution, *Econ. Bot.,* 26, 32, 1972.
14. Evans, L. T., *Crop Evolution, Adaptation and Yield,* Cambridge University Press, Cambridge, U.K., 1993, 500.
15. Harlan, J. R., Genetic resources in wild relatives of crops, *Crop Sci.,* 16, 329, 1976.
16. Oka, H.-I., Adaptive evolution of crop plants and cropping systems: a review and perspectives, in *Frontiers of Research in Agriculture,* Roman & Littlefield, Calcutta, India, 1982.
17. Salik, J., and Merrick, L. C., Use and maintenance of genetic resources: crops and their wild relatives, in *Agroecology,* Carroll, C. R., Vandermeer, J. H., and Rossett, P. M., Eds., McGraw-Hill, New York, 1990, 817.
18. Schwanitz, F., *The Origin of Cultivated Plants,* Harvard University Press, Cambridge, MA, 1966, 1.
19. Karban, R., Induced plant responses to herbivory *Annu. Rev. Ecol. Syst.,* 20, 331, 1989.
20. Tallamy, D. W., and Raupp, M. J., Eds. *Phytochemical Induction by Herbivores,* Wiley, New York, 1991, 431.
21. Hutchinson, J. B., India: local and introduced crops *Phil. Tras. R. Soc. Lond.,* B275, 129, 1976.
22. Narain, A., Castor, in *Evolutionary Studies in World Crops, Diversity, and Change in the Indian Subcontinent,* Hutchinson, J. B., Ed., University Press, Cambridge, U.K., 1974, 71.
23. Stephenson, R. A., Brown, R. A., and Ashley, D. A., Some observations on the photoperiodism and the development of annual forms of domesticated cottons *Econ. Bot.,* 30, 409, 1976.
24. Lush, W. M., and Evans, L. T., The domestication and improvement of cowpeas (*Vigna unguiculata* (L.) Walp. *Euphytica,* 30, 579, 1981.
25. Harlan, J. R., deWet, J. M. J., and Prince, E. G., Comparative evolution of cereals, *Evolution,* 27, 311, 1973.
26. Evans, L. T., and Dunstone, R. L., Some physiological aspects of evolution in wheat *Aust. J. Biol. Sci.,* 23, 725, 1970.
27. Hanson, P. R., Jenkins, G., and Westcott, B., Early generation selection in a cross of spring barley *Z. Pflanzenzüch.,* 83, 64, 1981.
28. Riggs, T. J., and Hayter, A. M., A study of the inheritance and interrelationships of some agronomically important characters of spring barley *Theor. Appl. Genetics.,* 46, 257, 1975.
29. Hay, R. K. M., and Kirby, E. J. M., Convergence and synchrony—a review of the coordination of development in wheat *Aust. J. Agric. Res.,* 42, 661, 1991.
30. Hayes, P. M., and Strucker, R. E., Selection for heading date synchrony in wild rice *Crop Sci.,* 27, 653, 1987.
31. Edmeades, G., and Tollenaar, M., Genetic and cultural improvements in maize production, in *Intl. Congr. Plant Physiol.,* New Delhi, India: Indian Agricultural Research Institute, 1990.
32. Lathrap, D. W., Our father, the cayman, our mother the gourd: Spinden revisited, or a unitary model for the emergence of agriculture in the New World, in *Origins of Agriculture,* Reed, C. A., Ed., Mouton, The Hague, 1977, 713.

33. Donald, C. M., Competitive plants, communal plants, and yield in wheat crops, in *Wheat Science—Today and Tomorrow,* Evans, L. T., and Peacock, W. J., Eds., Cambridge University Press, Cambridge, U.K. 1981, 223.

34. Kadkol, G. P., Beilharz, V. C., Halloran, V. M., and Macmillan, R. H., Anatomical basis of shatter-resistance in the oilseed Brassicas, *Aust. J. Bot.,* 34, 595, 1986.

35. Lush, W. M., and Evans L. T., The seedcoats of cowpeas and other grain legumes: structure in relation to function, *Field Crops Res.,* 3, 267, 1980.

36. Chapin, F. S. I., Groves, R. H., and Evans, L. T., Physiological determinants of growth rate in response to phosphorus supply in wild and cultivated *Hordeum* spp., *Oecologia,* 79, 96, 1989.

37. Ku, M. S. B., Agarie, S., Nomura, M., Fukayama, H, Tsuchida, H., Ono, K., Hirose, S., Toki, S., Miyao, M., and Matsuoka, M., High-level expression of maize phosphoenolpyruvate carboxylase in transgenic rice plants, *Nat. Biotechnol.,* 17, 76, 1999.

38. Matsuoka, M., Nomura, M., Agarie, S., Miyao-Tokutomi, M., and Ku, M. S. B., Evolution of C4 photosynthetic genes and overexpression of maize C4 genes in rice, *J. Plant Res.,* 111, 333, 1998.

39. United States Department of Agriculture, *Agricultural resources and environmental indicators, 1996–1997,* Economic Research Service, Natural Resources and Environment Division, Agricultural Handbook No. 712, 1997.

40. Crawley, M. J., Insect herbivores and plant population dynamics, *Annu. Rev. Entomol.,* 34, 531, 1989.

41. Pedigo, L. P., Hutchins, S. H., and Higley, L. G., Economic injury levels in theory and practice, *Annu. Rev. Entomol.,* 31, 341, 1986.

42. Southwood, T. R. E., and Norton G. A., Economic aspects of pest management strategies and decisions, in *Insects: Studies in Population Management,* Geier, P., Ed., Ecological Studies of Australia, Canberra, 1973, 168.

43. Higley, L. G., and Pedigo, L. P., Eds., *Economic Thresholds for Integrated Pest Management,* University of Nebraska Press, Lincoln, 1996.

44. Welter, S. C., Thresholds for interseasonal management, in *Economic Thresholds for Integrated Pest Management,* Higley, L. G., and Pedigo, L. P., Eds., University of Nebraska Press, Lincoln, 1996, 227.

45. Hedin, P. A., Ed., *Plant Resistance to Insects,* American Chemical Society, Washington D.C., 1983.

46. Maxwell, F. G., and Jennings, P. R., *Breeding Plants Resistant to Insects,* John Wiley & Sons, New York, 1980, 683.

47. Smith, C. M., *Plant Resistance to Insects : A Fundamental Approach,* John Wiley & Sons, New York, 1989, 286.

48. Russell, G. E., *Plant Breeding for Pest and Disease Resistance,* Butterworths, Boston, 1978, 485.

49. Smith, C. M., and Quisenberry, S. S., The value and use of plant resistance to insects in integrated crop management, *J. Agricult. Entomol.,* 11, 189, 1994.

50. Smith, C. M., Khan, Z. R., and Pathak, M. D., *Techniques for Evaluating Insect Resistance in Crop Plants,* CRC Press, Boca Raton, 1994, 320.

51. Ho, D. T., Heinrichs, E. A., and Medrano, F., Tolerance of the rice variety Triveni to the brown planthopper, *Nilaparvata lugens, Environ. Entomol.,* 11, 598, 1982.

52. Jung-Tsung, W., Heinrichs, E. A., and Medrano, F. G., Resistance of wild rice, *Oryza* spp., to the brown planthopper, *Nilaparvata lugens* (Homoptera: Delphacidae), *Environ. Entomol.,* 15, 648, 1986.

53. Panda, N., and Heinrichs, E. A., Levels of tolerance and antibiosis in rice varieties have moderate resistance to the brown planthopper, *Nilparvata lugens* (Stal) (Hemiptera: Delphacidae), *Environ. Entomol.,* 12, 1204, 1983.

54. Cole, L. C., The population consequences of life history phenomena, *Q. Rev. Biol.,* 29, 103, 1954.

55. Gadgil, M., and Solbrig, O. T., The concept of r- and K-selection: evidence from wild flowers and some theoretical considerations, *Am. Nat.,* 102, 14, 1972.

56. Jaremo, J., Nilsson, P., and Tuomi, J., Plant compensatory growth: herbivory or competition?, *Oikos,* 77, 238, 1996.

57. Jaremo, J., Tuomi, J., Nilsson, P., and Lennartsson, T., Plant adaptations to herbivory: mutualistic versus antagonistic coevolution, *Oikos,* 84, 313, 1999.

58. Shen, C. S., and Bach, C. E., Genetic variation in resistance and tolerance to insect herbivory in *Salix cordata, Ecol. Entomol.,* 22, 335, 1997.

59. Tiffin, P., and Rausher, M. D., Genetic constraints and selection acting on tolerance to herbivory in the common morning glory *Ipomoea purpurea, Am. Nat.,* 154, 700, 1999.

60. Stowe, K. A., Experimental evolution of resistance in *Brassica rapa*: correlated response of tolerance in lines selected for glucosinolate content, *Evolution,* 52, 703, 1998.

61. Mole, S., Trade-offs and constraints in plant-herbivore defense theory: a life-history perspective, *Oikos,* 71, 3, 1994.

62. Mauricio, R., Rausher, M. D., and Burdick, D. S., Variation in the defense strategies of plants: Are resistance and tolerance mutually exclusive?, *Ecology,* 78, 1301, 1997.

63. Simms, E. L., and Triplett, J., Costs and benefits of plant-responses to disease—resistance and tolerance, *Evolution,* 48, 1973, 1994.

64. Rosenthal, J. P., and Kotanen, P. M., Terrestrial plant tolerance to herbivory, *Trends Ecol. Evol.,* 9, 145, 1994.

65. Rosenthal, J. P., and Dirzo, R., Effects of life history, domestication and agronomic selection on plant defence against insects: evidence from maizes and wild relatives, *Evol. Ecol.,* 11, 337, 1997.

66. Bazzaz, F. A., Chiariello, N. R., Coley, P. D., and Pitelka, L. F., Allocating resources to reproduction and defense, *Bioscience,* 37, 58, 1987.

67. Bloom, A. J., Chapin, F. S., and Mooney, H. A., Resource limitation in plants—an economic analogy, *Annu. Rev. Ecol. Syst.,* 16, 363, 1985.

68. Chapin, F. S., III, The cost of tundra plant structures: evaluation of concepts and currencies, *Am. Nat.,* 133, 1, 1989.

69. Abrahamson, W. G., and Caswell, H., On the comparative allocation of biomass, energy, and nutrients in plants, *Ecology,* 63, 982, 1982.

70. Hickman, J. C., and Pitelka, L. F., Dry weight indicates energy allocation in ecological strategy analysis of plants, *Oecologia,* 21, 117, 1975.

71. Hole, C. C., and Scott, P. A., Effect of number and configuration of fruits, photon flux density and age on the growth and dry matter distribution of fruits of *Pisum sativum* L., *Plant Cell Environ.,* 6, 31, 1983.

72. Watson, M. A., and Casper, B. B., Morphogenetic constraints on patterns of carbon distribution in plants, *Annu. Rev. Ecol. Syst.,* 15, 233, 1984.

73. Tissue, D. T., and Nobel, P. S., Carbon relations of flowering in a semelparous clonal desert perennial, *Ecology,* 71, 273, 1990.

74. Reekie, E. G., and Bazzaz, F. A., Reproductive effort in plants. III. Effect of reproduction on vegetative activity, *Am. Nat.,* 129, 907, 1987.

75. Reekie, E. G., and Bazzaz, F. A., Reproductive effort in plants. I. Carbon allocation to reproduction, *Am. Nat.*, 129, 876, 1987.

76. Snow, A. A., and Wigwam, D. F., Costs of flower and fruit production in *Tipaularia discolor* (Orchidaceae), *Ecology*, 70, 1286, 1989.

77. Reekie, E. G., and Bazzaz, F. A., Reproductive effort in plants. II. Does carbon reflect the allocation of other resources?, *Am. Nat.*, 129, 897, 1987.

78. Bryant, J. P., Chapin, F. S. III, and Klein, D. R., Carbon/nutrient balance of boreal plants in relation to vertebrate herbivory, *Oikos*, 40, 357, 1983.

79. Chapin, F. S. III., Integrated responses of plants to stress a centralized system of physiological responses, *Bioscience*, 41, 29, 1991.

80. Coley, P. D., Bryant, J. P., and Chapin, F. S. III, Resource availability and plant antiherbivore defense, *Science*, 230, 895, 1985.

81. Gifford, R. M., and Evans, L. T., Photosynthesis, carbon partitioning, and yield, *Annu. Rev. Plant Physiol.*, 32, 485, 1981.

82. Gifford, R. M., Thorne, J. H., Hitz, W. D., and Giaquinta, T., Crop productivity and photoassimilate partitioning, *Science*, 225, 801, 1984.

83. Mooney, H. A., The carbon balance of plants, *Annu. Rev. Ecol. Syst.*, 3, 315, 1972.

84. Mooney, H. A., and Chu, C., Carbon allocation to *Heteromeles arbutifolia*, a California evergreen shrub, *Oecologia*, 14, 295, 1974.

85. Mooney, H. A., and Gulmon, S. L., The determinants of plant productivity—natural versus man-modified communities, in *Disturbance and Ecosystems: Components of Response*, Mooney, H.A., and Godron, M., Eds., Springer-Verlag, Berlin, 1983, 146.

86. Mooney, H. A., Gulmon, S. L., and Johnson, N. D., Physiological constraints on plant chemical defenses, in *Plant Resistance to Insects*, Hedin, P. A., Ed., American Chemical Society, Washington D.C., 1983, 21.

87. Björkman, C., and Anderson, D. B., Trade-off among antiherbivore defences in a South American blackberry (*Rhus bogotensis*), *Oecologia*, 85, 247, 1990.

88. del Moral, R., On the variability of chlorogenic acid concentration, *Oecologia*, 9, 289, 1972.

89. Hanover, J. W., Experimental variation in the monoterpenes of the *Pinus monticola* Dougl., *Phytochemistry*, 5, 713, 1966.

90. Krischik, V. A., and Denno, R. F., Individual, populations and geographic patterns in plant defense, in *Variable Plants and Herbivores in Natural and Managed Systems*, Denno, R. F., and McClure, M. S., Eds., Academic Press, New York, 1983, 463.

91. Larsson, S., Wirén, A., Lundgren, L., and Ericsson, T., Effects of light and nutrient stress on leaf phenolic chemistry in *Salix dasyclados* and susceptibility to *Galerucella lineola* (Coleoptera), *Oikos*, 47, 205, 1986.

92. Lightfoot, D. C., and Whitford, W. G., Interplant variation in creosotebush foliage characteristics and canopy arthropods, *Oecologia*, 81, 166, 1989.

93. Lightfoot, D. C., and Whitford, W. G., Variation in insect densities on desert creosotebush: Is nitrogen a factor?, *Ecology*, 68, 547, 1987.

94. Mihaliak, C. A., and Lincoln, D. E., Growth pattern and carbon allocation to volatile leaf terpenes under nitrogen-limiting conditions in *Heterotheca subaxillaris* (Asteraceae), *Oecologia*, 66, 423, 1985.

95. Simms, E. L., Costs of plant resistance to herbivory, in *Plant Resistance to Herbivores and Pathogens: Ecology, Evolution and Genetics*, Fritz, R. S., and Simms, E. L., Eds., University of Chicago Press, Chicago, 1992, 392.

96. Sohn, J. J., and Policansky, D., The costs of reproduction in the mayapple *Podophyllum peltatum* (Berberidaceae), *Ecology*, 58, 1366, 1977.

97. Gross, K. L., and Soule, J. D., Differences in biomass allocation to reproductive and vegetative structures of male and female plants of a dioecious perennial herb, *Silene alba* (Miller) Kraus, *Am. J. Bot.,* 68, 801, 1981.

98. Lovett Doust, J., and Lovett Doust, L., Leaf demography and clonal growth in female and male *Rumex acetosella, Ecology,* 68, 2056, 1987.

99. Onwekwelu, S. S., and Harper, J. L., Sex ratio and niche differentiation in spinach (*Spinacea oleracea* L.), *Nature,* 282, 609, 1979.

100. Saulnier, T. P., and Reekie, E. G., Effect of reproduction on nitrogen allocation and carbon gain in *Oenothera biennis, J. Ecol.,* 83, 23, 1995.

101. Reekie, E. G., and Bazzaz, F. A., Cost of reproduction as reduced growth in genotypes of two congeneric species with contrasting life histories, *Oecologia,* 90, 21, 1992.

102. Rosenthal, J. P., and Welter, S. C., Tolerance to herbivory by a stemboring caterpillar in architecturally distinct maizes and wild relatives, *Oecologia,* 1995.

103. Welter, S. C., and Steggall, J. W., Contrasting the tolerance of wild and domesticated tomatoes to herbivory: agro-ecological implications, *Ecol. Appl.,* 3, 271, 1993.

104. Jackson, L. L., and Dewald, C. L., Predicting evolutionary consequences of greater reproductive effort in *Tripsacum dactyloides,* a perennial grass, *Ecology,* 75, 627, 1994.

105. Haukioja, E., and Hakala, T., Life-history evolution in *Anodonta piscinalis*: correlation of parameters, *Oecologia,* 35, 253, 1978.

106. Lovett Doust, J., Plant reproductive strategies and resource allocation, *Trends Ecol. Evol.,* 4, 230, 1989.

107. Boeken, B., Life histories of desert geophytes—the demographic consequences of reproductive biomass partitioning patterns, *Oecologia,* 80, 278, 1989.

108. Bergström, R., and Danell, K., Effects of simulated winter browsing by moose on morphology and biomass of two birch species, *J. Ecol.,* 75, 533, 1987.

109. Roudy, B. A., and Ruyle, G. B., Effects of herbivory on twig dynamics on a Sonoran desert shrub *Simmondsia chinesis* (Link) Schn., *J. Appl. Ecol.,* 26, 1989.

110. Bilbrough, C. J., and Richards, J. H., Growth of sagebrush and bitterbrush following simulated winter browsing: mechanisms of tolerance, *Ecology,* 74, 481, 1993.

111. Marquis, R. J., Plant architecture, sectoriality and plant tolerance to herbivores, *Vegetatio,* 127, 85, 1996.

112. Gill, D. E., Individual plants as genetic mosaics: ecological organisms versus evolutionary individuals, in *Plant Ecology,* Crawley, M., Ed., Blackwell Scientific, Oxford, U.K., 1986, 321.

113. Loomis, R. S., Lou, Y., and Kooman P. L., Integration of activity in the higher plant, in *Theoretical Production Ecology: Reflections and Prospects,* Rabinge, R., Ed., Pudoc, Wageningen, 1990, 105.

114. Doebley, J., Stec, A., and Hubbard, L., The evaluation of apical dominance in maize, *Nature,* 386, 485, 1997.

115. Aarssen, L. W., Hypotheses for the evolution of apical dominance in plants—implications for the interpretation of overcompensation, *Oikos,* 74, 149, 1995.

116. Bonser, S. P., and Aarssen, L. W., Meristem allocation: a new classification theory for adaptive strategies in herbaceous plants, *Oikos,* 77, 347, 1996.

117. Coughenour, M. B., Graminoid responses to grazing by large herbivores: adaptations, exaptations, and interacting processes, *Ann. Missouri Bot. Garden,* 72, 852, 1985.

118. Lubbers, A. E., and Lechowicz, M. J., Effects of leaf removal on reproduction vs. belowground storage in *Trillium grandiflorum, Ecology,* 70, 85, 1989.

119. Gange, A. C., and Brown, V. K., Effects of root herbivory by an insect on a foliar-feeding species, mediated through changes in the host plant, *Oecologia,* 81, 38, 1989.

120. Archer, S., and Tiezen, L. L., Growth and physiological responses of tundra plants to defoliation, *Arctic and Alpine Res.*, 12, 531, 1980.

121. Richards, J. H., and Caldwell, M. M., Soluble carbohydrates, concurrent photosynthesis and efficiency in regrowth following defoliation: a field study with *Agropyron* species, *J. Appl. Ecol.*, 22, 907, 1985.

122. Gold, W. G., and Caldwell, M. M., The effects of the spatial pattern of defoliation on regrowth of a tussock grass. 1. Growth responses, *Oecologia*, 80, 289, 1989.

123. Donald, C. M., The breeding of crop ideotypes, *Euphytica*, 17, 385, 1968.

124. Janssens, M. J. J., Neumann, I. F., and Froidevaux, L., Low-input ideotypes, in *Agroecology: Researching the Ecological Basis for Sustainable Agriculture*, Gliessman, S.R., Ed., Springer-Verlag, New York, 1990, 130.

125. Butler, J. L., and Briske, D. D., Population structure and tiller demography of the bunchgrass *Schizachyrium scoparium* in response to herbivory, *Oikos*, 51, 306, 1988.

126. Olson, B. E., and Richards, J. H., Annual replacement of the tillers of *Agropyron desertorum* following grazing, *Oecologia*, 76, 1, 1988.

127. Richards, J. H., Mueller, R. J., and Mott, J. J., Tillering in tussock grasses in relation to defoliation and apical bud removal, *Ann. Bot.*, 62, 173, 1988.

128. Welker, J. M., Briske, D. D., and Weaver, R. W., Nitrogen-15 partitioning within a three generation tiller sequence of the bunchgrass *Schizachyrium scoparium*: response to selective defoliation, *Oecologia*, 74, 330, 1987.

129. Gifford, R. M., and Marshall, C., Photosynthesis and assimilate distribution following differential tiller defoliation, *Aust. J. Bot. Sci.*, 26, 517, 1973.

130. Miller, R. F., and Rose, J. A., Growth and carbon allocation of *Agropyron desortorum* following autumn defoliation, *Oecologia*, 89, 482, 1992.

131. Doak, D. F., Lifetime impacts of herbivory for a perennial plant, *Ecology*, 73, 2086, 1992.

132. Doak, D., The consequences of herbivory for dwarf fireweed: different times scales, different morphological scales, *Ecology*, 72, 1397, 1991.

133. Boyer, J. S., Plant productivity and environment, *Science*, 218, 443, 1982.

134. Mooney, H. A., and Gulmon, S. L., Constraints on leaf structure and function in reference to herbivory, *Bioscience*, 32, 198, 1982.

135. Chapin, F. S. III, The mineral nutrition of wild plants, *Annu. Rev. Ecol. Syst.*, 11, 233, 1980.

136. Tingey, W. M., and Singh, S. R., Environmental factors influencing the magnitude and expression of resistance, in *Breeding Plants Resistant to Insects*, Maxwell, F. G., and Jennings, P. R., Eds., John Wiley & Sons, New York, 1980, 87.

137. Sousa, O. J., and Foster, J. E., Temperature and the expression of resistance in wheat to the Hessian fly, *Environ. Entomol.*, 5, 333, 1976.

138. McNaughton, S. J., Wallace, L. L., and Coughenour, M. B., Plant adaptation in an ecosystem context: effects of defoliation, nitrogen, and water on growth of an African C4 sedge, *Ecology*, 64, 307, 1983.

139. Stafford, R. A., Allocation responses of *Abutilon theophrasti* to carbon and nutrient stress, *Am. Midl. Nat.*, 121, 225, 1989.

140. Ruess, R. W., Nutrient movement and grazing: experimental effects of clipping and nitrogen source on nutrient uptake in *Kyllinga nervosa*, *Oikos*, 43, 183, 1984.

141. Clement, C. R., Hopper, M. J., Jones, L. H. P., and Leafe, E. L., The uptake of nitrate by *Lolium perenne* from flowing nutrient solutions: II. Effects of light, defoliation, and relationship to CO2 flux, *J. Exp. Bot.*, 29, 1173, 1978.

142. Chapin, F. S. III, Nutrient allocation and responses to defoliation in tundra plants, *Arct. Alp. Res.*, 12, 553, 1980.

143. Davidson, J. L., and Milthorpe, F. L., Leaf growth in *Dactylis glomerata* following defoliation, *Ann. Bot.,* 30, 173, 1966.

144. Richards, J. H., Plant response to grazing: the role of photosynthetic capacity and stored carbon reserves, in *Rangelands: A Resource Under Siege,* Joss, P. J., Lynch, P. W., and Williams, O. B., Eds., Cambridge University Press, Cambridge, U.K., 1986, 428.

145. Ryle, G. J. A., and Powell, C. E., Defoliation and regrowth in the graminaceous plant: the role of current assimilate, *Ann. Bot.,* 39, 297, 1975.

146. Trlica, M. J., Distribution and utilization of carbohydrate reserves in range plants, in *Rangeland Plant Physiology,* Sosebee, R. E., Ed., Soc. Range Management, Range Sci. Ser. No. 4, Denver, CO, 73, 1977.

147. Schmid, B., Miao, S. L., and Bazzaz, F. A., Effects of simulated root herbivory and fertilizer application on growth and biomass allocation in the clonal perennial *Solidago canadensis, Oecologia,* 84, 9, 1990.

148. Godfrey, L. D., Meinke, L. J., and Wright, R. J., Vegetative and reproductive biomass accumulation in field corn: response to root injury by western corn rootworm (Coleoptera: Chrysomelidae), *J. Econ. Entomol.,* 86, 1557, 1993.

149. Godfrey, L. D., Holtzer, T. O., Spomer, S. M., and Norman, J. M., European corn borer (Lepidoptera: Pyralidae) tunneling and drought stress: effects on corn yield, *J. Econ. Entomol.,* 84, 1850, 1991.

150. Mott, J. J., Ludlow, M. M., Richards, J. H., and Parsons, A. D., Effects of moisture supply in the dry season and subsequent defoliation of persistence of the Savanna grasses *Themeda triandra, Heteropogon contortus* and *Panicum maximum, Aust. J. Agric. Res.,* 43, 241, 1992.

151. Wallace, L. L., McNaughton, S. J., and Coughenour, M. B., Compensatory photosynthetic responses of three African graminoids to different fertilization, watering, and clipping regimes, *Bot. Gaz.,* 145, 151, 1984.

152. Haile, F. J., Higley, L. G., Ni, X. Z., and Quisenberry, S. S., Physiological and growth tolerance in wheat to Russian wheat aphid (Homoptera: Aphididae) injury, *Environ. Entomol.,* 28, 787, 1999.

153. Maschinski, J., and Whitham, T. G., The continuum of plant responses to herbivory: the influence of plant association, nutrient availability, and timing, *Am. Nat.,* 134, 1, 1989.

154. Reader, R. J., Wilson, S. D., Belcher, J. W., Wisheu, I., Keddy, P. A., Tilman, D., Morris, E. C., Grace, J. B., and McGraw, J. B., Plant competition in relation to neighbor biomass: an intercontinental study with *Poa pratensis, Ecology,* 75, 1753, 1994.

155. Grace, J. B., and Tilman, D., *Perspectives on Plant Competition,* Academic Press, San Diego, 1990, 484.

156. Caldwell, M. M., Plant requirements for prudent grazing, in *Developing Strategies for Rangeland Management,* Westview Press, Boulder, CO, 1984, 117.

157. Louda, S. M., Keeler, K. H., and Holt, R. D., Herbivore influences on plant performance and competitive interactions, in *Perspectives on Plant Competition,* Grace, J. B., and Tilman, D., Eds., Academic Press, San Diego, 1990, 413.

158. Weiner, J., Asymmetric competition in plant populations, *Trends Ecol. Evol.,* 5, 360, 1990.

159. Mutikainen, P., Walls, M., and Ojala, A., Effects of simulated herbivory on tillering and reproduction in an annual ryegrass, *Lolium remotum, Oecologia,* 95, 54, 1993.

160. Painter, E. L., Detling, J. K., and Steingraeber, D. A., Grazing history, defoliation, and frequency-dependent competition: effects on two North American grasses, *Am. J. Bot.,* 76, 1368, 1989.

161. Swank, S., and Oechel, W. C., Interactions among the effects of herbivory, competition, and resource limitation on chaparral herbs, *Ecology,* 72, 104, 1991.
162. Lee, T. D., and Bazzaz, F. A., Effects of defoliation and competition on growth and reproduction in the annual plant *Abutilon theophrasti, J. Anim. Ecol.,* 68, 813, 1980.
163. Muller-Scharer, H., The impact of root herbivory as a function of plant density and competition: survival, growth and fecundity of *Centaurea maculosa* in field plots, *J. Appl. Ecol.,* 28, 759, 1991.
164. Fowler, S. V., and Rauscher, M. D., Joint effects of competitors and herbivores on growth and reproduction of *Aristolochia reticulata, Ecology,* 66, 1580, 1985.
165. Crockett, R. P., and Crookston, R. K., Tillering of sweet corn reduced by clipping of early leaves, *J. Amer. Soc. Hort. Sci.,* 105, 565, 1980.
166. Russell, W. A., Evaluations for plant, ear, and grain traits of maize cultivars representing different eras of breeding, *Maydica,* 29, 375, 1985.
167. Caldwell, M. M., Richards, J. H., Manwaring, J. H., and Eissenstat, D. M., Rapid shifts in phosphate acquisition show direct competition between neighbouring plants, *Nature,* 327, 615, 1987.
168. Marquis, R. J., Evolution of resistance in plants to herbivores, *Evol. Trends Plants,* 5, 23, 1991.
169. Nault, B. A., and Kennedy, G. G., Limitations of using regression and mean separation analyses for describing the response of crop yield to defoliation: a case study of the Colorado potato beetle (Coleoptera: Chrysomelidae) on potato, *J. Econ. Entomol.,* 91, 7, 1998.
170. Huntly, N., Herbivores and the dynamics of communities and ecosystems, *Ann. Rev. Ecol. Syst.,* 22, 477, 1991.
171. Velausamy, R., and Heinrichs, E. A., Tolerance in crop plants to insect pests, *Insect Sci. Appl.,* 7, 689, 1986.
172. Clement, S. L., and Quisenberry, S. S., Eds., *Global Plant Genetic Resources for Insect-Resistant Crops,* CRC Press, Boca Raton, 2000, 295.
173. Rosenthal, J. P., *Effects of Life History, Domestication, and Agronomic Selection on Plant Resistance to Insects: Maizes and Wild Relatives,* University of California, Berkeley, 1993.
174. Simms, E. L., and Fritz, R. S., The ecology and evolution of host-plant resistance to insects, *Trends Ecol. Evol.,* 5, 356, 1990.
175. Tingey, W. M., The environmental control of insects using plant resistance, in *CRC Handbook of Pest Management in Agriculture,* Pimentel, D., Ed., Boca Raton, 1981.
176. Duvick, D. N., Genetic contibutions to yield gains of U.S. maize hybrids, 1930–1980, in *Genetic Contributions to Yield Gains of Five Major Crop Plants,* Fehr, W. R., Ed., CSSA, Madison, WI, 1984, 15.

11 Crop Disease and Yield Loss

Brian D. Olson

CONTENTS

11.1 INTRODUCTION

Crop loss caused by plant pathogens has been reviewed extensively in a number of review articles over the past 50 years,[1,2] beginning with Chester[3] in 1950. This work established the rationale for complete and thorough assessments of plant disease and its impact on crop production. For simplicity, I will refer to the effects of plant diseases on crop production as "crop loss." With many plant pathogens, Cook[4] states, "diseases that affect the growing plant and thereby limit the ability of the plant to yield do not cause 'crop loss' nor can they 'reduce yields.' " Cook recommends the use of terminology such as "yield limiting factor" or "constraint to yield." This is true for crops where pests (weed, insect, or pathogen) reduce photosynthetic processes of the plant. But, in crops where growers must allow the grain or harvestable portion of the crop to mature before harvest, pests may directly infect, infest, or contaminate the crop, resulting in crop loss. The pests downgrade the quality of the crop, which makes it less valuable for commercial sale.

11.1.1 DIRECT DISEASES

There are many diseases that directly affect the fruit or harvestable portion of a crop. One well-known direct disease is apple scab caused by *Venturia inaequalis*. The pathogen infects leaves, causing lesions that reduce the photosynthetic capabilities of the apple tree. But more importantly the pathogen infects and produces lesions on fruit, rendering the fruit unmarketable. Consequently, damage thresholds for apples grown for fresh fruit are low.

Damaged fruit on an apple tree or any crop produce more ethylene than non-damaged fruit, accelerating the ripening of the damaged fruit as well as adjacent fruit on the same plant and adjacent plants. This process causes uneven fruit ripening in a field, making it difficult for growers to harvest all the fruit under optimal conditions. When damaged fruit are harvested and placed in storage for a few days, weeks, or months, the production of ethylene from the damaged fruit accelerates ripening of all fruit in the storage area. Damaged fruit can be culled during harvest but this significantly increases the length of time and costs necessary to harvest the crop, affecting the total cost to produce the crop.

Pests that cause direct effects to fruit or the harvestable crop do not always reduce the quantity of biomass produced by the crop, but do significantly reduce the quality and value of the crop even with low levels of damage. The emphasis of this chapter will be on diseases that infect roots, stems, and foliage, causing an indirect effect on crop production. The main focus will be on the major global food crops such as potato, wheat, corn, and rice.

11.1.2 THE DISEASE TRIANGLE AND INDIRECT DISEASES

The disease triangle is the simplest model that describes the interrelationships between environment, pathogen, and host (crop) necessary for disease development and effects on yield. An example of this is early blight of potato caused by *Alternaria solani*. Dry warm weather conditions, suboptimal soil nitrogen levels, and drought conditions with intermittent rains provide optimal conditions for early blight epidemics of potato.[5] Given these conditions (a susceptible host, a source of inoculum, and no fungicide applications), the disease epidemic rapidly increases. The potato plants grow poorly because of both environment and leaf necrosis caused by the early blight disease. The reduction in healthy leaf tissue reduces the ability of the plant to photosynthesize and produce assimilates for growth. Because potato tuber bulking occurs late in the growing season, late early-blight epidemics cause greater reduction in yield than early epidemics.[6]

The conditions listed above are optimal for the expression of early blight of potato, but not late blight. Optimal environmental conditions for late blight are adequate soil nitrogen, cool evening temperatures (13 to 19°C), and prolonged periods of rainfall or heavy dews.[5] With these environmental conditions (a susceptible host, abundant foliage, and a source of inoculum), the pathogen infects the potato foliage and a late blight epidemic ensues. If these conditions continue for 10 to 14 days with no fungicide applications, all of the above-ground potato plant tissue in the field will be destroyed.

The common threat of both early and late blight diseases of potato is yield loss caused by the destruction of potato foliage. The destruction of foliage reduces biomass accumulation of the plant and indirectly reduces the yield of the crop compared to a pathogen-free crop.

To protect potato foliage from early and late blight, timely fungicide sprays are applied. Fungicides are not applied continually to a crop but are applied when conditions (environment, host, and pathogen) are favorable for disease development. This is frequently determined by environmental conditions, but sometimes predictive strategies are used, such as indicator plants, spore catches, and/or insect counts depending on the disease and crop.[7] Because late blight can rapidly destroy the potato foliage and *Phytophthora infestans* can infect tubers, growers tolerate little to no disease in their potato fields. The difference in percent of visual symptoms evaluated for economic thresholds (ET) and economic injury levels (EIL) is much less for foliar plant pathogens than for foliar insect pests. Generally, insecticide sprays provide immediate (curative) control of an insect, while fungicides generally are prophylactic, not curative. Some curative fungicides are available for some diseases, but must be applied soon after infection, before symptom development. Once symptoms have developed, it is difficult to eradicate the fungus or bacteria and prevent secondary spread of the infection propagules.

11.2 DISEASE ASSESSMENTS AND YIELD LOSS

Over the last 60 years, plant pathologists have worked on many different methods to determine the relationship between disease and crop yield. In laboratory and field studies, disease incidence and severity are assessed to determine differences in cultivars, inoculum source, tillage practices, irrigation schedules, environmental conditions, fungicides, and fungicide spray schedules. The first important factor that plant pathologists have considered is the standardization of assessments for different crops and diseases[2] and a complete understanding of the terminology.[8] While choosing an assessment method, it also is important to standardize the rating procedure to reduce variability within and between studies.[9, 10] If resources are limited or because of the type of trial, disease incidence or severity may be measured once, at a critical growth stage of the crop yield. When resources are not a constraint or because of the disease epidemic, disease incidence and severity are measured multiple times to quantify treatment differences with respect to the disease epidemic. Values from the single critical point or multiple point evaluations are then modeled with linear regression analysis to determine a relationship with yield.

11.2.1 SINGLE POINT EVALUATION MODELS

In cereals, assessments often are made at specific growth stages to determine relationships between treatment effects on disease incidence and severity and yield, using linear regression analysis.[11] The advantage of a single point model is the simplicity, but unless the assessment measures disease symptoms that affect biomass production, little to no correlation is measured between disease and yield. Backman and Crawford[12] measured disease severity of early and late leafspot and defoliation of peanut before harvest. Both disease and defoliation correlated (negatively) well with peanut yield within each year tested, but not between years. Within each year the early and late leafspot disease rating correlated positively with defoliation, which had a negative effect on photosynthesis and biomass accumulation. Backman noted poor correlation between years was most likely because of differences in growing conditions such as rainfall, fertility, and solar radiation.

11.2.2 AREA UNDER THE DISEASE PROGRESS CURVE (AUDPC) AND YIELD LOSS

In research studies, diseases such as early blight and late blight of potato are often assessed multiple times on a regular schedule to obtain a true picture of the epidemic. With multiple assessment of the disease (X_i) over time (t_i) the area under the disease progress curve (AUDPC) can be calculated and compared between treatments. Shaner and Finney[13] demonstrated that when comparing treatment differences for slow-mildewing of wheat, using AUDPC had a lower error of variances than logit transformations. AUDPC is calculated by multiplying the average disease assessment between two consecutive evaluation dates $[(X_i + X_{i+1})/2]$ by the difference in time between those dates $(t_{i+1} - t_i)$ and adding the values together for the entire epidemic period.

$$\text{AUDPC} = \sum^{n-1} [(X_i + X_{i+1})/2] \, (t_{i+1} - t_i) \qquad [11.1]$$

For a single epidemic, the AUDPC summarizes all the assessment data into one value for each treatment and differences between treatments are often more significant than comparisons made for any one assessment. Researchers then use linear regression analysis to correlate the AUDPC with differences in yield.[14, 15] Often the regression equations work only for the specific study and cannot be transferred from one study, location, or year to the next.[16] Shtienberg et al.[17] solved this problem for early and late blight of potato by associating the severity of the disease epidemics (AUDPC) with the effect of the disease on the relative bulking rate of the potato plant. This solved the problem where early blight epidemics before bulking and late blight epidemics after cessation of bulking had little to no effect on potato tuber bulking.

11.3 FOLIAR ASSESSMENTS AND YIELD LOSS

As in the work of Shteinberg et al.[17] many studies in the 1980s and early 1990s began to associate the effects of disease epidemics with physiological functions of the crop to improve predictions of yield loss. The assessment of early and late leafspot severity and defoliation two to three weeks before harvest measured the relative amount of defoliation caused by both diseases throughout the year.[12] For any single year, the regression lines provided good estimates of yield. The slopes were similar but could not be transferred from year to year. The data established a close relationship between yield and defoliation or infection. This established a relationship and effect between the foliar disease, healthy green leaf tissue, and yield. By indirectly quantifying the effect of plant disease on the relative quantity of foliage and function of the different growth stages of the crop, the studies began to account for photosynthesis, assimilate production, and biomass accumulation.

11.3.1 LEAF AREA INDEX (LAI)

To directly account for the assimilate production of healthy green foliage, the leaf area of healthy foliage must be measured. Measuring the leaf area would account for differences in abiotic and biotic stresses on plants from year to year. The stresses manifest themselves in healthy foliage, which accounts for biomass production and yield. Plants grown in unfavorable abiotic and biotic conditions have a lower leaf area, relative to plants grown in favorable conditions. Typically, plant leaf area is measured per square meter of soil surface and referred to as the leaf area index (LAI).

$$\text{LAI} = \text{leaf area M}^2 \, / \, 1 \, \text{M}^2 \text{ of soil surface} \qquad [11.2]$$

11.3.2 BEER'S LAW

Knowing the LAI of a crop, the amount of radiation intercepted by the crop can be estimated if the incident solar radiation also is measured. Beer's Law predicts the amount of solar radiation not intercepted by plant foliage at any level in a crop assuming the crop canopy is uniformly distributed.[18] This means that the crop canopy is homogeneous through all layers of the canopy. The average irradiance not intercepted by foliage decreases exponentially with increasing depth in the canopy. Beer's Law predicts the amount of sunlight to reach the soil surface plane as:

$$I = I_0(e^{-LAI}) \qquad [11.3]$$

where I is the amount of sunlight that reaches the soil surface and I_0 is the amount of irradiance immediately above the plant canopy. The algebraic transformation of Beer's Law into the amount of radiation intercepted (RI) by the crop foliage is

$$RI = I_0 - I \qquad [11.4]$$

If we substitute $I_0(e^{-LAI})$ for I then:

$$\text{RI} = I_0 - I_0(e^{-LAI}) \qquad [11.5]$$

then:

$$RI = I_o(1 - e^{-LAI}) \qquad [11.6]$$

However, no crop has a homogenous canopy. The equation is corrected with an extinction coefficient for the crop canopy. This value is the amount of shadowed leaf area projected on a square meter of soil surface divided by the LAI. Therefore:

$$I = I_o(e^{-kLAI}) \qquad [11.7]$$

and:

$$RI = I_o(1 - e^{-kLAI}) \qquad [11.8]$$

Khurana and McLaren[19] observed a positive correlation between LAI and RI in potato, where no foliar pests were present. The amount of radiation intercepted increased linearly from 0 to 70%, between LAI of zero to two. Between LAI of two and four, the amount of radiation intercepted increased from 70 to only 95%. Leaf area indexes above four generally intercept a constant 95% of the radiation. Given a constant light extinction coefficient k (= 0.72), the Beer's Law equation explained 88% of the variance. When the extinction coefficient k was replaced with a quadratic relationship where k changed with increasing LAI, then the Beer's Law equation explained 92% of the variances in the data. In this case, the data indicate as LAI increases over time, the plant canopy architecture and/or leaf angles change, causing changes in the extinction coefficient used in the Beer's Law equation. This general relationship between LAI and the amount of RI by the crop holds true for all crops (Figure 11.1).

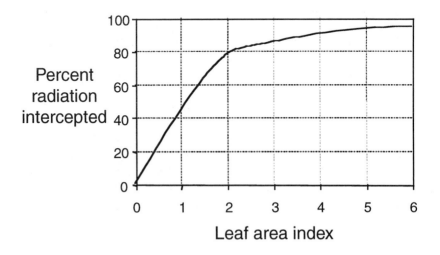

FIGURE 11.1   The relationship between the leaf area index (LAI) of a crop canopy and the percent radiation intercepted.

11.3.3 RADIATION USE EFFICIENCY

Watson[20, 21] demonstrated that the biomass production of different crops could be explained by measuring the LAI over time. When the LAI was integrated over the entire growing season, the leaf area duration (LAD) correlated well with yield. Monteith[22] took this one step further and suggested that biomass production of a crop was directly proportional to photosynthetically active radiation (PAR) by green plant tissue or LAI and the radiation use efficiency (RUE) of the plant could be calculated as:

$$RUE = (Total\ dry\ matter\ /\ M^2)\ /\ \smallint RI\ /\ M^2 \qquad [11.9]$$

The RI is the amount of irradiant energy intercepted by the plant. The plant transforms this energy by means of photosynthesis to produce assimilates for growth and production of seed. From this equation, Monteith showed that barley, potato, sugar beet, and apple produced about 1.4 g of carbohydrate per MJ of solar energy intercepted by healthy foliage. In the previously mentioned work of Khurana and McLaren,[19] when the PAR or RI was integrated over the entire growing season and plotted against the total dry weight a significant positive linear relationship was observed. The RUE for potato was 3.4 g of carbohydrate per MJ of solar energy. In studies where potatoes were grown in dry conditions, the RUE was lower compared to potatoes grown in a year with adequate rainfall, resulting in lower yields for the same RI as observed in other studies.[23]

11.4 EFFECTS OF DISEASES ON RI AND RUE

From the insights of Boote et al.[24] and Johnson[25] and summarization by Madden and Nutter,[11] foliar plant pathogens have been categorized into two groups: (1) those that reduce the amount of foliage or RI by a plant, and (2) those that reduce RUE of the foliage (Figure 11.2). Those pathogens that reduce the amount of foliage or RI do so by consuming foliage, accelerating leaf senescence, reducing plant stands, and/or stealing light. Then there are pathogens that interfere with the RUE of the photosynthetic process in the leaf by consuming assimilates, reducing the photosynthesis rate, and reducing plant turgor. Both categories of pathogens ultimately reduce the net photosynthetic ability of an infected plant compared to a healthy noninfected plant.

11.4.1 RADIATION INTERCEPTED

Until the late 1980s, the majority of studies investigating the relationship between diseases and crop yield attempted to correlate only disease severity or incidence at a single point or multiple points in time with yield or AUDPC and yield.[12, 26] Most of the correlations only had limited accuracy and did not predict yields at different locations or years. In 1987, Waggoner and Berger[27] helped clarify the relationship between disease and yield. They explained that disease symptoms reduce the healthy LAI of a plant. Consequently, the plant assimilates less biomass compared to a healthy plant. They used two equations to analyze data previously published in

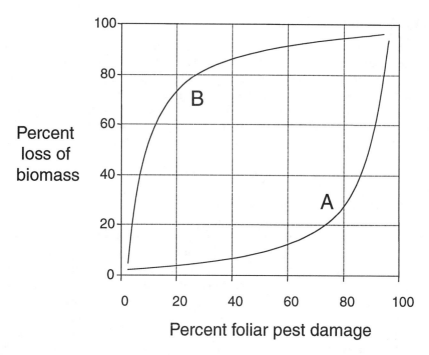

FIGURE 11.2   Foliar effects of pest damage caused by: A. reduction of leaf area indices (LAI), and B. reduced radiation use efficiencies (RUE).

refereed journals. The first equation integrates the effects of the disease on the healthy (disease free) LAI throughout the growing season (Equation 11.10). The second equation integrates the effects of the disease on the LAI and radiation absorption throughout the growing season (Equation 11.11), similar to that used by Khurana and McLaren.[19]

$$HAD = \sum [LAI_i (1 - X_i) + LAI_{i+1} (1 - X_{i+1})/2] (t_{i+1} - t_i) \qquad [11.10]$$

$$HAA = \sum I_o\{[(1 - X_i)(1 - e^{-kLAI_i}) + (1 - X_{i+1})(1 - e^{-kLAI_{i+1}})]/2\} (t_{i+1} - t_i) \qquad [11.11]$$

Values from each equation were correlated with yield. When HAD was plotted against yield, the two spring crops and one autumn crop each had different regression lines.[28, 29] The slopes of the regressions for the two spring crops were similar but much steeper than the autumn crop. The HAD equation does not consider any differences in incident radiation and RI. Solar radiation is significantly less in autumn compared to spring and early summer months. When the same data were plotted using the HAA equation, a single regression line fit both spring crops and autumn crops (Figure 11.3). This clearly supports the use of the RI equation to calculate biomass production of crops and yields.

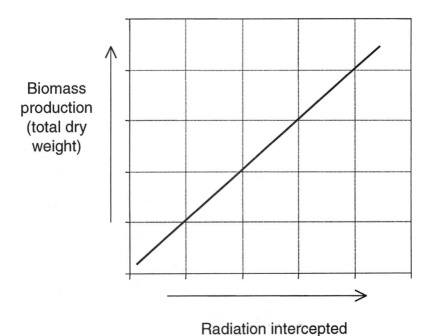

Radiation intercepted

FIGURE 11.3   The relationship between the total radiation intercepted by a crop canopy and the biomass production.

Johnson[30] pointed out the need to associate the effects of disease LAI with crop stage. This association would help define the role and effect of the different stages of the crop on crop yield. The information also might help define critical time periods to assess pest management strategies.

Many researchers have discussed the simplicity of the concept that radiation intercepted for a crop over an entire season will give an indication of the net assimilation or dry matter production of a crop. The amount of radiation intercepted is dependent on the changes in leaf area and incident radiation through the growing season.[31] For example, if cloudy conditions occurred in the beginning of the growing season of a potato crop, this would reduce the incident radiation portion of the equation. However, if the potato crop were limited in growth early in the season by poor crop emergence (dry conditions) and subsequent expression of early blight, the LAIs would be lower than those of a crop with good emergence and no early blight. If all other factors were equal then the net effect of shading on the LAI after six weeks may equal the net effect of stand loss and disease. In the case of shading, the plant is not able to assimilate and produce biomass at the maximum rate, resulting in a lower LAI than a non-shaded healthy plant. In the diseased crop the plant is able to assimilate and produce biomass only to have it destroyed by early blight. This hypothetical example demonstrates that the net result of plant biomass production is related to incident radiation, radiation intercepted, and the radiation use efficiency of the plant.

11.4.2 LAST'S FORGOTTEN CLASSIC

In the 1950s, Last[32] extensively investigated the effects of powdery mildew, caused by *Erysiphe graminis* D.C., on the growth of barley in a greenhouse study. Last's classic investigation[33] was cited in the 1960s and 1970s and infrequently in the 1980s and 1990s. He painstakingly collected data on total plant growth, root growth, foliar growth, and yield. The research demonstrated the effect powdery mildew has on the LAI of barley and its relationship with total plant growth, root growth, foliar growth, and yield.

Barley plants were inoculated with *Erysiphe graminis* soon after emergence and other noninoculated plants were protected from powdery mildew with sulfur every 7 to 10 days. The percentage of leaf tissue covered with powdery mildew for the inoculated plants ranged from 18 to 29% and 0 to 5% for the protected plants. As expected, LAI, total dry weight, plant height, and yield all were much greater for the protected plants than for the inoculated plants. Unexpectedly, the ratio of root dry weight to total plant dry weight or unit leaf area was much greater for the protected than for the inoculated plants. Last could not explain this difference, but fortunately he assessed the percent of leaf tissue infected with powdery mildew. In the calculation, Last used the entire leaf area, which included the powdery mildew infected leaf area. If the leaf area infected with powdery mildew is subtracted from the total leaf area and the root dry weight to leaf area ratio is recalculated, the difference between ratios for the inoculated and protected plants is much less than originally calculated by Last. Forty-one days after infection Last calculated the ratio of root dry weight to leaf area for protected and inoculated plants to be 1/7 and 1/13.5, respectively. When the values are recalculated after the leaf area infected with powdery mildew is subtracted from the total leaf area, the ratios for protected and inoculated plants are 1/6.9 and 1/10.7.

11.4.3 RADIATION USE EFFICIENCY

In Last's study the remaining difference in root dry weight to leaf area ratio between protected and inoculated plants could be caused by a reduction in RUE of the remaining healthy leaf area. When the incidence and severity of injury to foliage by a pest is measured, the amount of healthy LAI (HLAI) can be determined. The HLAI can account for the amount of radiation intercepted (RI) by the plant and potential net assimilation. But many biotic stresses may also reduce the radiation use efficiency (RUE) or potential net assimilation rate of the HLAI of the plant. If the relative RUE of a plant affected by the biotic stress is known, this value in addition to HLAI and insolation can determine the net assimilation of the plant. If all of the values are recorded regularly throughout the growing season, the values can be integrated to determine the net biomass production of the plant (Equation 11.12).

$$Y = \int RUE(t)RI(t)[1 - X]dt \qquad [11.12]$$

where $RUE(t)$ equals the radiation use efficiency, $RI(t)$ equals the amount of radiation intercepted by the plant, and $[1-X]$ equals the amount of healthy leaf area at any point in time (t)dt. If Equation 11.8 is substituted for RI then biomass production or yield can be calculated with Equation 11.13.

$$Y = \int RUE(t) \, ((I_0(1 - e^{-kLAI(t)})) \, (1 - X(t))) \, (t)dt \qquad [11.13]$$

Powdery mildew infested leaf area reduces the RI, but what remains uncertain is whether powdery mildew reduces the RUE of the remaining healthy leaf area. Studies have shown a linear negative correlation between powdery mildew infected small grain crops and the healthy leaf area.[34–39] However, Rabbinge et al.[40] reported that at maximum light intensities the net assimilation rate of powdery mildew infested leaves was less than what could be accounted for by the percentage of leaf area infested. These data indicate that powdery mildew of winter wheat has an effect on RI but also on RUE. Haigh et al.[41] did not observe the same dramatic reduction in net assimilation when oats were infected with *Erysiphe graminis* f. sp. *avenae,* nor did Balkema-Boomstra et al.[42] when spring barley, *Hordeum vulgare,* was infected with *Erysiphe graminis* f. sp. *hordei.* Consequently, the different effects of powdery mildew on the crop RUE probably are species-specific between pathogen and small grains host.

Necrotic lesions of late blight of potato on leaves and stems often destroy the entire leaf or stem. Haverkort and Bicamumpaka[43] observed a linear relationship between potato tuber yields and radiation intercepted over the entire growing season, regardless of the severity of late blight symptoms observed. Measurements of net photosynthetic rates of green potato leaf tissue from healthy and infected potato plants also supported this relationship.[44] There was no effect of the RUE of leaves with necrotic lesions on the stems. These studies provide evidence that *Phytophora infestans* does not reduce the RUE of the plant, but only limits the net assimilation of the plant by reducing the LAI.

Infection of rice with *Pyricularia oryzae* not only reduces the RI with the development of lesions on the leaves, but also causes a reduction in the RUE of the remaining healthy leaf tissue.[45, 46] Similarly, *Alternaria alternata* infections of cotton reduced the assimilation and transpiration rates more than what could be explained by infected leaf area.[47] *Pyricularia oryzae* and *Alternaria alternata* are both pathogens known to produce toxins that lead to plant stress, leaf chlorosis, and early senescence of infected leaves. If the amount of time between infection and leaf senescence is short, the ultimate effect on the plant is loss of RI. If the process takes longer, then part of the outcome should be a reduction in RUE as well as RI. Plant viruses also have a significant effect on photosynthetic metabolism and probably lower the RUE of the leaves.[48]

11.4.4 DISEASES OF ROOT AND STEM

Pathogens that infect plant roots and stems usually restrict the flow of water to the leaves, and therefore are expected to reduce RUE compared to noninfected plants.[1] In whole-plant field studies Gent et al.[49] demonstrated that *Verticillium* wilt of eggplants caused by *Verticillium dahliae* reduced the leaf LAI and RUE of the entire plant. Leaves on the infected plant wilted, became chlorotic, and abscised prematurely from the plant. As the infection progressed, the plant compensated for the reduction in transpiration flow with fewer and smaller leaves. This resulted in a lower LAI per infected eggplant plant compared to noninfected plants. Net assimilation per plant was less for the infected plants compared to the noninfected plants. However, when the CO_2 exchange was expressed in terms of rate per unit of leaf area, there was no difference between the infected and noninfected plants. Consequently, this study demonstrates that *Verticillium dahliae* infection of eggplant reduces the LAI, reducing the RI, but does not reduce the RUE of the remaining healthy leaf tissue.

Bowden and Rouse[50, 51] demonstrated that potato plants, when infected with *Verticillium dahliae,* not only lowered the RI compared to noninoculated plants but also lowered the RUE of most of the remaining leaf area. This study as well as other studies of vascular wilt diseases[52, 53] clearly shows that transpiration rates and RUE of the older leaves are lower in inoculated vs. noninoculated plants. The studies demonstrate that before symptom expression of the disease on older leaves is observed, the transpiration rate and CO_2 exchange rate begin to decrease in the inoculated vs. non-inoculated plants. The drought-stress like conditions caused by these vascular wilt diseases are most likely caused by plugged xylem vessels that reduce the transpiration stream in the plant, causing subsequent premature senescence of the older leaves. Not only do the older leaves prematurely drop off the plant, but the newly emerged leaves are often stunted in the inoculated vs. noninoculated plants. The premature leaf senescence and emergence of smaller leaves in the infected plant reduce the LAI and RI of the plant compared to a non-infected plant. What remains in question is whether the RUE of the remaining healthy leaf tissue has changed in an inoculated vs. noninoculated plant if the net assimilation rate is based on leaf area.

11.5 MODELING

To understand the interrelationships between RUE, RI, and plant growth, the different systems have been combined with mechanistic models. In a recent review, Madden and Nutter[11] provide some good insights into modeling crop losses from plant diseases. Not unlike Waggoner and Berger,[27] they observed that the most successful models describe the relationship between the disease and the effects on HLAI and RI. If these values are integrated over the entire growing season they correlate well with the total biomass production and yield of the different cropping systems tested.

In field studies investigating solitary and interaction effects of potato leafhopper, *Verticillium* wilt, and early blight on potato, Johnson et al.[26] observed that the effects of all three pests on potato yield and defoliation were not additive when compared to the effect of each pest alone. These and other results[54] demonstrate that when more than one foliar pest is present the pests compete for the same healthy green leaf tissue. The most accurate prediction of yield occurred when the area under the green leaf area curve (AUGLAC) was regressed with the area under the hopper burn curve (AUHBC).[26] The AUGLAC is essentially the integral of the effects early blight, *Verticillium* wilt, and potato leafhopper have on green leaf tissue or HLAI and RI.

The AUHBC accounts for the effect potato leafhopper has on the RUE of the remaining green leaf tissue HLAI. In subsequent work, Johnson[25] modeled the effects of early blight, *Verticillium* wilt, and potato leafhopper on potato foliage and yield. He found by measuring the effects of the multiple pests on HLAI, RI, and RUE, the model provided a basic framework to understand the interaction of the pests tested and effects of all pests on potato foliage and yield.

Luo et al.[54] modeled the effects of rice blast on rice growth and yield. They linked a rice leaf blast and rice growth model at the point where leaf blast affects leaf photosynthesis and biomass production. In the rice blast model the effects of the disease on RUE of healthy leaf tissue was accounted for using data from Bastiaans[46] where the visual lesion was correlated with the larger "virtual lesion." Again, the impacts of a disease on yield were measured by studying the effects of the disease on HLAI and RI.

11.6   BIOMASS PARTITIONING

The measurement of HLAI and integration of RI begin to account for the biomass production of the plant but do not account for the partitioning of the biomass within the plant. When different reproductive growth periods of peanut[55] were shaded (75% shade) for 14 to 21 days, shaded plants had significantly fewer flowers during flowering and pods during podding, and lower seed weight during maturation than nonshaded plants. This type of study indicates how the biomass partitioning or accumulation changes through the maturation of a crop. During the shading event, the shading is similar to the effect a disease has on RI and RUE of the foliage. But unlike the effects of disease that cannot be instantly removed, the shading was removed at the designated times. Therefore, to simulate the effects of foliar diseases on a plant and understand the effects on biomass partitioning it is probably best to remove plant tissue. Nutter[56] verified the removal of leaves from peanut plants simulated the effects of early and late leafspot on peanut yields.

In the studies of early and late blight of potato where the AUDPC was coupled with the potato bulking period, this greatly improved the ability to predict the effects of the epidemic on yields.[17] During the bulking period of potato, assimilates are accumulated as biomass in the tubers. In the early blight studies conducted by Johnson,[6] early season blight epidemics had much less effect on yield than late epidemics when the crop was in the bulking period.

The relative effects of foliar diseases on total biomass production can be accounted for by the measurement of HLAI and RI, but where the biomass is accumulated is dependent on the crop, crop stage, and pre-existing stresses. A multidisciplinary team most likely is necessary to develop a growth model for the crop.

11.7   REMOTE SENSING

The development of crop growth models that include the integration of the HLAI throughout the growing season provides the best estimate of the net assimilate production of the plant. This also is an excellent indicator of final yield of the crop. Collecting LAI data on a weekly or biweekly schedule on research plots is costly, but provides the necessary data to develop plant growth models. Collecting LAI data on a large scale, county and statewide, is prohibitive because of the time and expense involved. Consequently, many researchers have investigated the use of remote sensing imaging from satellites (LANDSAT) of crop canopies to supplement actual LAI measurements in the field. This area of study is still in its infancy even though it has been investigated for more than 30 years. The image sensitivity increases continually with improved technology.

Numerous studies have been conducted correlating measured vegetative indices from LANDSAT images of sorghum with actual ground LAI.[57, 58] The vegetative indices provide a good estimation of the LAI. When this information was coupled with crop growth models and meteorological data, a good estimation of yield also was observed. Maas[59] suggested that the satellite images could be used to initialize crop growth models for estimating crop yields. Other studies have been conducted with ground-based hand-held multispectral radiometers measuring reflectance to determine the vegetative indices.[60, 61] These data were correlated with measured LAI of the crop and photosynthetically active radiation measurements above, within, and below the crop canopy. The average of several vegetative indices recorded late in the growth stages of wheat and corn were positively (linearly) correlated with crops' dry matter, yield, and harvest index (yield/dry matter).

Satellite multispectral scanners and hand-held radiometers certainly have their place quantifying LAI and growth of a crop, but can they predict yield loss caused by diseases? Nilsson[62] and Nutter[63] review the use of remote sensing and image analysis in plant pathology and crop loss assessment. Nilsson and Johnsson[64] demonstrated that spectral reflectance measurements with a hand-held radiometer of barley infected with varying levels of barley stripe, *Pyrenophora graminea,* correlated positively with visual disease assessments and yield. On July 1 the least and heavily infested plots of barley strip had 2 and 49% of the plants infested, respectively. On July 5 the reflectance ratios for the near infrared and red wavebands for the same plots were 4.9 and 6.8, respectively. The final yields for the least and heavily infested plots were 3720 and 5409 kg/ha, respectively. The 28% reduction in reflectance ratio better described the 31% reduction in yield compared to 47% increase in disease incidence. The spectral reflectance partially measures the LAI of the crop while the disease infestation only measures the percent incidence of the disease. If disease severity had been rated, this may have revealed a better relationship between disease, spectral reflectance and yield. If the spectral reflectance in this study were associated with LAI and if the RI were integrated for the 10 observation dates, these data might better correlate to final yields.

In a similar study, Nutter[65] observed an association between spectral reflectance of peanut infected with late leafspot and yield. In this system the percent reflectance at 800 nm negatively correlated with both mechanical and disease-induced defoliation. Two weeks before harvest, the percent reflectance positively correlated with yields from four separate studies. The reflectance data were not analyzed collectively across all four studies because there were dramatic differences in yield between studies. If percent reflectance data were collected periodically throughout the growing season in the four separate peanut studies and related to LAI, then the RI integrated over all the observation dates may have allowed all the data to be analyzed together. Theoretically, this would demonstrate that RI as affected by late leafspot defoliation and different environmental conditions could explain most of the yield differences. Additionally, if disease severity was rated each time the spectral reflectance was measured, then the effects of leafspot on spectral reflectance and LAI could be correlated.

Dudka[66] used a hand-held radiometer to measure the reflectance of soybeans with varying levels of sclerotinia stem rot caused by *Sclerotinia sclerotiorum*. Disease incidence correlated positively with reflectance at 706 nm and negatively at 760 nm. It was inferred that yield data negatively correlated the incidence of sclerotinia stem rot.

The studies using the handheld radiometers demonstrate that remote sensing and image analysis have utility in quantifying the effects of disease on yield loss. Because the remote sensing was limited to a single finite point in time, the data collected only correlated well with each study and the relationships could not be integrated across studies. Instead, if the spectral measurements were collected over the entire growing season and related to LAI, then the RI by the crop could be integrated and calculated for the entire season and correlated to yield. This would then allow data from different studies to be collectively analyzed.

11.8   SUMMARY

Whether a pest attacks the roots, stem, or leaves, the effects ultimately will manifest themselves with a reduction in leaf area. The concept is simple, but often overlooked. Using Beer's Law and LAI, if a crop had an LAI of six, it could lose one half the LAI before the plant would see a dramatic reduction in RI (Figure 11.1). High LAI provides protection against loss biomass production and yield when leaves are destroyed by diseases or biotic stress, because of the relationship between LAI and RI. In the early growth stages of an annual plant, the LAI is less than three, and the loss of LAI has a significant effect on RI and net assimilate production (Figure 11.1). Early season disease epidemics that affect HLAI and RI may only delay the crop maturity, if control practices are implemented to stop the epidemic. Delayed maturity might reduce crop yields if the incident radiation between the normal maturation time and the delayed time is dramatically less due to weather and angle of the sun. For this reason it is important to quantify the severity of infestation and on what portion(s) of the crop, to record the growth stage, and measure the LAI if the effects will be correlated with crop yields.

Integrating RI and RUE over a growing season provides an accurate measurement of total biomass production of a crop as outlined by Monteith[22] and Goudriaan.[67] Differences in abiotic and biotic effects on the crop are then reflected in HLAI, RI, and RUE, accounting for differences in net assimilation, biomass, and yield of the crop. With these known variables, data from different locations and years can be collectively analyzed and summarized. The major difficulty with this concept is the length of time necessary to measure LAI data.[68] To help alleviate this problem in the future, limited LAI sampling could be integrated with remote sensing to save time and resources.

This approach should help us to better understand how single or multiple biotic stresses affect the net biomass production of a crop. This would allow the creation of models that would predict crop yields throughout the season, and be site specific, given historic environmental parameters. As the growing season progresses and the environmental conditions are updated into the model along with the relative RI and RUE, growth stage, predicted weather conditions, presence of pathogen, price of commodity, and cost of sprays, pest management decisions would be made as economic thresholds are reached.

REFERENCES

1. Gaunt, R. E., The relationship between plant disease severity and yield, *Annu. Rev. Phytopathol.,* 33, 119, 1995.
2. James, W. C., Assessment of plant diseases and losses, *Annu. Rev. Phytopathol.,* 12, 27, 1974.
3. Chester, K. S., Plant disease losses: their appraisal and interpretation, *Plant Dis. Rep. Suppl.* 193, 189, 1950.
4. Cook, R. J., Use of the term "crop loss," *Plant Dis.,* 69, 95, 1985.
5. Stevenson, W. S., Management of early and late blight, in *Potato Health Management,* Rowe, R. C., Ed., APS Press, St. Paul, MN, 1993, chap. 16.
6. Johnson, K. B., and Teng, P. S., Coupling a disease progress model for early blight to a model of potato growth, *Phytopathology,* 80, 416, 1990.
7. Backman, P. A., and Jacobi, J. C., Thresholds for plant-disease management, in *Economic Thresholds for Integrated Pest Management,* Higley, L. G., and Pedigo, L. P., Eds., University of Nebraska Press, Lincoln, 1996, chap. 8.
8. Nutter, F. W., Jr., Teng, P. S., and Royer, M. H., Terms and concepts for yield, crop losses and disease thresholds, *Plant Dis.,* 77, 211, 1993.
9. Nutter, F. W., Jr., Disease severity assessment training, in *Exercises in Plant Disease Epidemiology,* Francl, L. J., and Neher, D. A., Eds., APS Press, St. Paul, MN, 1997, chap. 1.
10. Tomerlin, J. R., and Howell, T. A., DISTRAIN: A computer program for training people to estimate disease severity on cereal leaves, *Plant Dis.,* 72, 455, 1988.
11. Madden, L. V., and Nutter, F. W., Jr., Modeling crop losses at the field scale, *Can. J. Plant Pathol.,* 17, 124, 1995.
12. Backman, P. A., and Crawford, M. A., Relationship between yield loss and severity of early and late leafspot diseases of peanut, *Phytopathology,* 74, 1101, 1984.
13. Shaner, G., and Finney, R. E., The effect of nitrogen fertilization on the expression of slow mildewing resistance in Knox wheat, *Phytopathology,* 67, 1051, 1977.
14. Broscious, S. C., Pataky, J. K., and Kirby, H. W., Quantitative relationships between yield and foliar diseases of alfalfa, *Phytopathology,* 77, 887, 1987.
15. Spitters, C. J. T., Van Roermund, H. J. W., Van Nassau, H. G. M. G., Schepers, J., and Kesdag, J., Genetic variation in partial resistance to leaf rust in winter wheat: disease progress, foliage senescence and yield reduction, *Neth. J. Plant Pathol.,* 96, 3, 1990.

16. Bryson, R. J., Sylvester-Bradley, R., Scott, R. K., and Pavely, N. D., Reconciling the effects of yellow rust on yield of winter wheat through measurements of green leaf area and radiation interception, *Aspects Appl. Biol.*, 42, 9, 1995.

17. Shtienberg, D., Bergeron, S. N., Nicholoson, A. G., Fry, W. E., and Ewing, E. E., Development and evaluation of a general model for yield loss assessment in potatoes, *Phytopathology*, 80, 466, 1990.

18. Jones, H. G., *Plants and Microclimate,* Cambridge University Press, London, 1983, chap. 2.

19. Khurana, S. C., and McLaren, J. S., The influence of leaf area, light interception and season on potato growth, *Potato Res.*, 25, 329, 1982.

20. Watson, D. J., Comparative physiological studies on the growth of field crops. I. Variation in net assimilation rate and leaf area between species and varieties, and within and between years, *Ann. Bot.*, 11, 41, 1947.

21. Watson, D. J., The dependence of net assimilation rate on leaf area index, *Ann. Bot.*, 22, 37, 1958.

22. Monteith, J. L., Climate and the efficiency of crop production in Britain, *Philos. Trans. R. Soc. London,* B281, 277, 1977.

23. Gallagher, J. N., and Biscoe, P. V., Radiation absorption, growth and yield of cereals, *J. Agric. Sci. Cambridge,* 91, 47, 1978.

24. Boote, K. J., Jones, J. W., Mishoe, J. W., and Berger, R. D., Coupling pests to crop growth simulators to predict yield reductions, *Phytopathology,* 73, 1581, 1983.

25. Johnson, K. B., Evaluation of a mechanistic model that describes potato crop losses caused by multiple pests, *Phytopathology,* 82, 363, 1992.

26. Johnson, K. B., Teng, P. S., and Radcliffe, E. B., Analysis of potato foliage losses caused by interacting infestations of early blight, Verticillium wilt, and potato leafhopper, and the relationship to yield, *J. Plant Dis. Prot.,* 94, 22, 1987.

27. Waggoner, P. E., and Berger, R. D., Defoliation, disease, and growth, *Phytopathology,* 77, 393, 1987.

28. Rotem, J., Bashi, E., and Kranz, J., Studies of crop loss in potato blight caused by *Phytophthora infestans, Plant Pathol.,* 32, 117, 1983.

29. Rotem, J., Kranz, J., and Bashi, E., Measurements of healthy and diseased haulm area for assessing late blight epidemics in potatoes, *Plant Pathol.,* 32, 109, 1983.

30. Johnson, K. B., Defoliation, disease, and growth: a reply, *Phytopathology,* 77, 1495, 1987.

31. Bergamin Filho, A., Carneiro, S. M. T. P. G., Godoy, C. V., Amorim, L., Berger, R. D., and Hau, B., Angular leaf spot of *Phaseolus* beans: relationships between disease, healthy leaf area, and yield, *Phytopathology,* 87, 506, 1997.

32. Last, F. T., Effect of powdery mildew on yield of spring-sown barley, *Plant Pathol.,* 4, 22, 1955.

33. Last, F. T., Analysis of the effects of *Erysiphe graminis* DC on the growth of barley, *Ann. Bot.,* 26, 279, 1962.

34. Carver, T. L. W., and Griffiths, E., Relationship between powdery mildew infection, green leaf area and grain yield of barley, *Ann. Appl. Biol.,* 99, 255, 1981.

35. Jenkyn, J. F., Effects of mildew (*Erysiphe graminis*) on green leaf area of Zephyr spring barley, 1973, *Ann. Appl. Biol.,* 82, 485, 1976.

36. Jenkyn, J. F., Effects of mildew on the growth and yield of spring barley: 1969–72, *Ann. Appl. Biol.,* 82, 485, 1976.

37. Lim, L. G., and Gaunt, R. E., Leaf area as a factor in disease assessment, *J. Agric. Sci. Cambridge,* 97, 481, 1981.

38. Wright, A. C., and Gaunt, R. E., Disease-yield relationships in barley. I. Yield, dry matter accumulation and yield loss models, *Plant Pathol.,* 41, 676, 1992.

39. Daamen, R. A., Assessment of the profile of powdery mildew and its damage function at low disease intensities in field experiments with winter wheat, *Neth. J. Plant Pathol.,* 95, 85, 1989.

40. Rabbinge, R., Jorritsma, I. T. M., and Schans, J., Damage components of powdery mildew in winter wheat, *Neth. J. Plant Pathol.,* 91, 235, 1985.

41. Haigh, G. R., Carver, T. L. W., Gay, A. P., and Farrar, J. F., Respiration and photosynthesis in oats exhibiting different levels of partial resistance to *Erysiphe graminis* D.C. ex Merat f. sp. *avenae* Marchal, *New Phytol.,* 119, 129, 1991.

42. Balkema-Boomstra, A. G., and Mastebroek, H. D., Effect of powdery mildew (*Erysiphe graminis* f.sp. *hordei*) on photosynthesis and grain yield of partially resistant geneotypes of spring barley (Hordeum vulgare L.), *Euphyticxa,* 126,1995.

43. Haverkort, A. J., and Bicamumpaka M., Correlation between intercepted radiation and yield of potato crops infested by *Phytophthora infestans* in central Africa, *Neth. J. Plant Pathol.,* 92, 239, 1986.

44. Van Oijen, M., Photosynthesis is not impaired in healthy tissue and blighted potato plants, *Neth. J. Plant Pathol.,* 96, 55, 1990.

45. Bastiaans, L., Effects of leaf blast on photosynthesis of rice. I. Leaf photosynthesis, *Neth. J. Plant Pathol.,* 99, 197, 1993.

46. Bastiaans, L., Ratio between virtual and visual lesion size as a measure to describe reduction in leaf photosynthesis of rice due to leaf blast, *Phytopathology,* 81, 611, 1991.

47. Ephrath, J. E., Shteinberg, D., Drieshpoun, J., Dinoor, A, and Marani, A., *Alternaria alternata* in cotton (*Gossypium hirsutum*) cv. Acala: effects on gas exchange, yield components and yield accumulation, *Neth. J. Plant Pathol.,* 95, 157, 1989.

48. Balachandran, S., Hurry, V. M., Kelly, S. E., Osmond, C. B., Robinson, S. A., Rohoszinski, J., Seaton, G. G. R., and Sims, D. A., Concepts of plant biotic stress. Some insights into the stress of virus-infected plants, from the perspective of photosynthesis, *Physiol. Plant.,* 100, 203, 1997.

49. Gent, M. P., Ferrandino, F. J., and Elmer, W. H., Effect of verticillium wilt on gas exchange of entire eggplants, *Can. J. Bot.,* 73, 557, 1995.

50. Bowden, R. L., and Rouse, D. I., Chronology of gas exchange effects and growth effects of infection by *Verticillium dahliae* in potato, *Phytopathology,* 81, 301, 1991.

51. Bowden, R. L., and Rouse, D. I., Effects *Verticillium dahliae* on gas exchange of potato, *Phytopathology,* 81, 293, 1991.

52. Haverkort, A. J., Rouse, D. I., and Turkensteen, L. J., The influence of *Verticillium dahliae* and drought on potato crop growth. I. Effects on gas exchange and stomatal behaviour of individual leaves and crop canopies, *Neth. J. Plant Pathol.,* 96, 273, 1990.

53. Lorenzini, G., Guidi, L., Nali, C., Ciompi, S., and Soldatini, G. F., Photosynthetic response of tomato plants to vascular wilt diseases, *Plant Sci.,* 124, 143, 1997.

54. Luo, Y., Teng, P. S., Fabellar, N. G., and TeBeest, D. O., A rice-leaf blast combined model for simulation of epidemics and yield loss, *Agric. Syst.,* 53, 27, 1997.

55. Hang, A. N., McCloud, D. E., Boote, K. J., and Duncan, W. G., Shade effects on growth, partitioning, and yield components of peanuts, *Crop Sci.,* 24, 109, 1984.

56. Nutter, F. W., Jr., and Littrell, R. H., Relationship between defoliation, canopy reflectance and pod yield in the peanut-late leafspot pathosytem, *Crop Prot.,* 15, 135, 1996.

57. Richardson, A. J., Wiegand, C. L., Arkin, G. F., Nixon, P. R., and Gerbermann, A. H., Remotely-sensed spectral indicators of sorghum development and their use in growth modeling, *Agric. Meteorol.,* 26, 11, 1982.

58. Wiegand, C. L., and Richardson, A. J., Leaf area, light interception, and yield estimates from spectral components analysis, *Agron. J.*, 76, 543, 1984.
59. Maas, S. J., Using satellite data to improve model estimates of crop yield, *Agron. J.*, 80, 655, 1988.
60. Gallo, K. P., and Daughtry, C. S. T., Techniques for measuring intercepted and absorbed photosynthetically active radiation in corn canopies, *Agron. J.*, 78, 752, 1986.
61. Wiegand, C. L., and Richardson, A. J., Use of spectral vegetation indices to infer leaf area, evapotranspiration and yield, *Agron. J.*, 82, 630, 1990.
62. Nilsson, H. E., Remote sensing and image analysis in plant pathology, epidemiology, crop loss assessment, phytopathometry, *Can. J. Plant Pathol.*, 17, 154, 1995.
63. Nutter, F. W., Jr., and Gaunt, R. E., Recent developments in methods for assessing disease losses in forage/pasture crops, in *Pasture and Forage Crop Pathology*, Chakraborty, S., Leath, K. T., Skipp R. A., Pederson, G. A., Bray, R. A., Latch, C. M., and Nutter, F. W., Jr., Eds., Soil Science Society of America, American Society of Agronomy, Crop Science Society of America, Madison, WI, 1996, 93.
64. Nilsson, H., and Johnsson, L., Hand-held radiometry of barley infected by barley stripe disease in a field experiment, *J. Plant Dis. Prot.*, 103, 517, 1996.
65. Nutter, F. W., Jr., Detection and measurement of plant disease gradients in peanut using a multispectral radiometer, *Phytopathology*, 79, 814, 1989.
66. Dudka, M., Langton, S., Shuler, R., Kurle, J., and Grau, C. R., Use of digital imagery to evaluate disease incidence and yield loss caused by Sclerotinia stem rot of soybeans, in *Proceedings of the Fourth International Conference on Precision Agriculture*, Robert, P. C., Rust, R. H., and Larson, W. E., Eds., American Society of Agronomy, Madison, WI, 1549, 1998.
67. Goudriaan, J., and Monteith, J. L., A mathematical function for crop growth based on light interception and leaf area expansion, *Ann. Bot.*, 66, 695, 1990.
68. Gaunt, R. E., and Bryson, R. J., Plant and crop yield potential and response to disease, *Aspects Appl. Biol.*, 42, 1, 1995.

12 Quantifying Crop Yield Response to Weed Populations: Applications and Limitations

John L. Lindquist and Stevan Z. Knezevic

CONTENTS

12.1 INTRODUCTION

Losses in crop yield and quality from the interactions between weeds and crops provide the basis for modern weed management.[1] Crop producers have relied upon herbicides since the early 1950s as the primary method for controlling weeds because they were cheap, convenient, and effective.[2] However, public concern about food safety and the environment has caused considerable debate about the impact of herbicides on agroecosystems. Excessive or inappropriate use of herbicides could be avoided if they were applied only when weed control is justified both from the biological and economic perspectives. Thus, the development of decision-support software that farmers, consultants, extension personnel, and other agronomists could use as a part of an integrated weed management (IWM) program is needed.[2]

IWM became a commonly used term in the early 1970s[3] and has since been defined in a number of ways.[4–6] Buchanan[7] described IWM as a combination of mutually supportive technologies in order to control weeds. Swanton and Weise[5] described it as a multidisciplinary approach to weed control utilizing the application of numerous alternative management practices. Management practices useful in an IWM program include tillage systems, inter-row cultivation, cover crops, biological control, competitive crops, crop rotation, and herbicide application.[4, 5] Concepts important in IWM include the critical period of weed interference, weed population biology, economic thresholds, and crop loss assessment. Because there are many weed species, each exhibiting unique population dynamics, a single IWM program will not be appropriate for all weed problems.[4] However, if IWM concepts are implemented in a systematic manner, significant advances in weed management can be achieved.[5] The number of concepts and management strategies involved in IWM present the grower with a large volume of information to evaluate before reaching a weed management decision.[8] Computers are well suited for evaluating this information and, therefore, may facilitate weed management decision making.

A basic requirement for decision making in weed management would be models for crop loss assessment that are based on observations of weed infestations early in the growing season.[9] Quantifying interactions among plants is complicated and continues to be an area of debate, particularly among ecologists. Much of the debate can be alleviated if the research objectives are clear.[10] There seem to be two main goals of conducting crop–weed interference research. The first is to quantify the effects of commonly observed weed populations on the yield of conventionally managed crops. For example, 20 years ago, Zimdahl[11] reported more than 500 citations of research conducted primarily with this goal in mind. However, relatively few of those citations provided data useful for quantifying crop loss across a range of weed infestations—a necessity for weed management decision support. The second goal is to understand the mechanisms of competition so that variation in observed effects of weeds on crop yield can be quantified.

In quantifying the effects of a weed species on crop yield, most researchers focus on crop yield of an individual or of a population as the dependent variable, and on weed (and sometimes crop) population density as the independent variable. In this chapter, we provide a brief overview of (1) the theory for using weed population as a predictor of crop yield reduction (loss), (2) how crop yield in relation to weed density is commonly used in weed management decision making, and (3) the limitations of these approaches for weed management. Detailed analysis of the causes (mechanisms) of these effects is discussed in Chapter 13.

12.2 CONCEPTS FROM POPULATION BIOLOGY

Because population density will be used as a predictor of crop-weed interactions, it is useful to look to population biology for some of the principles that form the theoretical foundation for quantifying crop-weed interference. A population is a group of individuals of the same species that have a high probability of interacting with each other.[12] The variable of interest in population biology is the number of individuals, and the goal is to understand and predict the dynamics of populations. Hastings[12] provided a very concise summary of the principles behind density-independent and dependent growth, part of which is summarized here.

12.2.1 DENSITY-INDEPENDENT GROWTH

The simplest model of population growth is one where we assume density independence. Density independence means that growth does not depend on the number of individuals; rather, per capita growth is constant regardless of the size of the population. In other words, we assume that the rate of births and deaths is proportional to the number of individuals present.

In modeling population growth, one could take a discrete or a continuous time approach. The discrete time models assume that each generation is unique (no overlap), whereas the continuous time model accounts for overlap. Using the continuous time model means we can count the population size at all times. For simplicity, and because we will later be interested in growth in biomass over time (a continuous variable), we will focus on the continuous time model.

As mentioned, birth and death rates are proportional to population density (N), so if b = per capita birth rate and m = per capita death rate, then rate of change in population density (dN/dt) can be written using:

$$\frac{dN}{dt} = bN - mN \qquad [12.1]$$

which we rewrite as:

$$\frac{dN}{dt} = rN$$

where $r = b - m$ = the intrinsic rate of increase of the population. If we separate N from t in Equation 12.1, integrate over time, and solve for N_t, we obtain:[12]

$$N_t = N_0 \exp(rt) \qquad [12.2]$$

where N_t and N_0 are population size at time t and $t = 0$, respectively. Note that our assumptions about the independence of birth and death rates are critical in evaluating the utility of this model. If we allow exponential growth to occur for any substantial length of time, we will soon calculate absurdly large population densities. Hastings[12] pointed out that the fundamental question of population ecology was to determine the causes and consequences of the deviation from exponential growth. Or simply stated, what regulates populations?

12.2.2 DENSITY-DEPENDENT GROWTH

Ignoring for the moment the mechanisms of population regulation, it is probably safe to assume that birth and death rates do depend on the number of individuals in the population. Any farmer who has experimented with optimizing crop population knows that size of the seed head on each plant differs depending on seeding rate. Therefore, we need to make an assumption about the relationship between the rate of growth per unit density. The simplest approach is to assume that the relationship between per capita growth rate ($f(N)$) and population size is linear (Figure 12.1):

$$f(N) = \frac{dN}{Ndt} = r\left(1 - \frac{N}{K}\right)$$
[12.3]

From Figure 12.1 we see that when $N = 0, f(0) = r =$ intrinsic rate of increase of the population, and when $N = K, f(N) = 0$. Note from Equation 12.2 that when $r = 0$, the population remains constant. Hence, when $N = K$ and $f(N) = 0$, the population no longer increases in size. Therefore, K defines the greatest number of individuals the environment can sustain (the carrying capacity).

Equation 12.3 is a simple logistic model, which can be solved explicitly for N_t. First, rate of change in the population can be written as:

$$\frac{dN}{dt} = rN\left(1 - \frac{N}{K}\right)$$
[12.4]

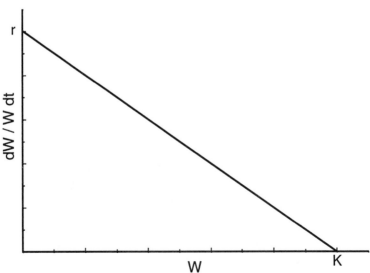

FIGURE 12.1 Per capita growth rate (f(N) = dN/Ndt) of a population as a function of population density (N).

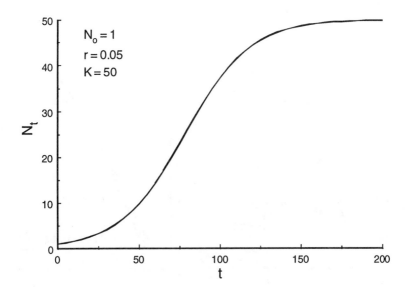

FIGURE 12.2   Logistic growth of the number of individuals in a population over time.

$$N_t = \frac{N_0 e^{rt}}{1 + \dfrac{N_0(e^{rt} - 1)}{K}} \qquad [12.5]$$

and using similar mathematical arguments used to derive Equation 12.2, we can calculate population size at any time using the equivalents listed in Figure 12.2.

Equation 12.5 results in a symmetric sigmoid type relationship as depicted in Figure 12.2. It is clear from this relationship that as population size increases, either the rate of births decreases or the rate of deaths increases. In either case, dN/Ndt decreases.

As agronomists interested in yield of our crop within a specific growing season, we are not particularly interested in the effects of competition on birth and death rates (except those that influence grain yield), but on the rates of growth and senescence of plants within a growing season. In the next section, we use many of the concepts discussed above to explore the effects of competition on growth and yield.

12.3   YIELD-DENSITY RELATIONSHIPS

Justification for weed–crop interference experimental designs has long been based on the observed relationships between yield and density of a single species. Two kinds of relationship have generally been observed,[13] the asymptotic and the parabolic (Figure 12.3). These relationships are observed because intraspecific competition (competition between neighboring individuals of the same species) becomes more intense as density increases. The decline in grain yield observed at high densities (Figure 12.3b) is typically the result of density-dependent floral abortion, or

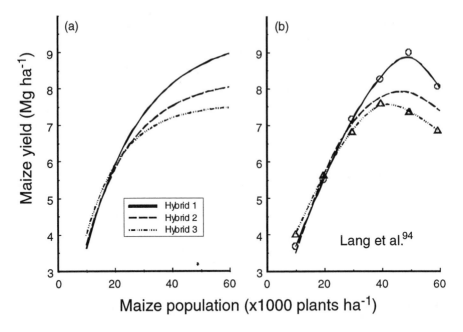

FIGURE 12.3 Examples of the asymptotic (a) and parabolic (b) yield–density relationship for maize. (Data in (a) are hypothetical; those in (b) are redrawn from Lang et al.[95])

barrenness. The observed relationship between total biomass per unit area and density is rarely parabolic,[13] instead reaching some asymptote defined by the limitations (carrying capacity) of the environment.

Willey and Heath[13] provided a review of the functional forms used through the 1960s to quantify yield–density relationships. They concluded that the reciprocal equations were best suited for explaining these relationships because they are the only type of equations that could best explain both forms of the yield–density relationship and because their parameters have some biological meaning. Remarkably, little theoretical research has been conducted over the past 40 years to improve upon these equations.

Shinozaki and Kira[14] derived the reciprocal equations to explain the yield density relationships. They made three very important assumptions. First, the growth of a plant can be described using a general logistic curve:

$$\frac{dw}{wdt} = \lambda \left(1 - \frac{w}{W_{max}}\right)$$ [12.6]

where λ is the intrinsic rate of increase in biomass, w is biomass of an individual plant at time t, and W_{max} is the maximum attainable biomass of an individual plant. Note that the form of Equation 12.6 is identical to that of Equation 12.3. Following similar arguments to those used to derive the density-dependent growth equation, we can obtain:

$$w = \frac{W_{max}}{1 + c \exp(-\lambda t)}$$ [12.7]

where c is an integration constant. This equation is of a different form than Equation 12.5, but is virtually identical in its shape. Both W_{max} and λ were assumed constant and independent of time, and λ independent of density. Similar to the assumption made in Section 12.2.2, we explicitly assume that the relationship between per unit biomass rate of growth (dW/Wdt) and biomass of the individual is linear. To our knowledge, research designed to test this assumption has not been conducted.

The second assumption of Shinozaki and Kira[14] was critical. To this point we have made arguments about the growth of a population (numbers of individuals) or of the biomass of an individual plant. The elegance of their analysis was in the method used to describe crop yield (Y = biomass per unit area) in relation to population density. They assumed that final yield per unit area (Y) of a plant in monoculture is constant and independent of density:

$$Y = W_{max} N \qquad [12.8]$$

where N is density. This has become known as the law of constant final yield. Equation 12.8 implies that, at the time a plant reaches W_{max}, the quantity of biomass in an area does not depend upon the number of individuals. This raises an issue of scale in crop–weed interference studies. If we place a single maize (*Zea mays*) plant in a one ha field, it clearly will not produce as much biomass as a one ha field with 75,000 maize plants. Therefore, we must be careful when measuring and reporting population density to use units that are relevant to an individual of the species of interest. Densities used in crop–weed interference research are commonly reported in units of plants m^{-2}, which is probably close to a relevant unit for many row crop species. In other words, one maize plant in a one m^{-2} area may produce as much biomass as ten maize plants m^{-2}, thus satisfying the law of constant final yield.

The third assumption made by Shinozaki and Kira[14] was that all plants are seeded simultaneously at $t = 0$ and average seed weight is constant and independent of density. If plants were of different ages or had different initial growth rates owing to variation in seed size, emergence time, or microenvironment, it is likely that initially large individuals will continue to dominate smaller individuals in the population.[15, 16] Effects of this hierarchical structure of plant size on competition within populations have been a topic of interest in the ecological literature (e.g., Pacala and Weiner[17]), but are beyond the scope of this discussion.

Based on these three important assumptions, Shinozaki and Kira[14] solved Equation 12.7 for w when $t = 0$:

$$w_o = \frac{W_{max}}{1 + c} = \frac{\dfrac{Y}{N}}{1 + c}$$

and from this we obtain:

$$c = \frac{Y}{w_o N} - 1$$

Recall that c is a constant, so by substituting it back into Equation 12.7, the reciprocal yield equation can be derived:

$$\frac{1}{w} = a + bN \qquad\qquad [12.9]$$

where:

$$a = \frac{\exp(-\lambda t)}{w_o}$$

and

$$b = \frac{1 - \exp(-\lambda t)}{Y}$$

Equation 12.9 shows that the relationship between the reciprocal of per plant yield and population density is linear. This form of the reciprocal equation only explains the asymptotic relationship between yield and density in monoculture (Figure 12.3a). However, Kira et al.[18] argued that the parabolic relationship could be explained using additional assumptions about plant allometry. Bleasdale and Nelder[19] and Farazdaghi and Harris[20] developed alternative forms of Equation 12.9 that can be used to explain both asymptotic and parabolic relationships.[13]

Shinozaki and Kira[14] conducted further analyses where they relaxed the assumptions that λ and W_{max} are independent of time. They showed that even when λ or W_{max} were allowed to vary with time, the simple reciprocal yield equation (Equation 12.9) can be derived. However, under such conditions, a and b are redefined so that λ or W_{max} varies with time.

Equation 12.9 can be solved for yield (Y) in order to quantify yield per unit area as a function of plant density (N):

$$Y = \frac{N}{a + bN} \qquad\qquad [12.10]$$

where a and b are identical to that described above. From Equation 12.10 it can be seen that as density increases, Y approaches a value of $1/b$. Willey and Heath[13] argued that $1/b$ is a measure of the potential of a given environment. It can also be seen that if b is small, yield per plant approaches $1/a$ at a density of 1.0. At densities less than 1.0, yield per plant becomes very small. Willey and Heath[13] point out that this is somewhat unrealistic because yield per plant levels out at densities too low for interplant competition to occur. However, if $1/a$ provides an estimate of yield per plant in a competition-free situation, then a may be an indicator of the genetic potential of that genotype. Therefore, a and b have some biological meaning attributed to them. Yet, because genotypes and environments vary, estimates of these empirical coefficients also may vary. This will have important consequences for the utility of empirical interference relationships shown in the following section.

12.4 QUANTIFYING INTERFERENCE EFFECTS

Replacement series (RS) experiments were one of the first experimental designs used to study the effects of interspecific interference. In the RS design, the relative proportions of two competitors are varied while maintaining their combined density constant.[10, 21] It has been argued that this approach confounds the effects of density vs. proportion (e.g., Roush et al.[22]). Spitters[23] argued that use of the RS design for intercropping studies was inappropriate because it does not allow the experimenter to identify a density combination that optimizes intercrop yield.

Spitters[23] proposed an alternative approach using arguments based on the reciprocal yield relationship. He argued that plants compete for a number of potentially limiting growth factors (e.g., light, water, nutrients), and that biomass production is approximately linear in relation to uptake of the resource that is most limiting. Hence, the ability of each plant to obtain that resource is reflected in its biomass. Therefore, if an environment contains a constant supply of limiting resources, the yield of an individual is expected to decrease with increasing numbers of individuals (density) and the relationship between yield per plant and density can be explained using the reciprocal yield equation (Equation 12.9). The parameters a and b represent the reciprocal of the biomass of an isolated plant and how biomass per plant decreases with the addition of each additional plant, respectively. The ratio b/a represents the decrease in biomass per plant relative to its value without competition and is, therefore, a measure of intraspecific competition.

It is important to note that use of the reciprocal yield equation (Equation 12.9) requires the assumption that the effect on plant yield of adding plants is additive. Spitters[23] argued that if adding plants of the same species has an additive effect on $1/w$, then it is reasonable to assume that adding plants of a different species has an additive effect on $1/w$. Therefore, Equation 12.9 can be rewritten to account for the effects of a second species on the biomass of an individual of the first:

$$\frac{1}{w_1} = a_{1,0} + b_{1,1}N_1 + b_{1,2}N_2 \qquad [12.11]$$

where the first subscript represents the species of interest, and the second subscript represents another species competing with the first. The term $b_{1,2}$ is a measure of interspecific competition. From Equation 12.11 it can be seen that adding one plant of species 1 has an equal effect on $1/w_1$ as adding $b_{1,1}/b_{1,2}$ plants of species 2.[23] Thus, the effect of species 2 on species 1 is directly proportional to the effect of species 1 on species 1.

Spitters et al.[24] expanded on earlier work with intercrops[23] by showing that biomass of the crop can be related to crop and weed density:

$$\frac{1}{w_c} = a_{c,o} + b_{c,c}N_c + b_{c,w}N_w \qquad [12.12]$$

where the subscripts c and w represent crop and weed, respectively. Crop yield per unit area (Y_c) can be derived from Equation 12.12 to obtain:

$$Y_c = \frac{N_c}{a_{c,o} + b_{c,c} N_c + b_{c,w} N_w}$$

[12.13]

In many cases we are interested in quantifying the effect of adding weeds on the yield of a uniformly managed crop. Hence, crop–weed interference studies are commonly designed such that crop density is held constant, while weed density is varied. Spitters et al.[24] argued that, when using these additive series[25] or partial additive[26] designs, the reciprocal yield equation can be written solely as a function of weed density:

$$\frac{1}{w_c} = a_o + b_w N_w \text{ and } Y_c = \frac{N_c}{a_o + b_w N_w}$$

[12.14]

—where $a_o = a_{c,o} + b_{c,c} N_c$. Because we are essentially interested in the proportional reduction in yield resulting from weed interference (yield loss, Y_L):

$$Y_L = 1 - \frac{Y_c}{Y_{wf}}$$

[12,15]

where Y_{wf} is weed-free crop yield (N_c/a_o). Incorporating Equation 12.14 into Equation 12.15 results in:

$$Y_L = \frac{b_w N_w}{a_o + b_w N_w}$$

[12.16]

Equation 12.16 is a rectangular hyperbola where $b_w/(a_o + b_w)$ represents the fractional yield loss caused by the first weed plant added to the population, and the upper asymptote is forced to a value of 1.0.[24]

Using similar arguments to those of Spitters,[23] Firbank and Watkinson[27] proposed that in two-species mixtures, mean yield of each species is dependent upon the relative frequencies of the two species and upon overall density. They suggested using the following to quantify per-plant yield in response to density:

$$w = \frac{W_{max}}{(1 + c N)^d}$$

[12.17]

where W_{max} is maximum per plant yield (yield of an isolated plant), c is the area required to achieve a yield of W_{max}, and d describes the efficiency of resource utilization of the population. Inversion of this equation results in:

$$\frac{1}{w} = \frac{(1 + c N)^d}{W_m}$$

which, if d is equal to 1, then the law of constant final yield holds (Equation 12.8) and we obtain the reciprocal yield equation:

$$\frac{1}{w} = \frac{1}{W_m} + \frac{cN}{W_m}$$

where $1/W_m = a$ and $c/W_m = b$. Equation 12.17 was then expanded to include multi-species interactions:

$$w_c = \frac{W_{\text{max},c}}{(1 + c_c (N_c + \alpha N_w))^d}$$

where subscripts represent species and α is the competition coefficient. If the law of constant final yield holds, the value of d is 1.0, and:

$$Y_c = \frac{Y_{wf}}{1 + \dfrac{\alpha \cdot N_w}{N_c}}$$

where weed-free yield $(Y_{wf}) = W_{\text{max}}$, N_c/c_c. From this, proportional yield loss is

$$Y_L = \frac{\alpha N_w}{N_c + \alpha N_w} \qquad [12.18]$$

Equation 12.18 is a rectangular hyperbola where $\alpha/(N_c + \alpha)$ represents fractional yield loss resulting from the first weed added to the population and, like Equation 12.16, the upper asymptote is 1.0. Forcing the asymptote to a value of 1.0 (i.e., yield loss must approach 100% at very large weed density) is a problem with these equations because rarely are 100% yield losses observed in field experiments. Figure 12.4 provides an example where Equations 12.16 and 12.18 were fit to data obtained in a maize-velvetleaf, *Abutilon theophrasti*, mixture experiment.

Cousens[28] used similar concepts, but took a completely different approach to quantifying the effects of weeds on crop yield. His primary interest was to look at proportional yield loss, and he argued that when no weeds were present, there can be no yield reduction from weeds. Similar to the approaches discussed above,[23, 24, 27] he assumed that the effects of weeds are additive at low weed densities. In other words, yield loss (Y_L) as weed density approaches zero is linear:

$$Y_L = IN$$

He then assumed that yield loss can never exceed 100%, but typically approaches some asymptote below 100%. As weed density increases, intraspecific competition among weed plants reduces the effect of each weed on crop yield. Cousens[28] assumed

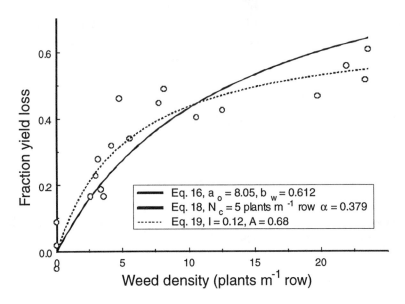

FIGURE 12.4 Maize yield loss in relation to velvetleaf (*Abutilon theophrasti*) population density. Data taken from Lindquist et al.[33]

that this reduction in the interspecific competitive effect is a linear function of weed density $(1+sN)$, so that:

$$Y_L = \frac{IN}{1 + sN}$$

where s defines the intraspecific competition among individuals of the weed species. As weed density approaches infinity, yield loss approaches I/s, which Cousens[28] defined as A. Hence, $s = I/A$, and:

$$Y_L = \frac{IN}{1 + \dfrac{IN}{A}} \qquad\qquad [12.19]$$

However, we can only measure yield, not yield loss. Because we know the shape of the expected yield loss–weed density relationship, then yield (Y) is equal to the product of yield in a weed-free situation (Y_{wf}) and $1-Y_L$ (where Y_L is the proportional yield loss in mixtures). Therefore, yield can be calculated using:

$$Y = Y_{wf}\left(1 - \frac{IN}{1 + \dfrac{IN}{A}}\right) \qquad\qquad [12.20]$$

Equations 12.19 and 12.20 are currently among the most commonly used equations in the analysis of crop–weed interference.[29–36]

Recall that one of the critical assumptions made by Shinozaki and Kira[14] was that all plants were seeded simultaneously at $t = 0$ and average seed weight was constant and independent of density. This assumption was consistent in all the analyses discussed above. In weed–crop mixtures, this assumption commonly fails because weeds emerge before or following crop emergence. Cousens et al.[37] expanded upon his earlier analysis to account for variation in the relative time (T) of crop and weed emergence (where T is negative if the weed emerges first and positive if the crop emerges first). They assumed that the value of I declines exponentially with T:

$$I = g \exp(-hT) \qquad [12.21]$$

where g is the value of I when $T = 0$ and h is the rate at which I decreases toward zero as T increases. Incorporation of Equation 12.21 into Equation 12.19 results in:

$$Y_L = \frac{gN}{\exp(hT) + \dfrac{gN}{A}} \qquad [12.22]$$

The relationship between Y_L and T for a given weed density is sigmoidal. Cousens et al.[37] suggested that the exponential relation between I and T may still be problematic because I will approach infinity as T becomes large and negative. They further suggested that a sigmoidal relationship between I and T may be more appropriate and presented two additional equations to account for this alternative approach. However, the simpler Equation 12.22 has been used in a number of studies to effectively quantify yield loss as a function of both density and relative time of emergence of weeds in maize,[31, 38] sorghum, *Sorghum bicolor*,[32] soybean, *Glycine max*,[35, 39] white beans, *Phaseolus vulgaris*,[36] and rice, *Oryza sativa*.[40]

It seems reasonable that if we had knowledge of how specific resources or management influenced the relative effects of weeds on crop yield, we could use Equation 12.22 to quantify crop yield loss in relation to weed density and the relative supply of that resource or management. For example, if we knew that a crop becomes more competitive relative to a weed as fertilizer nitrogen supply increases, then it may be reasonable to assume that the value of I decreases exponentially with increasing nitrogen supply. In this case, we could substitute nitrogen supply for T, and use Equation 12.22 to quantify yield loss in relation to weed density and nitrogen supply. Alternatively, it is expected that sublethal herbicide doses can reduce weed competitiveness. It is possible that this reduction is similar in effect to late emergence, in which case Equation 12.22 also could be used to relate crop yield loss to weed density and herbicide dose.

An alternative approach to dealing with the potential problem of varying density and time of emergence was proposed by Kropff and Spitters.[41] They argued that the competitive strength of a species is determined by its share in leaf area at the moment when the canopy closes and interspecific competition begins. Therefore, they proposed a method of weighting species density with the average leaf area of individual plants at the time of observation. The product of density and leaf area is, of course, the leaf area index (LAI). Incorporating the calculation of LAI into a yield loss equation derived from Equation 12.13, they presented an equation that expresses yield loss as a function of weed and crop LAI:

$$Y_L = \frac{qL_w}{1 + (q - 1)L_w}$$ [12.23]

where L_w is the relative leaf area of the weed species $(LAI_w/(LAI_c + LAI_w))$ and q is a relative damage coefficient, or a measure of the competitiveness of the weed with respect to the crop.[42] Because this equation depends on the development of leaf area before canopy closure, the value of q, theoretically, will change with the period between crop emergence and the time of observation of relative leaf area of the weed. Assuming that leaf area growth during this period is exponential, the value of q will change according to:

$$q = q_o \exp(RGRL_c - RGRL_w)t$$ [12.24]

where $RGRL_c$ and $RGRL_w$ are relative growth rates of crop and weed leaf area, respectively, t is time (expressed in degree-days, °C day^{-1}), and q_o is the value of q when L_w is observed at time $t = 0$ (time of observation for which q has been determined from experimental data). Unfortunately, Equation 12.23 has the same problem as Equations 12.16 and 12.18 in that the upper asymptote is

$$Y_L = \frac{qL_w}{1 + \left(\dfrac{q}{m} - 1\right)L_w}$$ [12.25]

forced to a value of 1.0. Therefore, Kropff et al.[43] modified the model to include a parameter to define maximum relative crop yield loss (m).

The potential advantage of Equation 12.25 is that it accounts for crop and weed density as well as the early leaf area growth of each species, which will depend upon relative time of emergence. Problems with this approach could occur if crop or weed growth as a function of degree-days is not exponential from emergence until canopy closure, or if species density or environmental factors modify exponential growth during this period.

12.5   USE OF CROP–WEED INTERFERENCE RELATIONSHIPS FOR MAKING WEED MANAGEMENT DECISIONS

Herbicides effectively reduce weed interference. However, cost, concern over the development of resistant weed populations, and environmental issues associated with specific herbicide chemistry require that herbicides be used judiciously. Use of bioeconomic models to aid growers in making cost-effective weed management decisions has been hailed as a method to achieve both reduced herbicide application and optimum economic return.[44–46] Bioeconomic models link the concepts of herbicide selection based on efficacy, cost of management, and crop–weed interference.[8]

Early development of decision support models focused on herbicide efficacy, often with consideration of treatment cost, crop rotation, and soil properties.[8] Efficacy-based models were developed to incorporate the large amounts of information commonly reported in herbicide use guides published by most agricultural universities. These models were typically designed to present herbicide treatments useful for a particular production system, and did not include any assessment of costs and benefits associated with the management options.

As agricultural economists became interested in the increasing costs of weed management, "bioeconomic models" were developed. These models take the earlier herbicide efficacy-based models a step further by recommending a herbicide treatment (or more broadly, any management tactic) only when economics justify it. The ultimate goal of the bioeconomic model is to maximize profit.[35] Decision rules in bioeconomic models are commonly based on a damage function and a profit function.[35, 47–49] The damage function is used to quantify the effects of a weed population on crop yield and is defined using any of the approaches described in Section 12.4. The profit function links cost of management with its effect on the weed population.

The bioeconomic model HERB was developed in 1986 to make economic calculations necessary for herbicide choice decisions in soybean.[47, 50] The WeedSOFT model[51] is a modified version of HERB that has become popular among crop consultants in Nebraska, and is currently being expanded for use across much of the maize and soybean production regions of the U.S. The profit function used to calculate net gain (G_n) in WeedSoft is

$$G_n = Y_{wf} P(Y_{L,o} - Y_{L,m}) - C \qquad [12.26]$$

where Y_{wf} is weed-free crop yield (or the yield goal), P is price obtained for the crop, $Y_{L,o}$ and $Y_{L,m}$ are the proportional loss in yield from a given weed density in the absence of and after management has removed $E_f N$ weeds (where E_f is efficacy of the management tactic, or proportion of weeds killed), respectively. C is the total cost of implementing the management tactic. Proportional yield reduction within WeedSoft is calculated using a function similar to Equation 12.19 with adjustments for the relative growth stages of the weed and crop at the time of management.[51] The model calculates net gain (G_n) for all management tactics available for controlling the weed species observed within the crop being planted. Potential tactics are listed in order of greatest to least economic gain, and the manager selects the tactic that best fits the operation.

Ultimately, Equation 12.26 relies on threshold theory. The concept of a threshold was used first in the fields of entomology and plant pathology.[52] Based on concern about insecticide resistance, nontarget effects and residue problems, Stern et al.[53] proposed an economic threshold as a "rational" basis for using insecticides. Economic thresholds were traditionally defined as "the level at which damage can no longer be tolerated and, therefore, the level at or before which it is desirable to initiate deliberate control activities."[52] Cousens[54] defined an economic threshold for weed management as "the weed density at which the cost of weed control equals the increased return on yield in the current year." Because they account for crop losses only in the current cropping season, economic thresholds are single-year measures of weed effects.[55]

Managing weeds according to the threshold concept means leaving some weeds unmanaged in the field. The number of weeds to remain in the field before a management tactic is required is the economic threshold weed density. Using the definition of Cousens,[54] the single-year economic threshold (T_e) is the weed density at which C in Equation 12.26 (cost) is equal to $Y_{wf}P(Y_{L,o} - Y_{L,m})$ (increased return from management). Therefore:

$$T_e = \frac{C}{Y_{wf}P(Y_{L,o} - Y_{L,m})} \qquad [12.27]$$

where C, Y_{wf}, P, $Y_{L,o}$, and $Y_{L,m}$ are defined as above. Substitution of Equation 12.19 into Equation 12.27 and rearrangement results in a quadratic equation:

$$0 = (1 - E_f)\left(T_e \frac{I}{A}\right)2 + \left(2 - E_f - Y_{wf}PA \frac{E_f}{C}\right)\left(T_e \frac{I}{A}\right) + 1$$

which can be solved algebraically for T_e.[54, 56] We can also substitute Equation 12.22 into Equation 12.27 to obtain:

$$0 = C((\exp(hT)A + gT_e)(\exp(hT)A + gT_e - gT_eE_f)) - gPY_{wf}gT_eE_f\exp(hT)A^2$$

which also can be solved algebraically for T_e. A similar approach could be used to identify an economic threshold relative leaf area (L_w) based on yield loss calculated from Equation 12.23.

12.6   LIMITATIONS OF EMPIRICAL CROP–WEED INTERFERENCE DATA FOR DECISION MAKING

There are a number of problems with the threshold concept that constrain its utility and the implementation of decision support models.[8, 54, 57–60] Biological constraints include: (1) concern about the future cost of allowing unmanaged weeds to produce seed, (2) the effects of management (especially herbicide) on growth, reproduction, and competitive ability of surviving weed plants, (3) limitations in data and methodologies needed to include the effects of multiple weed species, (4) the instability of crop–weed interference relationships, (5) field scouting to quantify weed densities, and (6) variation in spatial distribution of weeds within crop fields.[8]

12.6.1   SEEDS PRODUCED BY UNMANAGED WEEDS

A major opposition in accepting the threshold concept by farmers is the concern about the seeds that may have been produced by uncontrolled weeds.[8] Some weed scientists have long argued that the only feasible economic threshold weed density is zero (e.g., Norris[61]). A primary reason for this is a concern about the future cost of allowing even a few surviving weeds to produce seed. Norris[61] suggested that one barnyardgrass (*Echinochloa crus-galli*) plant per hectare is capable of reinfesting one hectare, assuming uniform distribution of seeds, at a density that would lead to a 5% sugarbeet (*Beta vulgaris*) yield loss if not controlled. Because of physical dormancy and highly plastic reproduction in velvetleaf, Sattin et al.[62] concluded that a zero threshold approach should be used for velvetleaf management in maize. These authors argue that even if current year economics do not justify treatment, the risk of having a weed population increase is unacceptable.

The approach to calculating economic thresholds presented in Section 12.5 does not include information on the reproductive output of weeds at or below the calculated threshold density,[61] or on other characteristics of the population dynamics of the weed. To adjust for this, Cousens[54] proposed that an economic optimum threshold (EOT) approach be adopted. The EOT weed density is determined by optimizing expected net returns over multiple years of simulated weed population dynamics, weed–crop interference, and weed management.[44] Several authors have shown that the single-year economic threshold is three to ten times greater than the EOT density.[44, 55, 63, 64] Lindquist et al.[44] determined that the EOT for velvetleaf in a maize–soybean rotation was 0.025 seedlings m^{-2}, which is 4 to 40 times smaller than the single year economic thresholds calculated by Lindquist et al.[65] However, by linking an EOT strategy with alternative management practices that reduce velvetleaf seed production by 80%, annualized economic return was increased by as much as 12% and the number of years that control was required could be reduced by 25%.[44] Teasdale[66] found that increasing the maize population from 60,000 to 90,000 plants ha^{-1} reduced velvetleaf seed production by 69 to 94% when velvetleaf emerged with the maize. Therefore, use of the EOT strategy to reduce herbicide application and increase long-term economic return for the grower appears promising. However, few decision support tools currently incorporate multi-year economics or the EOT approach in their decision rules. Improvements in this direction are needed.

12.6.2 EFFECTS OF MANAGEMENT ON WEED COMPETITIVENESS AND SEED PRODUCTION

Definition of efficacy (E_f) is a problem in weed management decision support. Most weed scientists evaluate efficacy of a management tactic by visually comparing weed pressure in a management treatment with that in an untreated control. This definition includes weed mortality and reduction in leaf area and biomass of the weed population.[67] The efficacy term used in Equation 12.26 only refers to proportion of plants killed. No account is taken of the effects of management on competitiveness and seed production of surviving weeds. In other words, it is assumed that weeds surviving management are equally competitive as those that were never exposed to management.

Weed growth and competitive ability are influenced by cultivation and herbicide. Steckel et al.[68] showed that cultivation resulted in up to 85% mortality, but also a 75%

reduction in seeds produced per surviving velvetleaf plant. Buhleret al.[69] showed that cultivation resulted in 35 to 59% mortality, and 11 to 28% reduction in biomass of surviving common cocklebur (*Xanthium strumarium*) plants. Cultivation plus herbicide resulted in 58 to 85% mortality and a 61 to 91% common cocklebur biomass reduction. Schmenk and Kells[70] showed that maize yield loss resulting from nine velvetleaf plants m^{-1} row was reduced from 30 to 35% to 10 to 15% when those nine weeds were survivors of the soil-applied herbicides atrazine or pendamethalin. Weaver[71] reported on a study to determine whether soybean yield loss varies for weeds that escaped the soil-applied herbicide metribuzin. She quantified soybean yield loss in response to the density of three weed species using Equation 12.19 and showed that weeds surviving metribuzin caused substantially smaller yield loss than unmanaged weeds. Although results from these studies show that management affects weed growth and competitive ability, they provide little information about the mechanisms of this effect.

Weed growth and competitive ability also are influenced by cultural practices. Lindquist and Mortensen[72] showed that four maize hybrids differed in their ability to tolerate competition from and suppress velvetleaf seed production. Traits that conferred tolerance and suppressive ability included optimum leaf area index, rate of canopy closure, and the height at which the greatest leaf area density occurs in the canopy.[73] These traits can be modified by varying cultural practices such as row spacing and population density, or through breeding. Tollenaar and Aguilera[74] showed that a modern hybrid (Pioneer 3902) had higher vertical leaf area distribution than an old hybrid (Pride 5), and that increasing maize density can result in leaves being distributed even higher in the canopy. In a follow-up study with the same hybrids, Tollenaar et al.[75] showed that 3902 also had greatest canopy LAI and its yield loss was reduced from 26% to 13%[76] under low vs. high maize density treatments. McLachlan et al.[77] showed that redroot pigweed, *Amaranthus retroflexus,* biomass was reduced by 89% under high vs. low maize density treatments, and Murphy et al.[78] showed that increased maize density and narrow row spacing reduced the biomass of late emerging weeds by up to 41%, and changing row spacing from 0.76 m to 0.5 m reduced yield loss from an average of 15 to 2%. Differences in yield or weed response in these studies were attributed to the influence of maize density and/or row spacing on weed-free maize LAI and its influence on photosynthetic photon flux density (PPFD) interception. Dingkuhn et al.[79] conducted crosses between a low-yielding African rice, *Oryza glaberrima,* and improved *O. sativa* tropical–japonica rices and identified traits that could result in improved rice yield potential and strong competitors against weeds.

Further research is needed to quantify the effects of weed management on growth, competitiveness, and seed production of weeds exposed to various management tactics. Of particular value would be data that show whether the response is due to (1) competition (and for which resource), (2) destruction of existing leaf tissue but not of the growing point, (3) a delay in development due to a temporary reduction in some metabolic process, or (4) a permanent reduction in some metabolic process that subsequently influences resource utilization. An understanding of how management influences weed growth, for how long, and what environmental factors influence the effect would help in the development of methods for adjusting crop–weed interference and weed seed production relationships in weed management decision support programs.

12.6.3   EFFECTS OF MULTIPLE WEED SPECIES

Most research on crop–weed interference is focused on the effects of a single weed species on crop yield. Decision-support models are only useful if they account for multiple weed species. A method of incorporating multi-species weed densities into Equation 12.19 has been utilized for a few weed species.[39, 80] However, estimation of the I and A coefficients using this method requires data from multi-species weed–crop interference research, which are not currently available for most species mixtures.

Assuming an additive effect of all weed species on crop yield reduction, Berti and Zanin[81] proposed a method to predict crop yield loss from multi-species weed infestations by transforming mean density of each species into a density equivalent (N_{eq}). This method adjusts actual mean weed density based upon the relative competitive effect of each species on crop yield. To obtain density equivalent, we first rewrite Equation 12.19 using:

$$YL_N = \frac{ACN}{1 + CN}$$

[12.28]

where N is mean weed density for a field, A represents YL_N as $N \to \infty$ and C is the ratio of I (dYL_N/dN as $N \to 0$) to A.[82] A hypothetical weed species with arbitrarily set values of A and C coefficients (redefined as A_{eq} and C_{eq}) is defined. Therefore, crop yield loss is redefined as:

$$YL_{N,eq} = \frac{A_{eq}C_{eq}N_{eq}}{1 + C_{eq}N_{eq}}$$

[12.29]

where N_{eq} is the adjusted density. Setting Equations 12.28 and 12.29 equal and isolating N_{eq} yields:

$$N_{eq} = \frac{ACN}{A_{eq}C_{eq} + C_{eq}CN(A_{eq} - A)}$$

[12.30]

Density equivalent is therefore obtained for any weed species based upon species-specific values of A, C, and N. Benefits of this method are that A and C can be obtained from the more common two-species mixture experiments, and that density equivalents are additive and a single equation can be used to describe the impact of all species present in the mixture:

$$YL_{n,j} = \frac{A_{eq} \sum_{i=1}^{j} N_{eq,j}}{1 + C_{eq} \sum_{i=1}^{j} N_{eq,j}}$$

[12.31]

where A_{eq} and C_{eq} are analogous to A and C in equation 28, but are constant for all $N_{eq,i}$, the subscript i is a species identifier, and j is the total number of species present.

Although this is a relatively convenient approach for incorporating the effects of multiple weed species on crop yield, it is limited by at least two factors. First, there remains some question as to whether the densities of multiple weed species are always additive. Second, estimates of A and C (I/A) are not available for every weed species in every crop. Moreover, several researchers have shown that estimates of A and C for a single weed vary among environments, even among years within the same field.[30–36, 83]

12.6.4 INSTABILITY OF CROP–WEED INTERFERENCE RELATIONSHIPS

If the relationship between measured yield loss and weed density varies among years and locations, then T_e will also vary. Estimates of I and A obtained for several weed species in mixture with maize have been shown to vary greatly across the north central U.S. and Canada.[30–36, 83] Lindquist et al.[33, 34] found that estimates of the parameter I in Equation 12.19 varied more than estimates of A across years and locations. This was unfortunate because yield loss at low weed density (I) is more important in determining economic thresholds than maximum yield loss (A). They found that estimated single-year economic thresholds for foxtails in maize ranged from 3.2 to 94 plants m^{-1} row between years at one location. Clearly, this instability in interference relationships undermines the utility of a T_e approach to improve weed management decisions. If a common yield loss relationship cannot be used to calculate a T_e across locations or among years within a field, what value of T_e should a grower use?

Variation in crop–weed interference relationships may result from variation in the relative time of emergence of the crop and weed, differential response of the crop and weed to different weather conditions, shifts in the resource that is most limiting, or variation in cultural practices.[34] Because the relative leaf area model (Equation 12.25) accounts for both density and time of emergence, it might be argued that estimates of q and m would vary less than I and A in Equation 12.19. Possible variation in q and m across years and locations was evaluated in experiments with maize,[31] white bean,[36] soybean,[35] sugarbeet, and spring wheat, *Triticum aestivum*.[42] In all cases, q and m were as unstable as or more unstable across years and locations than I and A. Moreover, the practical application of the relative leaf area models may still be limited due to the lack of a method to estimate leaf area index quickly and accurately.

These problems suggest that further research is needed to understand the mechanisms of crop–weed interference. Quantitative understanding of how crop and weed growth and competition respond to environmental factors and management could be used to modify simple crop–weed interference relationships, or to develop more comprehensive models for simulating crop–weed growth and competition. The comprehensive models may then be used with historic weather data to evaluate the probability of observing a specific crop–weed interference relationship. These probabilities would be useful for evaluating the amount of risk associated with a specific weed management decision.

12.6.5 FIELD SCOUTING TO DETERMINE WEED DENSITIES

Crop producers are well aware of the effect of high weed densities on crop yields. However, it is at low weed densities that they must make weed management decisions. Therefore, field scouting is an important part of decision making. Accurately determining weed species present, their density, and relative time of emergence in the field will help determine if management is necessary.

One of the major constraints to utilizing decision support models at the farm level is a lack of practical sampling methods for estimating weed density throughout a field. Growers generally have a good feel for where "weed pressure" is greatest. Gerowitt and Heitefuss[84] proposed sampling at least 20 to 30 randomly selected points within each 4 to 5 ha field. To obtain the required information for operation of HERB, Wilkerson et al.[46] suggested collecting at least one sample per 0.5 ha, or 10 samples per field. Thomas[85] suggested using a W pattern for sampling weed populations within fields, as depicted in Figure 12.5. A wooden or wire 'quadrat' that encloses an area of 0.25 m² (50 by 50 cm) is placed on the ground and the number of weeds of each species counted and recorded. A minimum of 20 sampling units at least 20 paces apart should be taken across the field, and the average weed density is

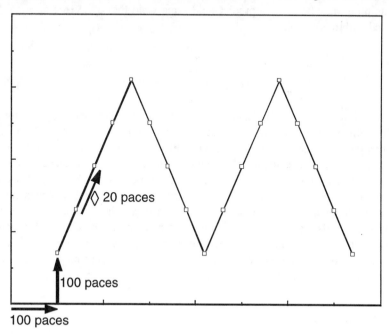

FIGURE 12.5 Possible pattern for scouting weed populations in farm fields (after Thomas[85]).

calculated for each species. Average weed density then can be used to calculate crop loss in any weed management decision support tool.

12.6.6 VARIATION IN SPATIAL DISTRIBUTION OF WEEDS

The A and C coefficients in Equations 12.28 and 12.30 are based on small plot weed–crop interference data with homogeneous weed densities. On a field scale, however, weed densities are not spatially homogeneous.[86–90] Field-scale mean weed density estimates may therefore be irrelevant considering the spatial diversity and density of weed populations across large areas. Use of field-scale mean density estimates in spatially heterogeneous weed populations results in under-prediction of yield loss at locations where weed density is high, and over-prediction in parts of the field where densities are low or weeds are absent. The net result of ignoring spatial heterogeneity is an over-prediction of whole-field yield loss.[91, 92] Spatial variation in weed density must therefore be accounted for to accurately predict crop yield loss.

Intensive sampling can be used to determine the spatial location of weed densities within fields.[93] Under an intensive sampling scheme, a grid system is imposed on a field and weed counts are made at each intersection of the grid, resulting in a discrete number of cells (d) per unit field area. Weed population density within each cell is commonly assumed homogeneous. Grid sampled data may be used to produce a spatial map of weed density within all cells. Maps of weed density produced from grid sampling may be used to predict field-scale yield loss by averaging yield loss predicted within each cell:

$$YL_{c,h} = \frac{\sum_{h=1}^{d} YL_{N,h}}{d} \qquad [12.32]$$

where $YL_{N,h}$ is predicted using Equation 12.30 for each cell (h) and d is total number of cells within the field. If the size of a cell on the sampling grid is on a scale at which field manipulations take place (e.g., the width of a spray boom), then a spatial map of weed density may be used to direct management decisions within each cell. This site-specific approach to weed management may be the most cost effective and result in greatest reduction of herbicide use.[93] A problem associated with this approach is the intensity with which sampling must take place. Grid sampling a field or farm can be expensive if weeds are counted at every grid intersection. Research is still needed to develop cost-efficient methods of accurately sampling weed populations in farmers' fields. Until this is accomplished, precision weed management may simply be too costly to benefit the grower.

12.7 SUMMARY

Interference from weed populations can reduce crop yield. Growers have used herbicides to reduce the impact of weed interference on crop yield for more than 50 years because they were cost effective. Currently, the U.S. government advocates implementation of IWM on at least 75% of the nation's farmland because of recent concern about the possible effects of pesticides on the environment. Initial attempts at implementing IWM focused on applying herbicides only when they were economically justified. Economic justification can only be evaluated if the effects of weed interference on crop yield can be quantified. In this chapter, we reviewed a number of different approaches to quantifying the effects of weed interference on crop yield. Emphasis was placed on developing the theoretical foundation for using these approaches, and showing how various quantitative relationships are related. It was shown that all population-based interference models rely on the assumption that plant growth through time follows a logistic function. A primary problem with population density-based theories of interplant competition is that they are useful only as *posteriori* descriptors that demonstrate the existence of competition and have little predictive power.[94]

A number of reasons for the inadequacy of density-based theories were discussed. First, economic justification of herbicide application based on a single-year crop–weed competition relationship does not account for the dynamics of weed seed return. Second, while management certainly may influence the competitiveness and seed production of weeds and crops, relatively little quantitative information is available to account for these effects. Third, little is known about the assumption of additivity when multiple weed species coexist within a grower's field, and coefficients for the appropriate equations are not available for every weed in all crops. Fourth, deviations from logistic growth within a growing season occur as a result of variation in weather and its influence on the availability of growth-limiting resources. Therefore, it should be of little surprise that many studies showed that crop–weed interference relationships vary among environments, limiting their utility for assisting in weed management decisions. Fifth, accurate estimates of weed density within farm fields are difficult and costly to obtain. Finally, growers typically apply a management tactic uniformly across their fields. A great deal of recent research shows that weed populations are not spatially homogeneous. It seems clear that the greatest reduction in pesticide use could be obtained if the compound was used only where it was needed (applied to existing weeds). Although the technology exists to apply management to specific areas of a field, sampling weed populations to obtain a spatial map of their presence is cost-limiting.

In the short term, quantitative relationships between crop yield and weed density may help growers make more informed weed management decisions. However, if weed scientists are to further improve weed management decision making, we must achieve an ability to predict the dynamics and outcome of crop–weed competition, both within and across growing seasons. The outcome of crop–weed interactions is driven by physiological mechanisms that regulate resource acquisition and plant growth. Therefore, further research is needed to understand how factors such as management and field position (edaphic factors) influence resource availability, and how external factors such as annual weather variation will influence the ability of crops and weeds to acquire resources that are available. Further discussion of these issues will be covered in the next chapter.

REFERENCES

1. Radosevich, S. R., Methods to study interactions among crops and weeds, *Weed Technol.,* 1, 190, 1987.
2. Schweizer, E. E., Lybecker, D. W., and Wiles, L. J., Important biological information needed for bioeconomic weed management models, in *Integrated Weed and Soil Management,* Hatfield, J. L., Buhler, D. D., and Stewart, B. A., Eds., Ann Arbor Press, Chelsea, MI, 1998.
3. Walker, R. H., and Buchanan, G. A., Crop manipulation in integrated weed management systems, *Weed Sci. Suppl.,* 30, 17, 1982.
4. Shaw, W. C., Integrated weed management systems technology for pest management, *Weed Sci. Suppl.,* 30, 2, 1982.
5. Swanton, C. J., and Weise, S. F., Integrated weed management: the rationale and approach, *Weed Technol.,* 5, 648, 1991.
6. Thill, D. C., Lish, J. M., Callihan, R. M., and Bechinski, E. J., Integrated weed management—a component of integrated pest management: a critical review, *Weed Technol.,* 5, 648, 1991.
7. Buchanan, G. A., Weed biology and competition, in *Research Methods in Science,* 2nd edition, Truelove, B., Ed., Southern Society of Weed Science, 1977, 25.
8. Martin, A. R., Mortensen, D. A., and Lindquist, J. L., Decision support models for weed management: in-field management tools, in *Integrated Weed and Soil Management,* Hatfield, J. L., Buhler, D. D., Stewart, B. A., Eds, Ann Arbor Press, Chelsea, MI, 1997, 363.
9. Kropff, M. J., Modelling the effects of weeds on crop production, *Weed Res.,* 28, 465, 1988.
10. Cousens, R. D., Design and interpretation of interference studies: Are some methods totally unacceptable? *N.Z. J. Forest. Sci.,* 26, 5, 1996.
11. Zimdahl, R. L., *Weed Crop Competition: A Review,* International Plant Protection Center, Corvallis, OR, 1980.
12. Hastings, A., *Population Biology: Concepts and Models,* Springer-Verlag, New York, 1997, chaps. 2, 4.
13. Willey, R. W., and Heath, S. B., The quantitative relationships between plant population and crop yield, *Adv. Agron.,* 21, 281, 1969.
14. Shinozaki, K., and Kira, T., Intraspecific competition among higher plants, vii, logistic theory of the C-D effect, *J. Inst. Polytech., Osaka City University ser.* D7, 35, 1956.
15. Weiner, J., A neighborhood model of annual–plant interference, *Ecology,* 63, 1237, 1982.
16. Weiner, J., Neighborhood interference amongst *Pinus rigida* individuals, *J. Ecol.,* 72, 183, 1984.
17. Pacala, S. W., and Weiner, J., Effects of competitive asymmetry on a local density model of plant interference, *J. Theor. Biol.,* 149, 165, 1991.
18. Kira, T., Ogawa, H., Hozumi, K., Koyama, H., and Yoda, K., Intraspecific competition among higher plants, v, Supplementary notes on the C-D effect, *J. Inst. Polytech., Osaka City University ser.* D7, 1, 1956.
19. Bleasdale, J. K. A., and Nelder, J. A., Plant population and crop yield, *Nature,* 188, 342, 1960.
20. Farazdaghi, H., and Harris, P. M., Plant population and crop yield, *Nature,* 217, 289, 1968.
21. de Wit, C. T., On competition, *Agric. Res. Report (Versl. Landbouwkd. Onderz.),* 66, 1, 1960.

22. Roush, M. L., Radosevich, S. R., Wagner, R. G., Maxwell, B. D., and Peterson, T. D., A comparison of methods for measuring effects of density and proportion in plant competition experiments, *Weed Sci.,* 37, 268, 1989.
23. Spitters, C. J. T., An alternative approach to the analysis of mixed cropping experiments. 1. Estimation of competition effects, *Neth. J. Agric. Sci.,* 31, 1, 1983.
24. Spitters, C. J. T., Kropff, M. J., and de Groot, W., Competition between maize and *Echinochloa crus-galli* analysed by a hyperbolic regression model, *Ann. Appl. Biol.,* 115, 541, 1989.
25. Snaydon, R. W., Replacement or additive designs for competition studies?, *J. Appl. Ecol.,* 28, 930, 1991.
26. Rejmanek, M., Robinson, G. R., and Rejmankova E., Weed crop competition: experimental designs and models for analysis, *Weed Sci.,* 37, 276, 1989.
27. Firbank, L. G., and Watkinson, A. R., On the analysis of competition within two-species mixtures of plants, *J. Appl. Ecol.,* 22, 503, 1985.
28. Cousens, R., A simple model relating yield loss to weed density, *Ann. Appl. Biol.,* 107, 239, 1985.
29. O'Donovan, J. T., and Blackshaw, R. E., Effect of volunteer barley (*Hordeum vulgare* L.) interference on field pea (*Pisum sativum* L.) yield and profitability, *Weed Sci.,* 45, 249, 1997.
30. Knezevic, Z. S., Weise, S. F., and Swanton, C. J., Interference of redroot pigweed (*Amaranthus retroflexus* L.) in corn (*Zea mays* L.), *Weed Sci.,* 42, 568, 1994.
31. Knezevic, Z. S., Weise, S. F., and Swanton, C. J., Comparison of empirical models depicting density of *Amaranthus retroflexus* L. and relative leaf area as predictors of yield loss in maize (*Zea mays* L.), *Weed Res.,* 35, 207, 1995.
32. Knezevic, Z. S., Horak, M. J., and Vanderlip, R. L., Relative time of redroot pigweed (*Amaranthus retroflexus*) emergence is critical in pigweed–sorghum (*Sorghum bicolor*) competition, *Weed Sci.,* 45, 502, 1997.
33. Lindquist, L. J., Mortensen, D. A., Clay, S. A., Schmenk, R., Kells, J. J., Howatt, K., and Westra, P., Stability of corn (*Zea mays*)–velvetleaf (*Abutilon theophrasti*) interference relationships, *Weed Sci.,* 44, 309, 1996.
34. Lindquist, J. L., Mortensen, D. A., Westra, P., Lambert, W. J., Bauman, T. T., Fausey, J. C., Kells, J. J., Langton, S. J., Harvey, R. G., Banken, K., Clay, S., and Bussler, B. H., Stability of corn (*Zea mays*)–foxtail (*Setaria* spp.) interference relationships, *Weed Sci.,* 47, 195, 1999.
35. Dieleman, A., Hamill, A. S., Weise, S. F., and Swanton, C. J., Empirical models of pigweed (*Amaranthus* spp.) interference in soybean (*Glycine max*), *Weed Sci.,* 43, 612, 1995.
36. Chikoye, D., Weise, S. F., and Swanton, C. J., Influence of common ragweed (*Ambrosia artemisifolia*) time of emergence and density on white bean (*Phaseolus vulgaris*), *Weed Sci.,* 43, 375, 1995.
37. Cousens, R., Brain, P., O'Donovan, J. T., and O'Sullivan, P. A., The use of biologically realistic equations to describe the effects of weed density and relative time of emergence on crop yield, *Weed Sci.,* 35, 720, 1987.
38. Boznic, A. C., and Swanton, C. J., Influence of barnyardgrass (*Echinochloa crus-galli*) time of emergence and density on corn (*Zea mays*), *Weed Sci.,* 45, 276, 1997.
39. Cowan, P., Weaver, S. E., and Swanton, C. J., Interference between pigweed (*Amaranthus* spp), barnyardgrass (*Echinochloa crus-galli*), and soybean (*Glycine max*), *Weed Sci.,* 46, 533, 1998.

40. Florez, J. A., Fischer, A. J., Ramirez, H., and Duque, M. C., Predicting rice yield losses caused by multispecies weed competition, *Agron. J.,* 91, 87, 1999.
41. Kropff, M. J., and Spitters, C. J. T., A simple model of crop loss by weed competition from early observations on relative leaf area of the weed, *Weed Res.,* 31, 97, 1991.
42. Lotz, L. A. P., Christensen, S., Cloutier, D., Fernandez-Quintanilla, C., Legere, A., Lemieux, C., Lutman, P. J. W., Pardo Iglesias, A., Salonen, J., Sattin, M., Stigliani, L., and Tei, F., Prediction of the competitive effects of weeds on crop yields based on the relative leaf area of weeds, *Weed Res.,* 36, 93, 1996.
43. Kropff, M. J., Lotz, L. A. P., Weaver, S. E., Bos, H. J., Wallinga, J., and Migo, T., A two parameter model for prediction of crop loss by weed competition from early observations of relative leaf area of the weeds, *Ann. Appl. Biol.,* 126, 329, 1995.
44. Lindquist, J. L., Maxwell, B. D., Buhler, D. D., and Gunsolus, J. L., Velvetleaf (*Abutilon theophrasti*) recruitment, survival, seed production and interference in soybean (*Glycine max*), *Weed Sci.,* 43, 226, 1995.
45. Swinton, S. M., and King, R. P., A bioeconomic model for weed management in corn and soybean, *Agric. Sys.,* 44, 313, 1994.
46. Wilkerson, G. G., Modena, S. A., and Coble, H. D., HERB: Decision model for postemergence weed control in soybean, *Agron. J.,* 83, 413, 1991.
47. Coble, H. D., and Mortensen, D. A., The threshold concept and its application to weed science, *Weed Technol.,* 6, 191, 1992.
48. Jordan, N., Weed demography and population dynamics: Implications for threshold management, *Weed Technol.,* 6, 184, 1992.
49. Maxwell, B. D., Weed thresholds: the space component and considerations for herbicide resistance, *Weed Technol.,* 6, 205, 1992.
50. Wilkerson, G. G., Modena, S. A., and Coble, H. D., HERB v2.0: Herbicide decision model for postemergence weed control in soybeans, Users manual. Bull. No. 113, Crop Science Department, North Carolina State University, Raleigh, 1988.
51. Neeser, C., Dieleman, J. A., Mortensen, D. A., and Martin A. R., WeedSOFT: state of the art weed management decision support system, *Weed Technol.,* (in review), 2000.
52. Headley, J. C., Defining the economic threshold, in *Pest Control Strategy for the Future,* National Academy Science, Washington D.C., 1972, 101.
53. Stern, V. M., Smith, R. F., van den Bosh, R. and Hagen, K. S., The integrated control concept, *Hilgardia,* 29, 81, 1959.
54. Cousens, R., Theory and reality of weed control thresholds, *Plant Prot. Quart.,* 2, 13, 1987.
55. Bauer, T. A., and Mortensen, D. A., A comparison of economic and economic optimum threshold for two annual weeds in soybeans, *Weed Technol.,* 6, 228, 1992.
56. Cardina, J., Regnier, E., and Sparrow, D., Velvetleaf (*Abutilon theophrasti*) competition and economic thresholds in conventional and no-tillage corn (*Zea mays*), *Weed Sci.,* 43, 81, 1995.
57. Auld, B. A., Menz, D. M., and Tisdell, C. A., *Weed Control Economics,* Academic Press, London, 1987, 177.
58. Norris, R. F., Weed thresholds in relation to long-term population dynamics, *Proc. West. Soc. Weed Sci.,* 37, 38, 1984.
59. Owen, M. D. K., Producer attitudes and weed management, in *Integrated Weed and Soil Management,* Hatfield, J. L., Buhler, D. D., and Stewart, B. A., Eds., Ann Arbor Press, Chelsea, MI, 1998, 43.
60. Weersink, A., Deen, W., and Weaver, S., Evaluation of alternative decision rules for postemergence herbicide treatments in soybean, *J. Prod. Agric.,* 5, 298, 1992.

61. Norris, R. F., Case history for weed competition/population ecology: barnyardgrass (*Echinochloa crus-galli*) in sugar beets (*Beta vulgaris*), *Weed Technol.*, 6, 220, 1992.
62. Sattin, M., Zanin, G., and Berti, A., Case history for weed competition/population ecology: velvetleaf (*Abutilon theophrasti*) in corn (*Zea mays*), *Weed Technol.*, 6, 213, 1992.
63. Doyle, C. J., Cousens, R., and Moss, S. R., Economics of controlling *Alopecurus myosuroides* in winter wheat, *Crop Prot.*, 5, 143, 1987.
64. Wallinga, J., Analysis of the rational long-term herbicide use: evidence for herbicide efficacy and critical weed kill rate as key factors, *Agric. Syst.*, 56, 323, 1998.
65. Lindquist, J. L., Mortensen, D. A., Westra, P., Kells, J. J., and Clay, S., Performance of INTERCOM for predicting the effects of maize-velvetleaf competition across the north central USA, *Weed Sci.*, (in review), 2000.
66. Teasdale, J. R., Influence of corn (*Zea mays*) population and row spacing on corn and velvetleaf (*Abutilon theophrasti*) yield, *Weed Sci.*, 46, 447, 1998.
67. Dieleman, J. A., and Mortensen, D. A., Influence of weed biology and ecology on development of reduced dose strategies for integrated weed management systems, in *Integrated Weed and Soil Management*, Hatfield, J. L., Buhler, D. D., and Stewart, B. A., Eds., Ann Arbor Press, Chelsea, MI, 1998, 333.
68. Steckel, L. E., DeFelice, M. S., and Sims, B. D., Integrating reduced rates of postemergence herbicides and cultivation for broadleaf weed control in soybeans (*Glycine max*), *Weed Sci.*, 38, 541, 1990.
69. Buhler, D. D., Gunsolus, J. L., and Ralston, D. F., Common cocklebur (*Xanthium strumarium*) control in soybean (*Glycine max*) with reduced bentazon rates and cultivation, *Weed Sci.*, 41, 447, 1993.
70. Schmenk, R., and Kells, J. J., Effect of soil-applied atrazine and pendimethalin on velvetleaf (*Abutilon theophrasti*) competitiveness in corn, *Weed Technol.*, 12, 47, 1998.
71. Weaver, S. E., Size-dependent economic thresholds for three broadleaf weed species in soybeans, *Weed Technol.*, 5, 674, 1991.
72. Lindquist, J. L., and Mortensen, D. A., Tolerance and velvetleaf (*Abutilon theophrasti*) suppressive ability of two old and two modern corn (*Zea mays*) hybrids, *Weed Sci.*, 46, 569, 1998.
73. Lindquist, J. L., Mortensen, D. A., and Johnson, B. E., Mechanisms of corn tolerance and velvetleaf suppressive ability, *Agron. J.*, 90, 787, 1998.
74. Tollenaar, M., and Aguilera, A., Radiation use efficiency of an old and a new maize hybrid, *Agron. J.*, 84, 536, 1992.
75. Tollenaar, M., Dibo, A. A., Aguilera, A., Wiese, S. F., and Swanton, C. J., Effect of crop density on weed interference in maize, *Agron. J.*, 86, 591, 1994.
76. Tollenaar, M., Nissanka, S. P., Aguilera, A., Weise, S. F., and Swanton, C. J., Effect of weed interference and soil nitrogen on four maize hybrids, *Agron. J.*, 46, 596, 1994.
77. McLachlan, S. M., Tollenaar, M., Swanton, C. J., and Weise, S. F., Effect of corn-induced shading on dry matter accumulation, distribution, and architecture of redroot pigweed (*Amaranthus retroflexus*), *Weed Sci.*, 41, 568, 1993.
78. Murphy, D. S., Yakubu, Y., Weise, S. F., and Swanton, C. J., Effects of planting patterns and inter-row cultivation on competition between corn (*Zea mays*) and late emerging weeds, *Weed Sci.*, 44, 856, 1996.
79. Dingkuhn, M., Johnson, D. E., Sow, A., and Audebert, A. Y., Relationships between upland rice canopy characteristics and weed competitiveness, *Field Crops Res.*, 61, 79, 1999.
80. Swinton, S. M., Buhler, D. D., Forcella, F., Gunsolus, J. L., and King, R. P., Estimation of crop yield loss due to interference by multiple weed species, *Weed Sci.*, 42, 103, 1994.

81. Berti, A., and Zanin, G., Density equivalent: a method for forecasting yield loss caused by mixed weed populations, *Weed Res.*, 34, 327, 1994.

82. Brain, P., and Cousens, R., The effect of weed distribution on predictions of yield loss, *J. Appl. Ecol.*, 27, 735, 1990.

83. Fischer, D. W., Harvey, R. G., Bauman, T. T., Hart, S. E., Johnson, G. A., Kells, J. J., Lindquist, J. L., and Westra, P., Effects of *Chenopodium album* competition with *Zea mays* in a regional study, *Weed Sci.* (in review), 2000.

84. Gerowitt, B., and Heitefuss, R., Weed economic thresholds in cereals in the Federal Republic of Germany, *Crop Prot.*, 9, 323, 1990.

85. Thomas G. A., Weed survey system used in Saskatchewan for cereal and oilseed crops, *Weed Sci.*, 33, 34, 1985.

86. van Groenendael, J. M., Patchy distribution of weeds and some implications for modelling population dynamics: a short literature review, *Weed Res.*, 28, 437, 1988.

87. Johnson, G. A., Mortensen, D. A., Young, L. Y., and Martin, A. R., The stability of weed seedling population models and parameters in eastern Nebraska corn (*Zea mays*) and soybean (*Glycine max*) fields, *Weed Sci.*, 43, 604, 1995.

88. Navas, M. L., Using plant population biology in weed research: a strategy to improve weed management, *Weed Res.*, 31, 171, 1991.

89. Thornton, P. K., Fawcett, R. H., Dent, J. B., and Perkins, T. J., Spatial weed distribution and economic thresholds for weed control, *Crop Prot.*, 9, 337, 1990.

90. Wiles, L. J., Oliver, G. W., York, A. C., Gold, H. J., and Wilkerson, G. G., Spatial distribution of broadleaf weeds in North Carolina soybean (*Glycine max*) fields. *Weed Sci.*, 40, 554, 1992.

91. Auld, B. A., and Tisdell, C. A., Influence of spatial distribution of weeds on crop yield loss, *Plant Prot. Quart.*, 3, 81, 1988.

92. Nordbo, E., and Christensen, S., Spatial variability of weeds, in *Proceedings of the Seminar on Site Specific Farming*, Olesen, S. E., Ed., Danish Institute of Plant and Soil Science, SP-report No. 26, 1995.

93. Lindquist, J. L., Dieleman, J. A., Mortensen, D. A., Johnson, G. A., and Pester-Wyse, D. Y., Economic importance of managing spatially heterogeneous weed populations, *Weed Technol.*, 12, 7, 1998.

94. Tilman, D., Mechanisms of plant competition for nutrients: the elements of a predictive theory of competition, in *Perspectives on Plant Competition*, Grace, J. B., and Tilman, D., Eds., Academic Press, San Diego, 1990, chap. 7.

95. Lang, A. L., Pendleton, J. W., and Dungan, G. H., Influence of population and nitrogen levels on yield and protein and oil contents of nine corn hybrids, *Agron. J.*, 48, 1956.

13 Mechanisms of Crop Loss Due to Weed Competition

John L. Lindquist

CONTENTS

13.1 INTRODUCTION

This book focuses on the effects of biotic *stress* on crop yield loss. In Chapter 12 we outlined a number of methods for quantifying the effects of weed interference on crop yield. In this context, it was important to use the term *interference* because experiments designed to show a relationship between crop yield and weed population density have not typically included methods of evaluating the *cause* of any observed yield reduction. In other words, although the yield reduction was likely caused by stress, the cause of that stress is unknown. It is possible that the loss was caused because the weed species under study acted as a trap crop for insects that subsequently caused stress through, for example, defoliation. Granted, most researchers interested in the effects of the weed would manage their experiment to eliminate such factors. However, the point is that if the stress was not quantified, it is impossible to say what actually caused the observed crop loss. Therefore, questions that need to be addressed include (1) how do weeds cause stress, and (2) how do we quantify the influence of this stress on crop yield?

    In this chapter, I focus on the mechanisms of interplant competition. The definition of competition used here is resource dependent, whereas others include phenomena such as allelopathy and some forms of symbiotic relationships.[1-3] These concepts are considered beyond the scope of this chapter. Several authors have discussed the importance of growth determining or limiting resources to crop–weed competition and weed management.[3–10] My aim here is to provide a quantitative perspective on how weeds influence the quantity of resource available to the crop (and vice versa), and how the crop responds to changes in the availability of those resources. In this context, I focus on the three resources that most commonly result in crop stress: light, soil water, and soil nitrogen.

13.1.1 INTERPLANT COMPETITION, THE STRESS FACTOR

The editors of this book define stress as a departure from optimal physiological conditions. Do weeds actually cause stress to the crop? Goldberg[2] argued that most interactions between plants occur through some intermediary such as resources. Weeds do not generally have a direct effect on the physiological status of a crop plant in the same way that an insect might (with the exception of parasitic weeds). However, both the weed and the crop have a direct effect on the resources available in their immediate environment. Moreover, both species have a unique response to the quantity of resources available within that environment. Therefore, *weeds cause crop yield loss indirectly through their influence on the resources required for crop growth.* The outcome of crop–weed competition is driven by the physiological mechanisms that regulate the effect of each species on a given resource, and their response to the quantity of that resource available to the plant.

In Chapter 12 we indicated that yield density relationships are built on the assumption that plant growth over time can be described using a general logistic function. This is typically true, but the exact shape of that curve commonly varies in different environments. For example, Figure 13.1 shows the growth of maize over time as influenced only by nitrogen fertilizer application.[11] The growth curve appears logistic, but the shape of the curve varies with nitrogen application. Since resource supply probably has a different effect on the growth of each species, it is no wonder that crop yield loss–weed density relationships vary among environments. Therefore, predicting the outcome of interplant competition requires accurate prediction of the quantity of resources available, the efficiency with which each species can acquire and utilize each resource, and the physiological consequences when the quantity of available resources does not meet demand.

13.1.2 THEORY OF INTERPLANT COMPETITION

A primary goal of many ecologists has been to predict the outcome of interplant competition.[12–14] Two highly debated theoretical approaches to predicting the effects of competition are those of Grime[14] and Tilman.[15–17] Although their concern was to develop a theory to describe the underlying mechanisms that define ecosystem structure (i.e., which species are dominant within a given ecosystem), I will give a brief overview of these two perspectives because they provide a useful theoretical foundation for the following sections.

Grime[14] classified plants based on their life history characteristics and adaptation to stress and disturbance. According to his system, plants adapted to low levels of both stress and disturbance are considered competitors, those adapted to high

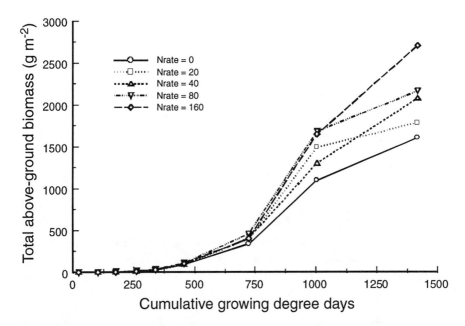

FIGURE 13.1 Maize total above-ground biomass (g m^{-2}) over time as influenced by nitrogen application rate (Nrate, kg N ha^{-1}, applied as ammonium nitrate) at the University of Nebraska Agricultural Research and Development Center near Mead, in 1994.

stress but low disturbance are considered stress tolerant, and those adapted to low stress and high disturbance are considered ruderal. Based on his classification, both weeds and crops would be considered ruderals. This classification tells us little about the dynamics of crop–weed competition within a particular cropping system. However, Grime defines competition as the tendency for neighboring plants to utilize the same resources and argues that success in competition is largely a reflection of the capacity for resource capture. Under Grime's theory, competitive ability is positively correlated with the maximum relative growth rate (RGR). Thus, he assumes that the ability to compete is determined by the ability to exploit resources rapidly rather than to tolerate resource depletion.[12]

Tilman[15] proposed a resource-based theory of competition. Given enough time, plants draw the concentration of resources down to a level (R*), below which the population is unable to maintain itself (i.e., growth rate is equal to death rate). According to his analysis, the species with the lowest R* will completely displace all other species *at equilibrium*.[12] Loosely, equilibrium occurs after a sufficient amount of time in which the resource has been depleted to its R* level. The issue of equilibrium is important in agronomic systems because resources are rarely if ever drawn down to the R* concentration and disturbance is managed to ensure the success of a specific class (ruderal) of plants. Therefore, crop–weed competition must be considered a more subtle and dynamic process.

Tilman[15-17] presented a system of two equations to account for the effects of resource supply on plant growth. The first states that the rate of change of biomass per unit biomass $(1/W_i)(dW_i/dt)$ of species i depends on the resource dependent net growth function $(f_i(R))$ and its loss rate (m_i):

$$\frac{1}{W_i}\frac{dW_i}{dt} = f_i(R) - m_i \qquad [13.1]$$

where $f_i(R)$ can be defined as a function of resource supply using a number of approaches.[17] The dynamics of the growth limiting resource (dR/dt) depend on the difference between the resource supply rate $(y(R))$ and resource consumption (C_i) summed over all species:

$$\frac{dR}{dt} = y(R) - \Sigma[RC_i(Rf_i(R))] \qquad [13.2]$$

where dR/dt is the rate of change in resource concentration and $C_i(f_i(R))$ indicates that resource consumption is itself resource dependent. Note that these equations are a simple representation of a mechanistic plant growth model. The beauty of their simplicity is that they are analytically tractable (an equation for R* can be obtained analytically). However, the net growth function $(f_i(R))$ depends not only on supply of the resource in question, but also on the past and present physiological status of the plant (e.g., tissue nutrient concentration, age, etc.), canopy morphology, and on environmental factors such as temperature and quantity of available radiation. The loss rate (respiration and senescence) will further depend on environmental factors and on stage of plant development. Resource supply rate will depend upon soil physical and chemical characteristics as well as environmental conditions. Resource consumption depends on the dynamics of resource supply and the demand of the plant, which is dependent upon previous growth and the partitioning of nutrients and photosynthate to different organs within the plant. All of these factors will vary temporally. Full understanding of the dynamics of crop and weed growth and competition requires that all of these factors be studied throughout the growth period and incorporated into our models.

Models that include these physiological components as well as the effects of dynamic environmental conditions are more appropriately dealt with through simulation, which involves stringing several mathematical functions together into algorithms that describe a particular process (e.g., soil water balance). Simulation models are excellent tools for gaining improved understanding of the mechanisms of interplant competition because they typically function on a daily time step, are responsive to edaphic factors and to daily inputs of weather data, and can be used to test hypotheses about the contribution of specific morphological and physiological factors to competitive outcome. A number of simulation models have been developed in which the mechanisms of interplant competition are described based on underlying physiological processes.[18] Although most of these models have focused on competition for light, a few authors have presented quantitative procedures for incorporating competition for water and soil nitrogen. Because many of the quantitative procedures have been presented by others, I will briefly outline some of these procedures and focus

my
discussion in the next sections on (1) the effects of limited resource supply on plant growth, (2) the effects of plant growth on the availability of resources, (3) the effects of limited resource supply on its uptake by different species, and (4) the importance of the linkage between resources in quantifying interplant competition. My goal is to highlight components of interplant competition that appear to be least understood and, therefore, represent the greatest need for quantitative research.

13.2   PLANT GROWTH AND COMPETITION FOR LIGHT

Light (photosynthetic photon flux) is necessary for the assimilation of carbon from atmospheric CO_2. The relationship between instantaneous CO_2 assimilation and the quantity of light available to a plant leaf (f_i(light) $= A_{c,i}$, μmol CO_2 m^{-2} s^{-1} = 1.7536 10^{-5} g CO_2 m^{-2} s^{-1}) has two components,[19] (1) a light limited region in which light-utilization efficiency (ε) is greatest, and (2) a light saturated region in which further increases in the quantity of light fail to increase photosynthesis ($A_{max,i}$). Optimal physiological conditions occur when all of the leaf area is exposed to enough light to saturate photosynthesis. However, in a dense plant canopy, saturation of all leaves is impossible because even leaves of the same plant will shade one another. If the leaves of a weed are displayed within a crop canopy they reduce the quantity of light available to a crop, thereby inducing stress. Weeds also may alter the physiology of crop growth by modifying the quality of light within the crop environment. Others have provided excellent reviews of the importance of both quantity and quality of light on weed growth and management.[3, 4, 7-10] The following discussion will focus on the effects of weeds on the quantity of light available to the crop.

Dry matter growth increase of species i under potential production conditions ($G_{p,i}$, g m^{-2} s^{-1}) may be calculated using:[20]

$$\frac{dW_i}{dt} = G_{p,i} = A_{c,i}\, 30/44 - R_{m,i} - R_{g,i} - S_i \qquad [13.3]$$

where 30/44 results from the conversion of CO_2 to dry weight, $R_{m,i}$ (g m^{-2} s^{-1}) is maintenance respiration,[21] $R_{g,i}$ (g m^{-2} s^{-1}) is growth respiration,[22] and S_i (g m^{-2} s^{-1}) is the rate of tissue loss, or senescence (which will depend primarily upon physiological age).

The dynamics of light availability at the canopy surface ($dR/dt = dI_o/dt$, where I_o is incident photosynthetic photon flux) is dependent solely on latitude and local atmospheric conditions (e.g., degree of cloudiness). Because light cannot be stored within the system, competition for light is always instantaneous.[20] A photon is either absorbed and utilized for growth, or it is lost. The dynamics of the amount of light available to a given leaf within a canopy (required to integrate Equation 13.3 for the entire canopy), are dependent upon the amount and distribution of photosynthetic area within the canopy and the light extinction coefficient (a measure of the efficiency of light interception). Algorithms for predicting the quantity of light available to and absorbed within mixed canopies and the subsequent amount of carbon assimilated have been described in detail elsewhere.[20, 23–26]

Because the quantity of radiation intercepted by each species is a primary determinant of growth under potential production conditions, a critical component of crop–weed competition models is how competition for light in mixed canopies is quantified. Several authors assumed that the fraction of radiation absorbed by a species ($F_{a,i}$) can be quantified based on the fraction of the total canopy LAI that species i occupies weighted by its extinction coefficient (k_i).[26–29] Therefore:

$$F_{a,i} = \frac{k_i LAI_i}{\sum\limits_{i=1}^{n} (k_i LAI_i)}$$ [13.4]

where the denominator is the sum of LAI for all n species in the canopy weighted by their respective extinction coefficients. Others have used similar approaches using slightly different definitions of efficiency of interception.[23, 25] Although Equation 13.4 makes excellent theoretical sense, to my knowledge this relationship has never been tested.

A true test of Equation 13.4 may not be possible because measuring the quantity of light absorbed by each species within a mixed canopy would be extremely difficult, if not impossible. However, we may be able to obtain an indirect test by measuring the quantity of light not absorbed by each species. For example, we can measure the quantity of radiation incident above the canopy (I_o), at the soil surface (I_s, or at any point within the mixed canopy), at the soil surface following removal of the crop (I_w), and at the soil surface following removal of the weed (I_c). The quantity of radiation intercepted by the mixed canopy is then represented by the difference $I_o - I_s$. We may then estimate the fraction of radiation intercepted by the crop and the weed using:

$$F_{a,c} = \frac{I_c - I_s}{I_o - I_s}$$

and   [13.5]

$$F_{a,w} = \frac{I_w - I_s}{I_o - I_s}$$

where the subscripts c and w represent the crop and weed, respectively. Methodology for measuring incident light and estimating the extinction coefficient in monoculture crop stands is well established in the literature.[30] Therefore, if the leaf area index of each species within the mixed canopy is measured at the time radiation interception measurements are taken, estimates of $F_{a,i}$ obtained using Equation 13.5 can be directly compared with those obtained using Equation 13.4.

Under the assumption that Equation 13.4 is correct, a number of canopy characteristics will have a critical impact on the quantity of light absorbed and subsequent plant growth within mixed plant canopies. In a sensitivity analysis of the model INTERCOM, Lindquist and Mortensen[31] showed that parameters having the greatest influence on maize yield in monoculture include rate of development during reproductive and vegetative stages, the extinction coefficient, and relative height of maximum leaf area density (i.e., vertical leaf area distribution). Traits in maize resulting in the greatest reduction in yield loss in a mixed maize-*Abutilon theophrasti* (velvetleaf) canopy include maximum height, relative height of maximum leaf area density, rate of vegetative development, and thermal time from emergence to 50% maximum height and leaf area. Caton et al.[32] found similar results in their analysis of a rice-*Echinochloa oryzoides* model.

In an empirical field study with four diverse maize hybrids, Lindquist and Mortensen[33] found that maize yield loss-velvetleaf density relationships varied among hybrids. Lindquist et al.[34] determined that maize traits having strongest correlation to yield loss included maximum leaf area index and height, thermal time between emergence and 50% maximum LAI and height, and vertical leaf area distribution. Others have shown an increased LAI and subsequent reduction in light transmission within maize canopies in response to increased crop population and narrow row spacing, ultimately resulting in the reduction in weed productivity and their effects on crop yield.[35, 36] These results corroborate those of the model sensitivity analysis. Improving crop competitiveness with weeds is an important method of reducing our current reliance on herbicides. Therefore, modification of these traits to optimize crop yield and competitiveness needs to be an ongoing objective for plant breeders and agronomists.

13.3   COMPETITION FOR SOIL WATER

The CO_2 required for photosynthesis is acquired by diffusion through stomata on the surface of leaves followed by active uptake into the chloroplasts of the mesophyll cells. These stomata are essentially portholes that allow for the exchange of gases between the atmosphere and the intercellular spaces within the leaf. Diffusion occurs as the result of differences in concentration of gases in the air, the boundary layer of the leaf surface, and the intercellular spaces of the leaf. Although water is required for photosynthesis, a far greater quantity of water is lost through stomata as a result of diffusion.[37] In support of this process of transpiration, a continuous flow of water from the soil through the outer cells of the root, into the xylem, and out through the leaf is required. If soil water content becomes limited, stomata at the leaf surface will close, thereby reducing the quantity of CO_2 that can be assimilated for plant growth.

Kropff[38] assumed that the potential growth rate of a species ($dW_i/dt = G_{p,i}$ from Equation 13.3 where only competition for light is assumed) is reduced in proportion to the ratio of actual ($T_{a,i}$, cm(= kg m^{-2}10)) to potential ($T_{p,i}$, cm) transpiration. Hence:

$$G_{s,i} = \frac{T_{a,i}}{T_{p,i}} G_{p,i} \qquad\qquad [13.6]$$

where $G_{a,i}$ is the water limited plant growth rate ($dW_{w,i}/dt$, g m^{-2} s^{-1}). When soil water supply becomes limited, the plant will experience stress and growth will be reduced. However, of importance to the current discussion is that weeds induce a stress on the crop through their direct use of stored water in the root zone of both species. How then can we quantify the demand for water by both the crop and weed? How does current water use influence current and future supply of available soil water? What is the effect of limited water supply on water uptake? In a competitive situation, how do we partition the quantity of water available for uptake between the two species?

Interplant competition for water involves direct and indirect processes because water can be stored in the soil.[38] The direct process occurs during a drought period when water directly limits plant growth (i.e., supply does not meet the transpiration demand of all species). In this case, competition for water is instantaneous. The indirect process occurs during periods when water is in adequate supply to meet the demands of both species. In rainfed environments where precipitation may be heavy but sporadic, the supply of water over time may be lower than the cumulative loss due to evapotranspiration. Therefore, transpiration during periods when water is not limiting growth will influence the stored pool of available water and may strongly influence growth later in the season.

Because soil water is influenced by more than just transpiration, any approach to predicting the quantity of water available for uptake (R_w) must account for the overall soil water balance. A number of authors have developed procedures for simulating the quantity of soil water available for uptake within the root zone (R_w, cm):[38–40]

$$R_w = R_{w,t-1} + [PR + CR - E - P] - [T_a] \qquad [13.7]$$

where $R_{w,t-1}$ (cm) is soil water content in the previous time increment, PR (cm) is the quantity of water made available through precipitation (including irrigation), CR (cm) is capillary rise of water from below the rooted zone, E (cm) is the quantity of water evaporated from the soil surface, P is percolation of water below the root zone, calculated as the amount of water in excess of field capacity, and T_a is actual transpiration by the canopy (summed across species). Methods of simulating capillary rise, soil evaporation, and percolation have been discussed elsewhere.[38, 41] Of importance to this discussion is the water consumption term in Equation 13.7 (T_a, the quantity of water transpired by the plant canopy), which depends upon the overall canopy demand for water.

A common approach to quantifying the demand for water by a plant canopy is to calculate a reference evapotranspiration (E_r, J m^{-2} d^{-1}), which, according to the Penman[42] approach, is dependent upon the quantity of light incident at the canopy surface, temperature, wind speed, and the vapor pressure deficit. Kropff[38] assumed that potential transpiration by species i ($T_{p,i}$) is proportional to the reference evapotranspiration and the fraction of light intercepted by that species ($F_{a,i}$) in the mixed canopy:

$$T_{p,i} = \left(\frac{E_r F_{a,i} c_i}{L_w} \right) 10 \qquad [13.8]$$

where c_i is a proportionality factor (0.9 and 0.7 for C_3 and C_4 species, respectively),[38] L_w is the latent heat of vaporization of water (J kg^{-1}), and 10 is required to obtain appropriate units for $T_{p,i}$ (cm).

The ratio of actual ($T_{a,i}$) to potential ($T_{p,i}$) transpiration has been shown to decrease linearly with soil water:[43]

$$\frac{T_{a,i}}{T_{p,i}} = \frac{\theta_a - \theta_{wp,i}}{\theta_{cr,i} - \theta_{wp,i}} \quad \text{where } \theta_{wp,i} \leq \theta_a \leq \theta_{cr,i} \qquad [13.9]$$

where θ_a is actual volumetric soil water content (obtained from R_w and soil bulk density) and the subscripts wp and cr refer to species-specific water content at permanent wilting point, and the critical soil water content (below which $T_{a,i}/T_{p,i}$ begins to decline), respectively.[38]

Equation 13.8 estimates the potential demand for soil water by species i. Note that the only way leaf area influences potential transpiration is through its effect on light interception ($F_{a,i}$). To my knowledge, this relationship has not been tested. Equation 13.9 estimates the effects of actual soil water content on the quantity of water transpired through the canopy of a given species. Ray and Sinclair[44] reported data useful for obtaining an empirical estimate of θ_{cr} and θ_{wp} for maize, and Moreshet et al.[45] provided the only example of experiments designed to obtain side-by-side comparisons of this relationship for crop and weed species.

Equations 13.6 to 13.9 can be used to simulate indirect competition for soil water. In other words, during periods when water is not limiting, plants utilize as much water as is required using Equation 13.9, which subsequently reduces the pool of available soil water. As the growing season progresses, if precipitation does not replenish the pool of available soil water, eventually the water content will fall below the $\theta_{cr,i}$ and growth will be reduced. Direct competition is not accounted for using this approach because when soil water becomes limiting, there is no accounting for how much soil water is available and whether that quantity meets the transpiration demand of all species in the mixed canopy. In other words, Equation 13.8 simply indicates that if the volumetric soil water falls below a threshold ($\theta_{cr,i}$), transpiration is reduced. Under conditions of limiting soil water, it is assumed that the quantity of available water is not great enough to meet the demands of both (or all) species (e.g., ($T_{a,c} + T_{a,w}$) > R_w). Under such conditions, how should $T_{a,i}$ be modified? Actual water uptake by a species will depend in part upon root length density and the efficiency with which the roots actively take up water. Should the quantity of available water be divided among species based on the relative quantity of roots in the soil profile? What if the species differ in their uptake efficiency? What if the depth of the actual root zone of the two species differs? These are factors that substantially complicate our ability to simulate direct competition for soil water. Moreover, because root physiology research is difficult and costly, there are few useful examples reported in the literature. Therefore, algorithms for simulating interplant competition for soil water are not as refined as those for simulating competition for light. Further research is clearly needed.

13.4 COMPETITION FOR SOIL NITROGEN

A relationship between crop productivity and nitrogen (N) application rate has long been known to exist and is commonly used for fertilizer management schedules. Because 50 to 80% of the nitrogen in plant leaves is found in photosynthetic proteins,[46-48] the observed correlation between light saturated CO_2 assimilation ($A_{max,i}$, μmol CO_2 m^{-2} s^{-1} = 1.7536 10^{-5} g CO_2 m^{-2} s^{-1}) and leaf nitrogen content ($N_{L,i}$, g N m^{-2}),[49-56] should be no surprise. Dry matter growth increase of species i under nitrogen limited production conditions ($dW_{N,i}/dt$, g m^{-2} s^{-1}) can be determined by adjusting the light saturated CO_2 assimilation rate ($A_{max,i}$) used to determine $A_{c,i}$ in Equation 13.3 for leaf nitrogen content ($N_{L,i}$). Therefore, the nitrogen dependent net growth function ($f_i(N_{L,i})$) is:

$$f_i(N_{L,i}) = A_{c,i}(\varepsilon, A_{max,i}(N_{L,i}))$$

which indicates that CO_2 assimilation is dependent upon ε and $A_{max,i}$, and the latter is dependent upon leaf nitrogen content. Sinclair and Horie[49] proposed the following relationship for quantifying the light saturated CO_2 assimilation–$N_{L,i}$ relationship:

$$A_{max,i} = [A_{max,i}] \left[\frac{2}{(1 + \exp(-a(N_{L,i} - b))} - 1 \right] \qquad [13.10]$$

where $[A_{max,i}]$ represents an absolute maximum CO_2 assimilation rate under optimal light, soil water, and leaf nitrogen conditions, a is a shape coefficient, and b is the leaf nitrogen content at which CO_2 assimilation reaches zero. Estimates of $[A_{max,i}]$, a, and b can be obtained by regressing observed CO_2 assimilation (in full sunlight) on leaf nitrogen content for each species.[50-57] Since leaf nitrogen content may be critical for plant growth, nitrogen uptake and partitioning within the plant must be accurately predicted. When soil nitrogen supply becomes limited, nitrogen uptake will be reduced and nitrogen partitioning to leaves may be modified. Therefore, weeds induce a stress on the crop through their direct use of stored nitrogen in the root zone of both species. How can we quantify the demand for nitrogen by both the crop and weed? How does current nitrogen use influence current and future supply of available soil nitrogen? What is the effect of limited nitrogen supply on nitrogen uptake? In a competitive situation, how do we partition the quantity of nitrogen available for uptake between the two species?

Several algorithms have been developed to predict soil nitrogen uptake (U_i, g N m^{-2} d^{-1}).[58-61] Typically, nitrogen uptake is predicted as the minimum of (1) the daily nitrogen demand of the species ($N_{demand,i}$, g N m^{-2} d^{-1}) and (2) the quantity of nitrogen available for uptake (R_N, g N m^{-2}). Thus:

$$U = MIN(N_{demand}, R_N) \qquad [13.11]$$

ten Berge et al.[58] calculated daily nitrogen demand for a given species using:

$$N_{demand,i} = MIN(U_N, U_M, U_P) \qquad [13.12]$$

where U_N is maximum potential nitrogen uptake rate, U_M is uptake limited by the maximum of the observed ratio of daily nitrogen uptake to daily biomass production, and U_P is uptake rate limited by the difference between potential and actual quantity of N in existing biomass.

Maximum potential nitrogen uptake (U_N, g N m^{-2} d^{-1}) can be obtained empirically as the maximum observed (actual) uptake rate for the species when grown under potential production conditions. Actual nitrogen uptake rate is measured as $\Delta N_{a,i}/\Delta t$, where $\Delta N_{a,i}$ and Δt are the change in measured nitrogen content of the species (g N m^{-2}) and the time interval (d) between sampling dates, respectively. Uptake (U_M) limited by the maximum of the ratio of daily nitrogen uptake to daily biomass production (q_N, g N (g dw)$^{-1}$) is the product of the predicted growth rate and q_N. An estimate of q_N is the maximum observed $\Delta N_a/\Delta W_i$, where W_i (g m^{-2}) is total biomass of the species. Nitrogen uptake limited by the difference between potential and actual amount of N in existing biomass (U_P, g N m^{-2}) is

$$U_P = W_i[N] - N_a \qquad [13.13]$$

where [N] (g N g^{-1} dw) is potential nitrogen concentration and N_a is actual nitrogen content of the species (g N m^{-2}). Many crop species show a consistent relationship between potential nitrogen concentration [N] and total biomass (W_i) when grown under potential production conditions:[62]

$$[N] = cW_i^{-d} \qquad [13.14]$$

where c is maximum observed nitrogen concentration and d is a shape coefficient. Estimates of c and d for weeds are not available, although Coleman et al.[63] presented evidence indicating this relationship is accurate for *Abutilon theophrasti*.

The quantity of soil nitrogen available for uptake within the root zone (R_N, g N m^{-2}) can be estimated using:

$$R_N = [S_N + F_d + S_{min}] - [U] \qquad [13.15]$$

where S_N (g N m^{-2}) is the quantity of soil nitrate in the root zone at planting, F_d (g N m^{-2}) is the quantity of fertilizer added, S_{min} (g N m^{-2}) is the quantity of native soil nitrogen mineralized, and U is cumulative nitrogen uptake, summed across species. S_N is easily determined by soil sampling and S_{min} can be determined as a function of cumulative soil thermal units using:

$$S_{min} = N_{pm}(1 - e^{-kT}) \qquad [13.16]$$

where N_{pm} is potentially mineralizable N (g N m^{-2}), k is the first order rate constant (°T^{-1}) derived from laboratory incubations of surface soil vs. cumulative soil thermal units, and T is cumulative soil thermal units. N mineralization is measured in the laboratory using aerobic incubation where soil is kept at 25°C[64–66] and 60% relative water content, and periodically leached with 0.01 M CaCl$_2$[67] to determine net N mineralization. Estimates of N_{pm} and k can be determined by fitting cumulative mineralized N on cumulative soil thermal units using Equation 13.16.

As with soil water, Equations 13.10 to 13.16 can be used to simulate indirect competition for soil nitrogen. During periods when soil nitrogen is not limiting, plants utilize as much nitrogen as is required using Equation 13.12, which subsequently reduces the pool of available soil nitrogen. As the season progresses, if nitrogen supply is insufficient to meet demand, nitrogen uptake is reduced. A subsequent reduction in tissue N content or partitioning of carbon to above-ground growth will reduce plant growth. Equation 13.11 indicates that when nitrogen supply is less than the overall demand, uptake is limited to the quantity available. However, it provides no information as to how much is acquired by each species. Therefore, direct competition for nitrogen is not accounted for. Under conditions of limited soil nitrogen supply, should the quantity of available nitrogen be divided among species based on the relative quantity of roots in the soil profile? What if the species differ in their uptake efficiency? What if the depth of the actual root zone of the two species differs?

Smethurst and Comerford[68] presented a method of simulating direct competition for soil nitrate. Their approach required that nitrogen uptake be calculated using concepts from solute transport theory (e.g., Nye and Tinker[69]). Nitrate uptake (U, μmol cm^{-3} d^{-1}) of a given species within a rooted layer j can be modeled using the procedures of Baldwin et al.[70]

$$U_j = 2\pi r_o \alpha C_{lo,j} L_{v,j} t \qquad [13.17]$$

where r_o is mean root radius (cm), α is root absorbing power (cm s^{-1}, $I_{max}/(K_m + C_{lo} - C_{min})$, where I_{max} (μmol m^{-2} s^{-1}) is the maximum N uptake rate, K_m (μmol cm^{-3}) is the solute nitrate concentration at which uptake is 1/2 I_{max}, C_{lo} is defined below, and C_{min} (μmol cm^{-3}) is minimum solute concentration required for uptake to occur, $L_{v,j}$ (cm cm^{-3}) is the root length density within the jth layer, t is the time of integration (1 d = 86400 s), and $C_{lo,j}$ (μmol cm^{-3}) is nitrate concentration at the root surface within layer j:

$$C_{lo,j} = \left(\cfrac{C_{1,j}}{\dfrac{\alpha}{v_{o,j}} + \left(1 - \dfrac{\alpha}{v_{o,j}}\right)\left(\dfrac{2}{2 - \dfrac{r_o v_{o,j}}{D_e b}}\right)\left(\dfrac{\left(\left(\dfrac{r_{dz,j}}{r_o}\right)^{\left(2 - \frac{r_o v_{o,j}}{D_e b}\right)} - 1\right)}{\left(\left(\dfrac{r_{dz,j}}{r_o}\right)^2 - 1\right)}\right)} \right) \qquad [13.18]$$

where $C_{1,j}$ is average concentration of nitrate in soil solution (μmol cm^{-3}, $C_{1,j} = C_{li,j}$ at t = 0), $v_{o,j}$ is water flux at the root surface (cm s^{-1}, equal to $T_{a,i}$), $r_{dz,j}$ is the radius of the depletion zone (cm) around the root, b is buffer power, and D_e is the effective diffusion coefficient (cm^2 s^{-1}):

$$D_e = D_l \, \theta_{a,j}^{0.5} \qquad [13.19]$$

where D_l is the diffusion coefficient of nitrate in water (cm^2 s^{-1}) and $\theta_{a,j}$ is actual volumetric water content. Since nitrate is a non-adsorbing ion, buffer power is set equal to $\theta_{a,j}$.

When individual roots of different species compete for nutrients, mean radial distance between adjacent roots ($r_{1,j}$) can be determined using $(1/(\pi L_{v,j,T})^{0.5}$, where $L_{v,j,T}$ is $L_{v,j}$ summed across species. However, because root characteristics (e.g., r_o, α) of the competing species may differ, the true zone of influence of the root of species P may be larger than that of a species Q root. The width of the true zone of influence defines the no-transfer boundary (r_{ntb}). The position of the no-transfer boundary between two competing roots within a time step may be calculated as:[68]

$$\frac{C_{lo,j,P}}{C_{lo,j,Q}} = \frac{(\alpha \, r_o)_Q \, y^2}{(\alpha \, r_o)_P \, (IRD_j - y)^2} \qquad [13.20]$$

where IRD_j is mean interroot distance ($2 \, r_{1,j}$, cm), P and Q denote species, and y (cm) is distance from root P to the no-transfer boundary (r_{ntb}, the species-specific estimate of $r_{dz,j}$).

The method of predicting uptake and competition for soil nitrate described above separated the rooting zone into multiple layers. The approach described previously, as well as that for predicting water uptake and competition, assumed that nitrogen or water content in the rooted zone was homogenous. This may be a useful first approximation, but clearly is not realistic. When a layered model of uptake and competition is used, a root growth model is required. Within such a model, at least two factors must be dealt with: (1) the rate of vertical penetration of roots, and (2) root length density with depth.[71] Jones et al.[72] reviewed a number of useful approaches for simulating root growth.

Although Equation 13.14 can be used to determine the expected nitrogen concentration of the entire canopy, these equations do not account for the partitioning of nitrogen to leaves once it is taken up, or for the potential effect of nitrogen supply on the partitioning of new growth. Nitrogen content of the leaves (N_L, g N m^{-2} leaf) over time can be predicted if the fraction of nitrogen taken up that is partitioned to leaves (P_L) throughout the growing season is known:

$$P_L = \frac{\left(\dfrac{\Delta N_{L,i}}{\Delta t}\right) LAI}{U_i} \qquad [13.21]$$

where $\Delta N_L/\Delta t$ is change in nitrogen content of the leaves, LAI is leaf area index (m^2 leaf m^{-2} ground), and U_i is measured nitrogen uptake ($\Delta N_a/\Delta t$) during a given sampling interval. Once the P_L–time relationship is known, N_L can be calculated from predicted nitrogen uptake and LAI. Unfortunately, the relationship represented by Equation 13.21 is most likely to be obtained from experiments conducted under potential production conditions. It is possible that nitrogen supply will modify the partitioning of nitrogen to leaves. Similarly, what will be the influence of nitrogen supply on the partitioning of new biomass among plant organs?

In a recent study in rice, Hasegawa and Horie[73] suggested that the effect of nitrogen supply on leaf area development was more important to crop growth and yield than its effect on $N_{L,i}$ and subsequent CO_2 assimilation. Although maize and velvetleaf leaf nitrogen content clearly differed among nitrogen application treatments (Figure 13.2),[57] maximum observed LAI and height of maize and velvetleaf are strongly influenced by nitrogen application rate (Figure 13.3). This observation suggests that when nitrogen is in short supply, plants make a tradeoff when allocating new biomass between tissues necessary for acquiring nitrogen vs. light. In other words, if nitrogen uptake is reduced due to limited nitrogen supply, plants partition less new growth to leaves and more to roots. This provides the plant with greater root biomass and volume so it can acquire more nitrogen, and it allows the plant to maintain a relatively constant nitrogen concentration in existing leaves. However, it limits the production of leaf area and therefore potential growth rate. Because maize is a C_4 species and expected to have greater nitrogen and radiation use efficiency than velvetleaf, root growth will need to increase more in velvetleaf than maize to maintain

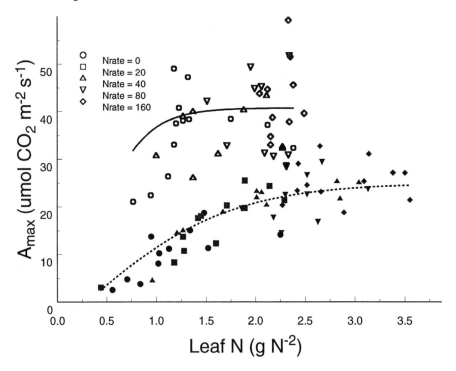

FIGURE 13.2 Maize (open symbols) and velvetleaf (closed symbols) net CO_2 assimilation rate (A_{max}, μmol CO_2 m^{-2} s^{-1} = 1.7536 10^{-5} g CO_2 m^{-2} s^{-1}) under full sun conditions in response to nitrogen content (N_L) of the most recent fully expanded leaf in plots with nitrogen application (Nrate) ranging from zero to 160 kg N ha^{-1}. Redrawn from Lindquist and Mortensen.[57]

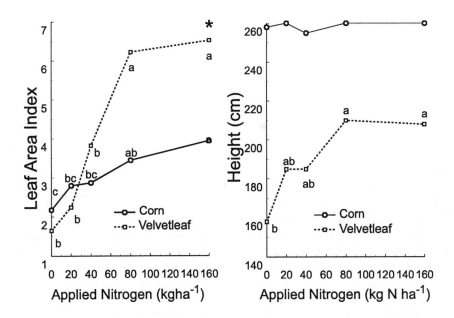

FIGURE 13.3 Maximum observed leaf area index (LAI) and height of maize (Pioneer 3379) and velvetleaf as influenced by amount of nitrogen applied (as ammonium nitrate). Means within a species that are associated with the same letter do not differ at the $p < 0.05$ level of significance. The asterisk indicates that maize and velvetleaf LAI differs at $p < 0.05$.

equivalent changes in nitrogen uptake. Consequently, the greater partitioning of new growth to roots in velvetleaf comes at a cost to its leaf area development under low nitrogen supply conditions. In contrast, when nitrogen is non-limiting, velvetleaf partitions most of its biomass to leaves and stems. The result is a taller plant with greater leaf area. Because velvetleaf leaf area is nearly completely concentrated in the upper 10% of its canopy height,[57] it becomes a better competitor for light when nitrogen is non-limiting. Figure 13.3 shows that velvetleaf leaf area index and height responded more strongly to nitrogen application than did maize.[74] The trade-off in nitrogen and carbon partitioning must be considered if our models are to account for the interactive effects of competition for both light and soil nitrogen.

During vegetative growth of annual plants, the fraction of new biomass that is partitioned to various organ groups depends both on stage of development and the degree of water or nitrogen stress imposed by the below-ground environment. Fraction of new biomass partitioned to an organ group (p_k) can be quantified using:

$$p_k = \frac{\Delta W_k}{\Delta W} \qquad [13.22]$$

where ΔW_k and ΔW represent change in organ and total plant biomass between sampling dates, respectively. It is assumed that the trade-off in partitioning discussed above occurs because plants functionally balance partitioning between roots, leaves, and stems to optimize photosynthetic area and the nitrogen content of leaf tissue.[75-77] Accomplishing both optimizes growth rate under nitrogen limited conditions. Under the assumption of balanced growth, the fraction of new biomass partitioned to the root (p_r) can be predicted using:[78]

$$p_r = \frac{\dfrac{[N]}{G_N}}{\dfrac{[C]}{G_c} - \dfrac{[N]}{G_N}} \qquad [13.23]$$

where [C] is carbon concentration of the plant, G_N, is daily gain in nitrogen for each unit root biomass (g N g root^{-1}), and G_C is daily gain in carbon for each unit above-ground biomass (g carbon g^{-1} shoot). G^N depends upon nitrogen uptake and previous root growth and G_C depends upon the amount of leaf area accumulated, environmental conditions (temperature, incident radiation), and leaf nitrogen content. The issue of partitioning new growth to optimize plant growth using this balanced growth hypothesis has been studied from a theoretical standpoint by many authors.[79-83] However, because root physiology research is difficult and costly, there are few data available in the literature to support or refute the theory (but see McConnaughay and Coleman[84]). Further research is clearly needed.

13.5   SUMMARY

Quantity of resources available within agricultural systems strongly influences growth and development of crops and weeds. Weeds cause crop loss primarily through their direct effect on the quantity of resources available to the crop. In most agricultural systems, light, soil water, and nitrogen are the most critical limiting resources. Quantifying the effects of each resource on plant growth is complex, but must be accomplished to accurately predict the effects of weeds on crop yield. I have outlined some of the approaches described in the literature for quantifying these effects, and pointed out many areas where further research would clearly benefit the current state of the science. In doing so, I focused my attention on the effects of resources on plant growth processes, ignoring for the most part respiration and senescence. There is evidence that plant stresses increase the rate of senescence,[85] but little theoretical work has been done to quantify respiration and senescence as a function of resource supply.

REFERENCES

1. Barbour, M. G., Burk, J. H., and Pitts, W. D., *Terrestrial Plant Ecology,* Benjamin Cummings, Menlo Park, 1987, chaps. 6 and 7.
2. Goldberg, D. E., Components of resource competition in plant communities, in *Perspectives on Plant Competition,* Grace, J. B., and Tilman D., Eds., Academic Press, San Diego, 1990, chap. 3.

3. Radosevich, S., Holt, J., and Ghersa, C., *Weed Ecology: Implications for Management,* John Wiley & Sons, New York, 1997, chaps. 5 and 6.

4. Holt, J. S., Plant responses to light: a potential for weed management, *Weed Sci.,* 43, 474, 1995.

5. DiTomaso, J. M., Approaches for improving crop competitiveness through manipulation of fertilization strategies, *Weed Sci.,* 43, 491, 1995.

6. Kropff, M. J., and van Laar, H. H., *Modelling Crop–Weed Interactions,* CAB International and the International Rice Research Institute, Wallingford, 1993, chaps. 4 to 6.

7. Patterson, D. T., Methodology and terminology for the measurement of light in weed studies—a review, *Weed Sci.,* 27, 437, 1979.

8. Patterson, D. T., Comparative ecophysiology of weeds and crops, in *Weed Physiology, Volume I, Reproduction and Ecophysiology,* Duke, S. O., Ed., CRC Press, Boca Raton, 1985, chap. 4.

9. Patterson, D. T., Effects of environmental stress on weed/crop interactions, *Weed Sci.,* 43, 483, 1995.

10. Aphalo, P. J., and Ballare, C. L., On the importance of information-acquiring systems in plant–plant interactions, *Functional Ecol.,* 9, 5, 1995.

11. Lindquist, J. L., unpublished data, 1994.

12. Grace, J. B., On the relationship between plant traits and competitive ability, in *Perspectives on Plant Competition,* Grace, J. B., and Tilman D., Eds., Academic Press, San Diego, 1990, chap. 4.

13. Grace, J. B., A clarification of the debate between Grime and Tilman, *Functional Ecol.,* 5, 583, 1991.

14. Grime, J. P., *Plant Strategies and Vegetation Processes,* Wiley, London, 1979.

15. Tilman, D., *Resource Competition and Community Structure,* Princeton University Press, Princeton, 1982.

16. Tilman, D., *Plant Strategies and the Dynamics and Structure of Plant Communities,* Princeton University Press, Princeton, 1988.

17. Tilman, D., Mechanisms of plant competition for nutrients: the elements of a predictive theory of competition, in *Perspectives on Plant Competition,* Grace, J. B., and Tilman D., Eds., Academic Press, San Diego, 1990, chap. 7.

18. Caldwell, R. M., Pachepsky, Y. A., and Timlin, D. J., Current research status on growth modeling in intercropping, in *Dynamics of Roots and Nitrogen in Cropping Systems of the Semi-arid Tropics,* Ito, O., Johansen, C., Adu-Gyamfi, J. J., Katayama, K., Kumar Rao, J. V. D. K., and Rego, T. J., Eds., Japan International Center for Agricultural Sciences, 1996, 617.

19. Boote, K. J., and Loomis, R. S., The prediction of canopy assimilation, in *Modeling Crop Photosynthesis — From Biochemistry to Canopy,* Crop Science Society of America Special Publication No., 19, Madison, WI, 1991, chap. 7.

20. Kropff, M. J., Mechanisms of competition for light, in *Modelling Crop-Weed Interactions,* Kropff, M. J., and van Laar, H. H., Eds., CAB International and the International Rice Research Institute, Wallingford, 1993, chap. 4.

21. Amthor, J. S., The role of maintenance respiration in plant growth, *Plant Cell Envir.,* 7, 561, 1984.

22. Penning deVries, F. W. T., Brunsting, A. H. M., and van Laar, H. H., Products, requirements, and efficiency of biosynthesis: a quantitative approach, *J. Theor. Biol.,* 45, 339, 1974.

23. Ryel, R., Barnes, P. W., Beyschlag, W., Caldwell, M. M., and Flint, S. D., Plant competition for light analyzed with a multispecies canopy model. I. Model development and

influence of enhanced UV-B conditions on photosynthesis in mixed wheat and wild oat canopies, *Oecologia*, 82, 304, 1990.

24. Graf, B., Gutierrez, A. P., Rakotobe, O., Zahner, P., and Delucchi, W., A simulation model for the dynamics of rice growth and development: Part I—The carbon balance, *Agric. Syst.*, 32, 341, 1990.

25. Rimmington, G. M., A model of the effect of interspecies competition for light on dry-matter production, *Aust. J. Plant Physiol.*, 11, 277, 1984.

26. Wiles, L. J., and Wilkerson, G. G., Modeling competition for light between soybeans and broadleaf weeds, *Agric. Syst.*, 35, 37, 1991.

27. Kropff, M. J., Modelling the effects of weeds on crop production, *Weed Res.*, 28, 465, 1988.

28. Spitters, C. J. T., and Aerts, R., Simulation of competition for light and water in crop-weed associations, *Aspects Appl. Biol.*, 4, 467, 1983.

29. Kiniry, J. R., Williams, J. R., Gassman, P. W., and Debaeke, P., A general, process-oriented model for two competing plant species, *Trans. ASAE*, 35, 801, 1992.

30. Flenet, R., Kiniry, J. R., Board, J. E., Westgate, M. E., and Reicosky, D. C., Row spacing effects on light extinction coefficients of corn, sorghum, soybean and sunflower, *Agron. J.*, 88, 185, 1996.

31. Lindquist, J. L., and Mortensen, D. A., A simulation approach to identifying the mechanisms of maize tolerance to velvetleaf competition for light, in *Proceedings 1997 Brighton Crop Protection Conference—Weeds*, Brighton, U.K., 1997, 503.

32. Caton, B. P., Foin, T. C., and Hill, J. E., A plant growth model for integrated weed management in direct-seeded rice, III. Interspecific competition for light, *Field Crops Res.*, 63, 47, 1999.

33. Lindquist, J. L., and Mortensen, D. A., Tolerance and velvetleaf (*Abutilon theophrasti*) suppressive ability of two old and two modern corn (*Zea mays*) hybrids, *Weed Sci.*, 46, 569, 1998.

34. Lindquist, J. L., Mortensen, D. A., and Johnson, B. E., Mechanisms of corn tolerance and velvetleaf suppressive ability, *Agron. J.*, 90, 787, 1998.

35. McLachlan, S. M., Tollenaar, M., Swanton, C. J., and Weise, S. F., Effect of corn-induced shading on dry matter accumulation, distribution, and architecture of redroot pigweed (*Amaranthus retroflexus*), *Weed Sci.*, 41, 568, 1993.

36. Murphy, S. D., Yakubu, Y., Weise, S. F., and Swanton, C. J., Effect of planting patterns and inter-row cultivation on competition between corn (*Zea mays*) and late emerging weeds, *Weed Sci.*, 44, 856, 1996.

37. Loomis, R. S., and Connor, D. J., *Crop Ecology: Productivity and Management in Agricultural Systems*, Cambridge University Press, Cambridge, U.K. 1992.

38. Kropff, M. J., Mechanisms of competition for water, in *Modelling Crop-Weed Interactions*, Kropff, M. J., and van Laar, H. H., Eds., CAB International and the International Rice Research Institute, Wallingford, 1993, chap. 5.

39. Campbell, G. S., Simulation of water uptake by plant roots, in *Modeling Plant and Soil Systems*, Hanks, J., and Ritchie, J. T., Eds., American Society of Agronomy Special Publication No., 31, Madison, WI, 1991, chap. 12.

40. Swaney, D. P., Jones, J. W., Boggess, W. G., Wilkerson, G. G., and Mishoe, J. W., Real-time irrigation decision analysis using simulation, *Trans. ASAE*, 26, 562, 1983.

41. Penning de Vries, F. W. T., Jansen, D. M., ten Berge, H. F. M., and Bakema, H., Simulation of ecophysiological processes of growth in several annual crops, Simulation monographs 29, International Rice Research Institute and Pudoc, Wageningen, 1989.

42. Penman, H. L., Natural evaporation from open water, bare soil and grass, Proc. R. Soc. London, Series A, 193, 120, 1948.

43. Doorenbos, J., and Kassam, A. H., Yield response to water, FAO Irrigation and Drainage Paper 33, FAO, Rome, 1979.

44. Ray, J. D., and Sinclair, T. R., Stomatal closure of maize hybrids in response to drying soil, *Crop Sci.*, 37, 803, 1998.

45. Moreshet, S., Bridges, D. C., NeSmith, D. S., and Huang, B., Effects of water deficit stress on competitive interaction of peanut and sicklepod, *Agron. J.*, 88, 636, 1996.

46. Makino, A., and Osmond, B., Effects of nitrogen nutrition on nitrogen partitioning between chloroplasts and mitochondria in pea and wheat, *Plant Physiol.*, 96, 355, 1991.

47. Sage, R. F., and Pearcy, R. W., The nitrogen use efficiency of C3 and C4 plants. II. Leaf nitrogen effects on the gas exchange characteristics of *Chenopodium album* and *Amaranthus retroflexus, Plant Physiol.*, 84, 958, 1987.

48. Hikosaka, K., and Terashima, I., A model of the acclimation of photosynthesis in the leaves of C3 plants to sun and shade with respect to nitrogen use, *Plant, Cell & Environ.*, 18, 605, 1995.

49. Sinclair, T. R., and Horie, T., Leaf nitrogen, photosynthesis, and crop radiation use efficiency: a review, *Crop Sci.*, 29, 90, 1989.

50. Muchow, R. C., Effects of leaf nitrogen and water regime on the photosynthetic capacity of kenaf (*Hibiscus cannabinus* L.) under field conditions, *Aust. J. Agric. Res.*, 41, 845, 1990.

51. Sage, R. F., and Pearcy, R. W., The nitrogen use efficiency of C3 and C4 plants. I. Leaf nitrogen, growth, and biomass partitioning in *Chenopodium album* and *Amaranthus retroflexus, Plant Physiol.*, 84, 954, 1987.

52. Sage, R. F., Sharkey, T. D., and Pearcy, R. W., The effect of leaf nitrogen and temperature on the CO_2 response of photosynthesis in the C3 dicot *Chenopodium album* L., *Aust. J. Plant Physiol.*, 17, 135, 1990.

53. Hilbert, D. W., Larigauderie, A., and Reynolds, J. F., The influence of carbon dioxide and daily photon flux density on optimal leaf nitrogen concentration and root:shoot ratio, *Ann. Bot.*, 68, 365, 1991.

54. Hunt, E. R. Jr., Weber, J. A., and Gates, D. M., Effects of nitrate application on *Amaranthus powelii* Wats. I. changes in photosynthesis, growth rates, and leaf area, *Plant Physiol.*, 79, 609, 1985.

55. Marshall, B., and Vos, J., The relation between the nitrogen concentration and photosynthetic capacity of potato (*Solanum tuberosum* L.) leaves, *Ann. Bot.*, 68, 33, 1991.

56. Ampong-Nyarko, K., and De Datta, S. K., Effects of light and nitrogen and their interaction on the dynamics of rice-weed competition, *Weed Res.*, 33, 1, 1993.

57. Lindquist, J. L., and Mortensen, D. A., Ecophysiological characteristics of four maize hybrids and *Abutilon theophrasti, Weed Res.*, 39, 271, 1999.

58. ten Berge, H. F. M., Wopereis, M. C. S., Riethoven, J. J. M., and Drenth, H., Description of the ORYZA_0 modules (version 2.0), in *ORYZA Simulation Modules for Potential and Nitrogen Limited Rice Production,* Drenth, H., ten Berge, H. F. M., and Riethoven, J. J. M., Eds., Simulation and Systems Analysis for Rice Production (SARP) Research Proceedings, DLO-Research Institute for Agrobiology and Soil Fertility, WAU-Department of Theoretical Production Ecology, Wageningen, and International Rice Research Institute, 1994, 43.

59. Hasegawa, T., and Horie, T., Modelling the effect of nitrogen on rice growth and development, in *Systems Approaches for Sustainable Agricultural Development: Applications of*

Systems Approaches at the Field Level, Kropff, M. J., Teng, P. S., Aggarwal, P. K., Bouma, J., Bouman, B. A. M., Jones, J. W., and van Laar, H. H., Eds., Kluwer Academic Publishers, Dordrecht, The Netherlands, 1997, 243.

60. Kropff, M. J., Mechanisms of competition for nitrogen, in *Modelling Crop-Weed Interactions,* in Kropff, M. J., and van Laar, H. H., Eds., CAB International and the International Rice Research Institute, Wallingford, 1993, chap. 6.

61. Sinclair, T. R., and Muchow, R. C., Effect of nitrogen supply on maize yield: I. Modeling physiological responses, *Agron. J.,* 87, 632, 1995.

62. Greenwood, D. J., Lemaire, G., Gosse, G., Cruz, P., Draycott, A., and Neeteson, J. J., Decline in percentage N of C3 and C4 crops with increasing plant mass, *Ann. Bot.,* 66, 425, 1990.

63. Coleman, J. S., McConnaughay, K. D. M., and Bazzaz, F. A., Elevated CO_2 and plant nitrogen-use: Is reduced tissue nitrogen concentration size-dependent?, *Oecologia,* 93, 195, 1993.

64. Honeycutt, C. W., Ziblinski, L. M., and Clapham W. M., Heat units for describing carbon mineralization and predicting net nitrogen mineralization, *Soil Sci. Soc. Am. J.,* 52, 1346, 1988.

65. Honeycutt, C. W., and Potero, L. J., Field evaluation of heat units for predicting crop residue carbon and nitrogen mineralization, *Plant and Soil,* 125, 213, 1990.

66. Honeycutt, C. W., Potero, L. J., and Halteman, W. A., Predicting nitrate formation from soil, fertilizer, crop residue, and sludge with thermal units, *J. Environ. Qual.,* 20, 850, 1991.

67. Cabrera, M. L., and Kissel, D. E., Evaluation of a method to predict nitrogen mineralized from soil organic matter under field conditions, *Soil Sci. Soc. Am. J.,* 52, 1027, 1988.

68. Smethurst, P. J., and Comerford, N. B., Simulating nutrient uptake by single or competing and contrasting root systems, *Soil Sci.,* 57,1361, 1993.

69. Nye, P. H., and Tinker, P. B., *Solute Movement in the Soil-root System,* University of California Press, Los Angeles, 1977.

70. Baldwin, J. P., Nye, P. H., and Tinker, P. B., Uptake of solutes by multiple root systems from soil, III, *Plant and Soil,* 38, 621, 1973.

71. Tardieu, F., Analysis of the spatial variability of maize root density. I. Effect of compactions on spatial arrangement of roots, *Plant and Soil,* 107, 259, 1988.

72. Jones, C. A., Bland, W. L., Ritchie, J. T., and Williams, J. R., Simulation of root growth, in *Modeling Plant and Soil Systems,* Agronomy Monograph No. 31, American Society of Agronomy, Madison, WI, 1991, chap. 6.

73. Hasegawa, T., and Horie, T., Leaf nitrogen, plant age and crop dry matter production in rice, *Field Crops Res.,* 47, 107, 1996.

74. Lindquist, J. L., unpublished data, 1994.

75. Agren, G. I., and Ingestad, T., Root:shoot ratio as a balance between nitrogen productivity and photosynthesis, *Plant, Cell & Environ.,* 10, 579, 1987.

76. Kastner-Maresch, A. E., and Mooney, H. A., Modelling optimal plant biomass partitioning, *Ecol. Modelling,* 75/76, 309, 1994.

77. Robinson, D., Compensatory changes in the partitioning of dry matter in relation to nitrogen uptake and optimal variations in growth, *Ann. Bot.,* 58, 841, 1986.

78. Reynolds, J. F., and Chen, J., Modelling whole-plant allocation in relation to carbon and nitrogen supply: coordination versus optimization, *Plant and Soil,* 185, 65, 1996.

79. Hilbert, D. W., Optimization of plant root:shoot ratios and internal nitrogen concentration, *Ann. Bot.,* 66, 91, 1990.

80. Hilbert, D. W., and Reynolds, J. F., A model allocating growth among leaf proteins, shoot structure, and root biomass to produce balanced activity, *Ann. Bot.,* 68, 417, 1991.
81. Ingestad, I., and Agren, G. I., Theories and methods on plant nutrition and growth, *Physiol. Plant.,* 84, 177, 1992.
82. Levin, S. A., Mooney, H. A., and Field, C., The dependence of plant root:shoot ratios on internal nitrogen concentration, *Ann. Bot.,* 64, 71, 1989.
83. Thornley, J. H. M., A balanced quantitative model for root:shoot ratios in vegetative plants, *Ann. Bot.,* 36, 431, 1972.
84. McConnaughay, K. D. M., and Coleman, J. S., Biomass allocation in plants: Ontogeny or optimality? A test along three resource gradients, *Ecology,* 80, 2581, 1999.
85. Girardin, P., Tollenaar, M., and Deltour, A., Effect of temporary N starvation in maize on leaf senescence, *Can. J. Plant Sci.,* 65, 819, 1985.

Index

1, 5-bisphosphate carboxylase/oxygenase, *see* Rubisco

A

Abiotic stress, 169
Abscisic acid, 124
Abutilon theophrasti, see Velvetleaf
Acalymma vittata, see Striped cucumber beetle
Acyrthosiphum pisum, 142
Acyrthosiphum condoi, 142
African rice, 222
Agriotes maneus, see Wireworms
Agrotis gladiaria, see Cutworms, claybacked
Agrotis ipsilon, see Black cutworm
Albert's squirrel, 145
Alfalfa, 16, 26, 27, 29, 75, 86, 88, 141
    canopy, 105
    cultivars, 110
    defoliation, 168
injury guild, 93, 94
regrowth, 31
Alfalfa caterpillar, 142
Alfalfa weevil, 29, 94
Alternaria alternata, 195
Alternaria solani, see Early blight
Altica subplicata, 151
Amblyseius cucumeris, 28
Antibiosis, 137
Anticarsia gemmatalis, see Velvetbean caterpillar
Antixenosis, 137
Apical dominance, 168
Apple, 25, 191
Apple scab, 186
Apterothrips apteris, 146
Area under the disease progress curve (AUDPC), 188
Architectural modifier, 92
Armyworm, 51-52
Arroyo willow, 144
Artemisia tridentata, 168, 173
Assimilate removal, 91, 92
Assimilate sappers, 6, 76, 92
ATP, 86, 87
Atropa acuminata, 141
AUDPC, *see* Area under the disease progress curve

B

Bacillus thuringiensis (Bt), 28, 29, 58
Barley, 26, 28, 54, 128, 195
    barley stripe, 198
    moisture stress, 122
    photosynthesis, 164
    powdery mildew, 194
    radiation use efficiency, 191
    tolerance to herbivory, 165
Barley stripe, 198
Barnyardgrass, 221, 239
Bean
    dry, 85, 87, 88, 126
    pinto, 26
    snap, 24
    white, 217, 224
Bean leaf beetle, 94
Beer's Law, 101, 189-190
Beet armyworm, 94, 142
Bemisia tabaci, 29, *see also* Whiteflies
Bermudagrass, 26
Beuvaria bassiana, 120
Bioeconomic model, 219
Biomass, 210
Birth rate, 207
Black box, 1-2, 73
Black cutworm, 31, 50
Blissus leucopterus leucopterus, see Chinch bug
Brassica rapa, see Canola
Brown planthopper, 119
Bt, *see* Bacillus thuringiensis
Bush lupine, 144
Busseola fusca, see Stem borer

C

C-3 plant species, 164
C-4 plant species, 164
Cabbage, 143
Cabbage butterfly, 148
Cabbage looper, 85
Cabbage *seed*pod weevil, 31
Canola, 26, 29, 31, 145, 148
Carbon exchange rate, *see* Photosynthesis and Plant gas exchange
Cassava, 143, 164
Centaurea maculosa, 173
Cephus cinctus, see Wheat stem sawfly